TI

Jack Cohen is an internationally known reproductive biologist, who was a university teacher for thirty years, and has published nearly a hundred research papers. His books include *Living Embryos, Reproduction; Spermatozoa, Antibodies and Infertility*; and *The Privileged Ape*, a rather different look at human evolution. He now works with the mathematician Ian Stewart with whom he has explored issues of complexity, chaos and simplicity, producing several joint papers. He acts as a consultant to top science-fiction writers, designing credible creatures and ecologies, and frequently appears on radio and television programmes. His hobbies include boomerang-throwing and keeping strange animals.

Ian Stewart was born in Folkestone in 1945. He graduated in Mathematics from Cambridge and obtained a Ph.D. from the University of Warwick, where he is now Professor of Mathematics. He is an active research mathematician with over 130 published papers, and he takes a particular interest in problems that lie in the gaps between pure and applied mathematics. Ian Stewart has written or co-authored over sixty books, including *Nature's Numbers*, shortlisted for the 1996 Rhône-Poulenc Prize for Science Books; *The Science of Discworld* (with Jack Cohen); *The Collapse of Chaos*; *Fearful Symmetry*; the bestselling *Does God Play Dice?*; *Figments of Reality* (also with Jack Cohen); and *Life's Other Secret* (several of which are published in Penguin). He is mathematics consultant for *New Scientist* and writes the 'Mathematical Recreations' column in *Scientific American*.

In 1995 the Royal Society awarded him the Michael Faraday Medal for the year's most significant contribution to the public understanding of science.

THE COLLAPSE OF CHAOS

Discovering Simplicity in a Complex World

JACK COHEN
&
IAN STEWART

PENGUIN BOOKS

PENGUIN BOOKS

Published by the Penguin Group
Penguin Books Ltd, 27 Wrights Lane, London W8 5TZ, England
Penguin Putnam Inc., 375 Hudson Street, New York, New York 10014, USA
Penguin Books Australia Ltd, Ringwood, Victoria, Australia
Penguin Books Canada Ltd, 10 Alcorn Avenue, Toronto, Ontario, Canada M4V 3B2
Penguin Books (NZ) Ltd, Private Bag 102902, NSMC, Auckland, New Zealand

Penguin Books Ltd, Registered Offices: Harmondsworth, Middlesex, England

First published in the United States of America by Viking Penguin, 1994
Published in Penguin Books 1995
Reissued 2000
1 3 5 7 9 10 8 6 4 2

Printed in England by Clays Ltd, St Ives plc

How exquisitely the individual mind to the external world
Is fitted—and how exquisitely, too
The external world is fitted to the mind
And the Creation *(by no lower name*
Can it be called) which they with blended might
Accomplish.

WILLIAM WORDSWORTH

The next great awakening of human intellect may well produce a method of understanding the qualitative content of equations. Today we cannot. Today we cannot see that the water-flow equations contain such things as the barber pole structure of turbulence that one sees between rotating cylinders. Today we cannot see whether Schrödinger's equation contains frogs, musical composers, or morality—or whether it does not.

RICHARD P. FEYNMAN

CONTENTS

THE COLLAPSE OF CHAOS

PREFACE

At the heart of this book lies a paradox. The more we learn about the universe, the more complicated it appears to be, but we have discovered that beneath those complexities lie deep simplicities, laws of nature. How can simple laws explain complex behavior? Where does the complexity "come from"? How can the enormous diversity of life on earth have arisen from simple chemicals? How can a structure as complicated as the brain evolve? Over the centuries science has developed an extensive system of answers to such questions, a working philosophy known as reductionism, which shows how complexities on one level of description can be traced back to the interaction of large numbers of simple elements on a lower level. Much of the complexity of living creatures, for example, can be traced to the presence within them of the substance DNA. This truly gigantic molecule encodes a huge amount of information that tells an organism what to do when it develops.

The universe does not always seem complex. In our daily lives, we experience the world as a simple place—in fact, we would be unable to function if we had to grapple with the complexities as such. So in order to comprehend our world and humanity's place within it, we must do more than just explain higher-level complexities in terms of lower-level simplicities. We must also explain why, on every level of existence, we can deal with the world as if it were simple.

Where do the *simplicities* of nature come from? The conventional answer is that deep down inside, nature is simple: It functions on the basis of simple laws. Any large-scale simplicities that we observe—such as the spiral form of galaxies, or the tendency of a flock of geese to string out in a V—are just the underlying simplicities becoming visible on a higher level. Unfortunately, this answer is no longer convincing. Chaos theory tells us that simple laws can have very complicated—indeed, unpredictable—consequences. Simple causes can produce complex effects. Complexity theory tells us the opposite: Complex causes can produce simple effects. And conventional reductionist science tells us that inside the great simplicities of the universe we find not simplicity but overwhelming complexity. A galaxy's spiral arms contain myriads of stars, dotted almost at random. One of the deepest simplicities of biology is that the genetic material of almost all life-forms is constructed using the same giant molecule; but its workings are based on an intricate chemical code, whose unraveling for humans alone will require an effort comparable to the entire American space program.

The Collapse of Chaos shows how simplicity in nature is generated from chaos and complexity. These twin themes lie at the frontiers of modern thinking and are commonly confused with each other. A spate of recent books has emphasized one or the other as a general principle for understanding the natural world. We ask how the great simplicities of nature can persist within a chaotic universe. Our story combines chaos and complexity, and derives simplicity from their interaction. We show that the same simple large-scale features occur in many different complex systems because patterns of that kind do not depend upon detailed substructure.

The book is in two parts. The first half is about what science knows; the second half is about how to *think* about what science knows—and what it doesn't. The first half is a guided tour of the Islands of Truth that have been mapped by conventional science; the second half is an adventurous and unorthodox dive into the Oceans of Ignorance that surround them. In the first half we can provide plenty of road maps to tell you in advance just where we plan to take you. We can't do that in the second half—because where we wish to travel, there are not only no road maps, but no roads. However, we do put up plenty of warning signs: Here Be Dragons.

In outline, the first half of *The Collapse of Chaos* explains how simple laws of physics lead, through the chemistry of DNA molecules, to the complexity of living creatures, evolution, intelligence, and human culture. This is the conventional reductionist view. Along the way we give a streamlined introduction to the central preoccupations of modern science, which we hope you will find accessible and well integrated. These areas include cosmology, quantum mechanics, the arrow of time, biological development, evolution, consciousness, intelligence, and culture. We present science not as a fixed body of established knowledge but as a developing body of ideas.

The pivotal chapter 7 argues that asking where complexity comes from is really the wrong question. A more important question is, Why is there any *simplicity*? The book's first half provides an accepted scientific database from which the second half can address this more subtle issue by viewing science in context. We argue that simplicities of form, function, or behavior emerge from complexities on lower levels because of the action of external constraints. The focus moves from things to rules that govern the behavior of things. We offer a novel and indeed sometimes heretical approach to the questions discussed in the first half. For example, we argue against the view of DNA as a genetic blueprint for organisms by demonstrating that in principle two totally distinct organisms might possess identical DNA. Does this happen in nature? Ah, that would be telling. . . . In place of the view that evolution is the direct result of chemical changes in DNA, we argue that most changes to the form of an organism occur without any genetic changes at all—and that even when the form stays constant, the genes may be changing substantially.

The final chapter combines content and context into two new concepts: simplexity and complicity. Simplexity is the tendency of simple rules to emerge from underlying disorder and complexity, in systems whose large-scale structure is independent of the fine details of their substructure. Complicity is the tendency of interacting systems to coevolve in a manner that changes both, leading to a growth of complexity from simple beginnings—complexity that is unpredictable in detail, but whose general course is comprehensible and foreseeable.

We don't want you to get the impression that because the second half exposes gaps in the orthodox theories of the first half we think the first half is wrong. On the contrary, we think it's right, as far as it goes. But it doesn't go far enough. The reductionist story is nowhere near as complete as it appears to be. However, we want to do more than just locate gaps; we want to explore how to fill them. We want to alert you to another, much more mysterious, world. We want to explain something that hasn't yet been formalized, so in place of formal definitions and descriptions we will offer illustrations, images, metaphors, examples. . . . Like thousands before us, we are trying to come to grips with "emergent phenomena"—collective behavior of a system that somehow transcends its components. Because it transcends them, it can't be "in" the components—so where is it? Tricky. Our collaboration on this book is, to us at least, an example of such a phenomenon. What has emerged from our collective deliberations includes many things that neither of us "knew" independently: We could only have written the book together.

If either of us were writing this on his own, he would be much surer he was right but (paradoxically) much more cautious in presenting his ideas. Instead, our joint voice knows that it is probably wrong all over the place but puts its ideas forward with immense confidence. We hope that even when we're wrong, our ideas will take your mind to new places—places well worth a visit. We believe that when we're wrong, we're constructively wrong—wrong in a more informative way than the orthodox story is right.

A preface is the traditional place to express thanks to friends and colleagues who were persuaded or bribed to read and criticize early drafts. They include Janet Brandon, Teri Bristol, Baron Mendes da Costa, Richard Craven, Dawn Ann Drzal, Alan Garfinkel, Brian Goodwin, Steve Gould (to whom special thanks for his sterling role as jokes consultant), Helen Haste, Kate Lyons, Robert Mash, Anne McCaffrey, Jacquie McGlade, Tim Poston, Terry Pratchett, Irving Rapaport, Lena Sarah, Rabbi Pete Tobias, and David and Helen Wake. We are also grateful to the Victorian computer scientist Augusta Ada Lovelace and the alien inhabitants of the planet Zarathustra, with whom we held many useful conversations, some of which have been recorded in the book. Although those conversations were entirely our own invention, we learned a lot by listening to what our characters were telling us. In particular we are grateful to Neeplphut for pointing out a serious misinterpretation of the image of an egg as a computer with a start-up disk.

This is the sort of book that has to have footnotes; but footnotes look so horribly academic. So we've put them all at the back, with enough information to indicate what they're about but with nothing at the front to show there's a footnote at all. The notes are a bonus, intended to be read once you've finished the main book. But if you want a precise reference to something, or think we ought to have qualified our statements a little more carefully, or think that we've said something stupid—take a look at the notes.

One of the most useful words in the English language is "despite." *Despite* the success of conventional science, we think there should be more to the scientific endeavor than just the study of ever more refined internal bits and pieces. *Despite* not having read all the great philosophers, we're invading their space and reinventing their wheels. And *despite* its many flaws, we hope you'll find *The Collapse of Chaos* a stimulating contrast to rigid orthodoxy and emerge from it just a little different.

Jack Cohen and Ian Stewart
Blackwell and Coventry
July 1993

1 SIMPLICITY AND COMPLEXITY

A yeshiva boy—a young man studying in a rabbinical college—took instruction from three rabbis. A friend asked him his reactions.

"The first I found very difficult, disorganized, and poorly explained, but I understood what he was saying. The second was a lot clearer, and much more clever. I understood part of that."

"And the third? They say he is very good."

"Oh, he was brilliant! Such a magnificent, resonant voice—it flowed as if from the heart. I was transported to realms beyond my imagining! So articulate, so lucid—and I didn't understand a word."

Do we live in a simple universe, or a complicated one? Common sense—the way we think when we go about our daily lives—treats the world as a simple collection of familiar objects bearing known relationships to one another. When we open the fridge, pour milk into a saucer, and put it down for the cat to drink, we don't need to grapple with the thermodynamics of refrigeration, the molecular structure of the chemicals that go together to make milk, the interatomic forces that hold ceramic materials together, or the patterns of electrical firing of cells in the cat's brain. Our own brains have found ways to structure our world and make it comprehensible without going down these ever more complex paths. If you look out of the window you will see about forty kinds of things—flowers, trees, birds, fences, cars, people, clouds—maybe two hundred if you look hard and make distinctions between daffodils and daisies. You transact most of your daily business using a basic vocabulary of some five hundred words.

Closer investigation, however, reveals complexities that seem too intricate for the human mind to comprehend. For example, the underlying complexity of the object that in commonsense mode we encapsulate by the term "flower" is staggering. First, think of all the funny bits and pieces that you find when you look in more detail, like stamens and pollen grains. Those are there so that the plant can reproduce by making seeds. To do this it often needs pollen grains from another plant of a similar kind. Bees are a common source of such grains, which they brush off when they visit one flower, and deposit on others. For their pains, bees are rewarded with nectar, which they make into honey. Flowers need bees and bees need flowers. How did this remarkable commerce between plants and insects come about?

Here's a second complexity of plants: They know a trick that animals don't. They can take carbon dioxide from the air around them, water from the soil beneath, and use light to pull these two chemicals to bits and recombine them to form more complicated chemicals such as sugars—a process known as photosynthesis. The trick is possible only because plants contain a rather complicated chemical, chlorophyll, which animals generally lack. How does chlorophyll work?

Yet another complexity, the most astonishing of all: Plants *grow* from tiny seeds. Is all the complexity of a mature plant somehow compressed inside the seed? Or does complexity arise spontaneously as the seed develops? Neither answer seems very satisfactory.

In order to answer this kind of question, we're going to have to dig deeper. What do we mean by "simplicity" and "complexity"? Can simple causes produce complicated effects, or must all of the complexity that we observe in the effect somehow be cryptically present in the cause? Is complexity a conserved quantity, or can you create complexity from nothing? The relation between simplicity and complexity in nature is one of the deepest and broadest questions that faces modern science.

COMPLEXITIES HAVE TEETH

The simplicities of our commonsense existence are the placid surface of a teeming ocean of complexity. This complexity is not just something outside us but hidden from view, like the ocean depths or the earth's geological strata. It is also, in a very real sense, inside us, just as chlorophyll is inside the plant. We have a comparably complicated chemical trick: Hemoglobin in our red blood cells transports oxygen around our bodies. And everything that we know about the world comes to us by courtesy of a major consumer of that oxygen, the most complex structure that we have yet encountered: the human brain. As the joke goes: "If our brains were simple enough for us to understand them, we'd be so simple that we couldn't".

Faced with the incredible hidden complexities of the universe, it is not surprising that many people take refuge in the commonsense simplicities and prefer not to dig into what lies beneath them. "Ignorance is bliss"; "What the eye doesn't see the heart doesn't grieve over." Unfortunately, the complexities of the real world have sharp teeth. Braking a car may seem a simple process: You just push the brake pedal and the car stops. However, if your mental picture of "brake" is no better than this—a kind of slogan, "Brakes are for stopping"—then you may get into terrible trouble on an icy road. If you understand a little bit more about how brakes work—they slow down the wheels' spin, so that friction between the tires and road can make the car slow down too—then you won't make that particular mistake. We are surrounded by technological gadgetry whose surface simplicity is becoming increasingly deceptive, as anyone who tries to program a video recorder or set a digital watch will discover. If you open up a Victorian steam engine you can get a fair idea of how it works. If you open the back of a television set nothing inside makes much immediate sense; indeed, most of the box is empty. Nature is like this too: Look at living creatures under a microscope and they make even less sense than a television. "No user-serviceable parts inside."

Is this just a problem of adopting the wrong point of view? The ordinary city-dweller finds the jungle complex and incomprehensible but is entirely comfortable when surrounded by the ordered simplicities of New York, such as department stores, subways, taxicabs, drug-dealers, and muggers. The jungle-dweller is baffled by New York but is entirely at home with the snakes and the spiders in the nice, simple jungle. *Crocodile Dundee* rested entirely on this simple premise, but with the Australian outback in place of the jungle. An apparent complexity that may actually be simple is the fact that most holly bushes have spiky leaves only near the bottom. If you've never noticed this, take a look next time you pass one. It's very striking.

Now, isn't it clever of the trees to save on the cost of making spikes and concentrate their efforts at the bottom, where animals might eat the leaves? How incredibly complex nature is! Except, of course, that we don't know whether it *does* cost the plant anything to make spikes. It may be harder to make all the leaves the same than it is to make them different, just as it's harder to make a flat piece of ground (all at the same level) than a sloping one (different heights). So the holly bush may be doing things the easy way, and it's just we who think it's complicated.

Even accepting that point, the universe still looks like a pretty complicated place if we remove our commonsense blinkers and look beneath our comfortable, illusory simplicities. If we don't want to be caught napping when that complexity decides to bite us, we must come to terms with it. There are two main approaches. Recondite professions (such as astrology or plumbing) claim to handle these hidden complexities in their own terms, but shy away from any attempt to explain their methods. The astrologer who casts your horoscope and predicts the approach of a tall dark stranger, and the plumber who produces an odd-shaped wrench to unscrew a nut that you didn't even know your sink possessed, are both keeping a lot up their sleeves. Science adopts a radically different approach. It claims to see beyond the apparent complexities to the underlying simplicities, which it calls laws of nature. By working with these simple laws, rather than trying to handle the complexities *as* complexities, science claims to render the world once more accessible to common sense. It is common sense on a more refined level, common sense with different intuitions; but when a physicist argues that perpetual motion machines are impossible because of the law of conservation of energy, the general line of thought is just as simple and transparent as the statement that the cat needs some milk because it's thirsty.

THIGHBONE EQUALS SPACESHIP?

Common sense, in short, has a great deal going for it—provided that it resembles the actual world sufficiently well for the purpose in hand. Common sense works when it is congruent to reality. When it is not, it can go horribly wrong—like, for example, throwing water onto a gasoline fire. The gasoline, still burning, floats on the water, and the fire just spreads. The slogan "Water puts out fires" *sounds* like common sense, but it goes wrong more often than it works. The trouble is, our brains mostly think in such slogans. The word "comprehend" originally meant "grasp." To understand something is to grasp it with your mind, to make it into an *object* that you can hold as a unit. As the human race has evolved, it has developed this technique into a way of life.

By reducing complexities to underlying simplicities, science has allowed our brains to grasp the hitherto ungraspable. And once you have grasped something, you can use it, as a tool—provided that you know you have grasped it, so that you can play the same trick consciously again. You can then grasp at what we shall call a meta-level, a level of greater generality, by grasping the concept "tool."

We will use the prefix "meta" frequently, to mean "a more general version of," or sometimes "a higher-level way of thinking about." For instance, meta-physics (a word that actually exists in dictionaries, in the form "metaphysics") is a higher-level theory of how physics works; it's not a part of physics itself. "Meta-plumbing" isn't in the dictionary, but if it were, it would mean "a higher-level theory of plumbing." Meta-plumbing would concentrate on such questions as how to look as though you really understand what you're doing when the living room is several feet deep in water—instead of how to make a good joint, which is ordinary plumbing.

Instead of a few special brain slogans such as "Rocks are for throwing at rabbits" or "Fire is for burning things," you can gear up a level to the meta-slogan "Tools are things that can enhance human abilities," and start thinking about developing new tools out of what you have already grasped. This process of research and development leads to technology. Stanley Kubrick's now classic science-fiction film *2001: A Space Odyssey* (with screenplay by Arthur C. Clarke and Kubrick) begins with a sequence of events whose star turn is the apeman Moon-Watcher. His investigations culminate in the discovery that you can beat a leopard's brains out with a bone. In a paroxysm of joy, Moon-Watcher tosses the thighbone tumbling end over end, high into the air—

—and it turns into a spaceship.

The symbolism may be trite, but the message is deep. When protohumanity learned how to generalize about the structure of the natural world, to classify similar objects under identical labels—in short, to exploit the power of metaphor—it latched onto a wonderful trick for simplifying what would otherwise be complex beyond human understanding. You can't track a snake through the jungle if every leaf, every insect, every broken branch is seen as a unique individual. You have to distinguish things-that-sting from things-that-are-harmless, things-that-break from things-that-block-your-path, things-that-make-noises-when-stepped-on from things-that-don't. It's not easy to get the idea of making a hut from sticks and mud unless you have a good grasp of the stickiness of mud and the muddlability of sticks. And you have to achieve all of this with a considerable degree of rapidity: Label thy neighbor before it labeleth thee. The mental computations must be in real time, so something quick and dirty is the order of the day. We have to think in slogans, because a really high degree of congruence with reality takes too long. So a flash of black and orange is labeled "tiger," when it might be just a funny-colored leaf—because tigers can bite. It's better to be safe than sorry.

WILD ELECTRONS

Once humanity grasped this trick—of attaching labels, generalizing, seeing the common simplicities in nature instead of the baffling complexities—then discovering the laws of nature, science, and technology was really just a lengthy exercise in research and development. The trick seems to work best when the underlying simplicities of the world are perceived on a mechanistic level: Physics is about electrons or atomic lattices rather than daffodils or prides of lions. Indeed, nature seems to simplify as we move toward subhuman or superhuman scales. On the ultramicroscopic scale, matter reveals itself as assemblages of huge quantities of particles of an extraordinarily limited number of types, everything being governed by the laws of quantum mechanics. On a pan-galactic level, there are just vast swaths of matter moving through empty space under the force of gravity, sliding gracefully and inevitably along the curved world-lines of general relativity. Modern physics claims to be on the verge of a final synthesis of the two theories, micro and macro, into a Theory of Everything.

On our human level, there seems to be a lot more going on: shoes and ships and sealing-wax, cabbages and kings. But science books tell us that this rich mixture, the stuff of which we build our lives, is no more than a logical, inevitable consequence of a few elegantly simple laws, those of quantum mechanics and relativity, or the presumably even more elegant laws of their long-sought synthesis. This claim is true in many respects, but the simplicity of physics is to some extent an illusion. We will explore both sides of the question as our story unfolds, but for the moment we fasten upon the possibly illusory nature of physics' simplicities. The point is not that physics deals in unrealities, but that its considerable successes are to a great extent dependent upon its choice of subject-matter.

The universe appears to simplify at nonhuman scales because we possess a very limited set of techniques for converting its behavior into human-scale effects, in both space and time. In order to observe nature on spatial scales markedly different from our own we must either amplify things (microscopes) or bring them closer (telescopes). We similarly alter time scales with high-speed or time-lapse photography, or by compiling records of observations for a very long time. We probably miss a lot of the fun by peering through glasses darkly. It takes a galaxy a million years to perform one stately revolution; how can we possibly appreciate the universe on the time scale of a galaxy? At the other extreme, subatomic particles go about their business at such a frenetic pace that we need the most expensive machines in the world just to catch a glimpse of them.

Physics takes a pragmatic and severely critical stance. It concentrates on simple, highly controlled systems; in return it expects impeccable agreement between experiment and theory. Thus quantum mechanics predicts, with spectacular success, the behavior of a single electron in the potential well of a proton; this is the hydrogen atom, whose theoretical analysis features prominently in the pages of any respectable physics text. The behavior of electrons "in the wild" is quite another matter, and while physicists generally *believe* that this is governed by the same laws that they have established experimentally for isolated electrons in a hydrogen atom, this belief is founded on extrapolation from the simple systems accessible to experiment (and—but only recently—on supercomputer calculations that simulate tiny quantities of bulk matter). By its nature, it cannot be tested directly, "in the wild." Physicists would argue, quite properly, that in the absence of evidence that wild electrons behave differently from tame ones, the onus of proof is upon the skeptic. Physics deals with an invented, simplified world. This is how it derives its strength, this is why it works so well: Its raw material is of a type that can be placed in simple settings. Sciences like biology are less fortunate.

HOW DO WE DISCOVER LAWS?

We're going to spend a lot of time talking about "laws of nature," and it will help if we can develop some appropriate mental imagery. The most suitable one for our purposes is Isaac Newton's law of gravity, which states that the gravitational force between any two bodies is inversely proportional to the square of the distance between them. (It also states that the force is directly proportional to their masses, but we won't worry about that bit.) Newton discovered his law by thinking about possible causes for the elliptical shape of the planetary orbits, previously discovered by Johannes Kepler. Because Newton was an unusually good mathematician—he's generally ranked among the top three of all time—it didn't take him too much thought to prove that a point particle obeying inverse-square-law attraction will pursue an elliptical orbit. He wasn't entirely impressed by this argument, though, because planets aren't points. Eventually he discovered that, on the assumption of an inverse-square law of gravity, spherical bodies produce exactly the same forces that they would if all their mass was concentrated at the central point. He was a lot happier after that. But the real clincher—dramatized in the almost certainly fictional story of the apple—was that the same law, correctly interpreted, also governs the motion of falling objects on earth. Apples and the moon both fall because they are at-

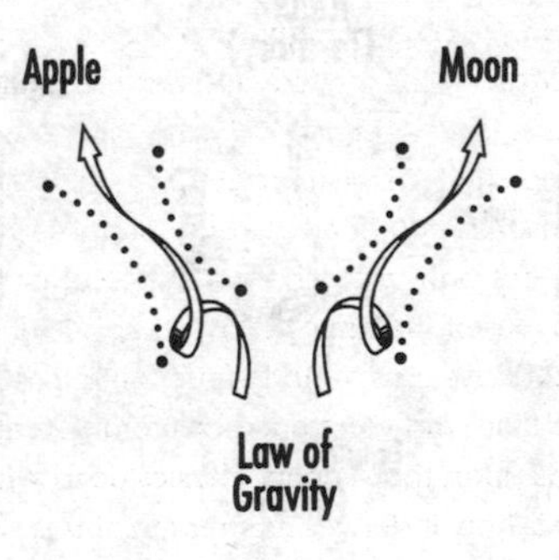

FIGURE 1
Anatomy of a law of nature

tracted to the center of the earth. We don't normally think of the moon falling, mind you. But really it's falling all the time; it moves sideways in its orbit so that the amount of falling is just enough to compensate for the earth's curvature. Although the moon perpetually falls, it remains much the same distance from the center of the earth. Newton invented a brilliant thought experiment: A cannon on top of a tower fires a projectile at ever higher speeds. As the projectile falls, it also moves forward. At sufficiently high speeds it travels all the way around the earth and hits the cannon. It's in orbit—and still falling.

On page 13 is a diagram of the kind of thought process that is involved in digging out the law of gravity (Fig. 1). At the top are two different real-world phenomena: the falling apple and the orbiting moon. Two funnels lead down to a deeper layer of explanation, where we find in each case the same principle: inverse-square-law attraction. Notice that in this image we look down the funnels, but the arrow of causality, which we assume must be present, runs upward. The direction of discovery is from top to bottom, peering into the funnel, but the direction of explanation is from bottom to top. In just the same way, a simple map explains complicated territory, but you explore the territory to draw the map. (See Fig. 2.)

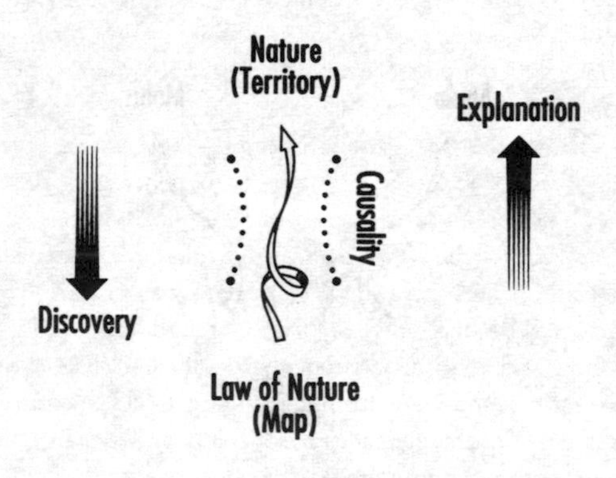

FIGURE 2

Discovery and explanation proceed in opposite directions.

Newton's discovery of the law of gravity required lots of funnels. At their tops were things like Mars, Jupiter's satellites, and the wobbling of the earth and moon on their axes. At the bottom of *every* funnel he found the same thing: inverse-square attraction. So he named his find the inverse-square *law* of *universal* gravitational attraction—"law" because he thought that it was always rigidly enforced, and "universal" because he was convinced that it applied to all pairs of bodies throughout the entire universe.

Newton discovered other laws of nature using similar methods. Starting from Galileo Galilei's observations on the motion of bodies, he dug out a number of "laws of motion," of which the most important is the statement that the acceleration of a body is proportional to the force acting on it. Especially in combination, Newton's laws were a spectacular success: They explained the "System of the World" to a remarkable degree of accuracy.

THEORIES DESTROY FACTS

As we've seen, you have to look hard and deeply at the world before you realize that it *is* complex. You have to look even harder to see that much of its complexity has simple causes. From Newton's time on, every scientist's dream was to discover a new law of nature—a new underlying simplicity that could be used to explain innumerable disconnected observational facts. But "law" is a word with curious and rather inappropriate overtones. Laws are invented by human societies to control individual or collective behavior; they proscribe certain activities and prescribe punishments for disobedience. But you can't break a natural law. Nevertheless, nature "obeys" simple laws. To the intense irritation of generations of schoolchildren, these laws seem to be mathematical. We human beings observe underlying simplicities in the way the universe works; we distinguish them by their mathematical elegance, and we call them laws of nature. This mathematical approach to the universe *works*; it underpins virtually all of science. "God," said Plato, "is a geometer." Paul Dirac called God a mathematician; James Jeans went further, defining God's specialty as pure mathematics. The best mathematics has a purity and elegance that somehow seem to capture the true essence of phenomena.

"Theories destroy facts," said Peter Medawar at a Mensa conference in the 1960s. He meant that a general theory can remove the need to record huge quantities of isolated facts. One way to specify where the planets in the solar system are (or will be) on different dates is to draw up huge tables of numbers. However, once you've got Newton's laws, all you need is a much shorter list saying where they are, and how fast they are moving, at *one instant of time*. All else is a matter of routine calculation. Some thirty years ago the mathematician René Thom deplored biology as "a cemetery of facts." Many biologists protested vehemently, thinking that Thom was deploring the facts. He wasn't. He was deploring the cemetery. He wanted a theory to organize all of the facts, and bring the cemetery to life.

In Newton's day, scientists (they called themselves natural philosophers then) held a rather strong view of laws of nature: They thought they were *true*. However, there has since been what Thomas Kuhn calls a paradigm shift—a change in worldview. Newton's law of gravitation has been replaced by something that predicts the observed motion more accurately: Albert Einstein's theory of general relativity. Einstein's theory, which views matter as the curvature of space-time, and gravity as the distortions in motion that result from this curvature, explains a number of oddities that do not fit the Newtonian picture. These include a slow drift ("precession") of the orbit of Mercury, and the bending of rays of starlight by the sun's gravitational field. Nowadays we can observe the light from a distant quasar, bent by an intervening galaxy that acts as a "gravitational lens." The result is that we see several images of the same quasar, a kind of cosmic mirage.

In a different paradigm shift, Newton's laws of motion were replaced by quantum mechanics, in which mass and energy come in tiny, indivisible

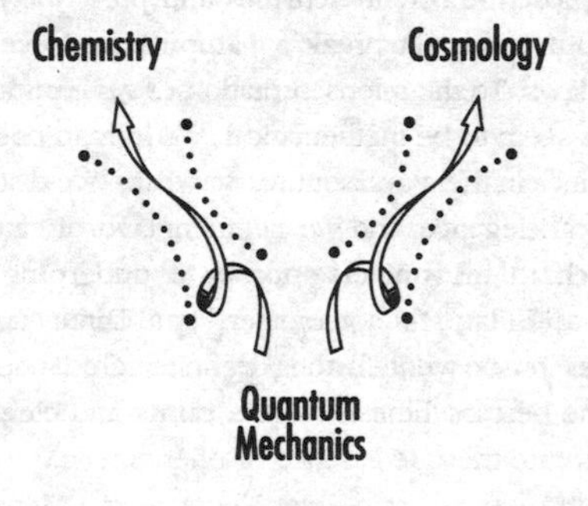

FIGURE 3
Funnel diagram for quantum mechanics

packages, and everything is as much a wave as a particle. Relativity and quantum mechanics are both systems of laws, in the same way that Newton's laws were. In particular, if you look down a funnel from either chemistry or cosmology, you see quantum mechanics at the bottom (Fig. 3).

Science's great generalizations are produced in just this manner—by finding similar laws, similar simplicities, inside different complexities. Systems that have the same deep similarities must obey the same simple rules.

As we'll describe at greater length in the next chapter, quantum mechanics is implicated in both the atomic structure of chemical elements and the Big Bang that gave rise to them. Cosmologists in particular are very impressed by this closing of the logical circle from the microscopic world to the macroscopic world. Like the scientists of Newton's time, many of them think that this proves that the law of nature expressed in the Big Bang is *true*. We don't even learn from history that we don't learn anything from history.

Even though relativity and quantum mechanics are better laws than Newton's—they fit more aspects of the real world more accurately—scientists still find Newton's laws extremely useful. The mathematics involved in Newton's laws is simpler; the mental picture of masses and forces is on a more human scale. Newton's laws break down badly only at the two extremes of the microscopically small or the astronomically large: When bodies move very fast, or have enormous densities, or are very small. For most human-scale phenomena, Newton's laws survive, essentially for reasons of convenience.

Biology too has "laws," but these are less frequently expressed in mathematical language. An example is chlorophyll. Millions of different plant species build carbohydrates from sunlight by the process known as photosynthesis, and down the funnel from each species lives the same chemical: chlorophyll. You don't need a different method for each different species of plant. The word "law" is not so appropriate here: Chlorophyll is an underlying universal. But the role of a generality is exactly the same as that of a law, and for our purposes we'll consider statements such as "Chlorophyll is responsible for photosynthesis" to be just as much a law of nature as Newton's law of gravity is.

IS THE UNIVERSE A COMPUTER?

Either because the laws of nature are couched in mathematical symbolism, or because science cannot progress safely in the presence of ambiguity and imprecision, scientists tend to express natural laws as mathematical statements. It would be wrong, however, to read too much into this. Consider a stone tumbling down a hillside, bouncing off rocks and molehills, until it reaches its final resting place at the bottom of the slope. If the stone really is implementing mathematical laws, then in a few seconds it will have performed a series of calculations beyond the capabilities of the fastest supercomputer. But is that really what the rock is doing? Measuring its own position to the hundreds of decimal places that we know are needed to guarantee the "correct answer"? Computing its way from collision to collision in an orgy of dynamical equations? Some physicists and philosophers think so; in their view, information, rather than matter, is the basic material of the universe. The universe itself then becomes a supercomputer of unprecedented speed and power, busily pursuing the consequences of its "program," its program being the laws of nature.

Alternatively, the simple laws that we consider fundamental may not be fundamental at all, but just approximations of how nature behaves, or consequences of that behavior. We now know that Newton's laws are not rigid rules that nature just obeys; they are excellent but sometimes inaccurate descriptions of what nature does. They are not nature's laws but human laws, and like all human laws they can be broken. Indeed, according to another human law, Murphy's, they always will be—an interesting case of self-reference. If nature breaks our laws, then our calculations will bear no relation to the way in which nature actually works. *We* may use the laws of dynamics to calculate where the stone will fall; but that's not how the stone does it. It certainly can't if we've got the *wrong laws*. The Newtonian stone has no choice; it is forced to fall wherever it does. In this view the universe is a machine rather than a computer; it is composed of matter, and it is *in the nature of matter* to behave in ways that happen, coincidentally, to mimic certain computations that appeal to humans.

That was a classical picture of a moving stone. The quantum picture is more subtle, and far stranger to human intuition. In a quantum view, the subatomic particles that make up the stone actually follow *all possible paths,* consistent with the laws of quantum mechanics. According to the quantum paradigm, what we see is the superposition of all of those potentialities. It just happens that the result of this strange process *looks* like a lump of rock moving under Newtonian laws. In this picture the Newtonian laws are viewed as mathematical consequences of the *real* quantum laws, valid for modest but bulky quantities of matter moving at moderate speeds.

In other areas of science, especially those where really accurate measurements or repeatable experiments aren't possible, people nowadays tend to speak of "models" rather than "laws." They look for underlying rules and regularities that explain a limited range of phenomena in simple, graspable terms. From that point of view, "laws" may be just spectacularly successful, very simple, models. The important thing is that, even though we can't be certain that what we think of as laws of nature are actually true, we do see a lot of patterns and regularities in the world, and we can use these patterns very effectively to bring certain aspects of the world under our control. For instance, the laws of aerodynamics work sufficiently well that airplanes designed using those laws stay up. The vast bulk of evidence, while not quite so conclusive, points to the flight of birds as a consequence of those same laws. However, we can't yet start with aerodynamics and end with a proof that a bird, too, will stay up; but despite such admitted uncertainties, there still seem to be simple laws at work. It's just that some operate further behind the scenes than others. Indeed, the further behind the scenes the laws are, the more we tend to think of them as being "fundamental."

CONSERVATION OF COMPLEXITY

Many people seem to have an intuition that can be most conveniently described as "conservation of complexity." They expect complicated effects to have complicated causes (and simple things to have simple causes). The eye is a very complicated organ; therefore, a simple explanation like Darwin's theory of evolution must be wrong. Economic inflation is a simple effect; therefore, there must be a simple cause, therefore an easy way to deal with it. Human consciousness is enormously complicated; therefore it cannot result from laws of nature: it must have some supernatural element. Early theories of biological development contended that inside every human sperm there is a homunculus, a tiny person perfect in every exquisite detail. The argument was that you can't make a person unless you've already got one, albeit in cryptic form.

We may tentatively define the complexity of a system as the quantity of information needed to describe it. One small whole number conveys very little information, and can describe only very simple things such as the position of a dot on a computer screen; but a sufficiently long list of numbers can describe far more complicated things. (A videotape is, at root, just such a list, and you can put a description of *anything* onto a long enough strip of videotape.) If you think that complexity—in this "bit-counting" sense—is conserved between cause and effect, then there is only one way in which simple laws can produce complex effects. Any complexity in nature, such as the intricate collisions of molecules in a gas, must arise because huge numbers of objects are interacting. When simple laws govern systems with a large number of variables, the underlying order is obscured by our inability to track every component, and it becomes inaccessible to our limited brainpower.

Within the last decade this view of the origin of complexity has been strongly challenged from a variety of directions. At the frontiers of today's mathematics are startling paradoxes about the way the world can change. In particular, we now know that rigid, predetermined, *simple* laws can lead to behavior so irregular that it is to all intents and purposes random. Simple rules can produce incredibly complex effects. An example is the infamous Mandelbrot set, which we describe in chapter 6; it is one of the most intricate geometric objects ever to have decorated a teenager's wall (Fig. 4). However, the computer program that generates it is only a few instructions long. This phenomenon—of vastly complex effects arising from simple causes—is known as chaos, and there is plenty of evidence that it is widespread. Stuart Kaufmann of the Santa Fe Institute has pioneered the study of a converse process, "antichaos," in which complex causes produce simple effects; this is also widespread. Complexity can get lost as well as being created.

We can still save the conservation principle, but only if we refine our concepts of simplicity and complexity to include the processes that generate phenomena as well as the phenomena themselves. Suppose we say that a system is simple if it can be *pre*scribed by simple rules, rather than *de*scribed by simple lists of numbers. This transfers the problem of complexity from results to processes, from effects to causes. It turns out that

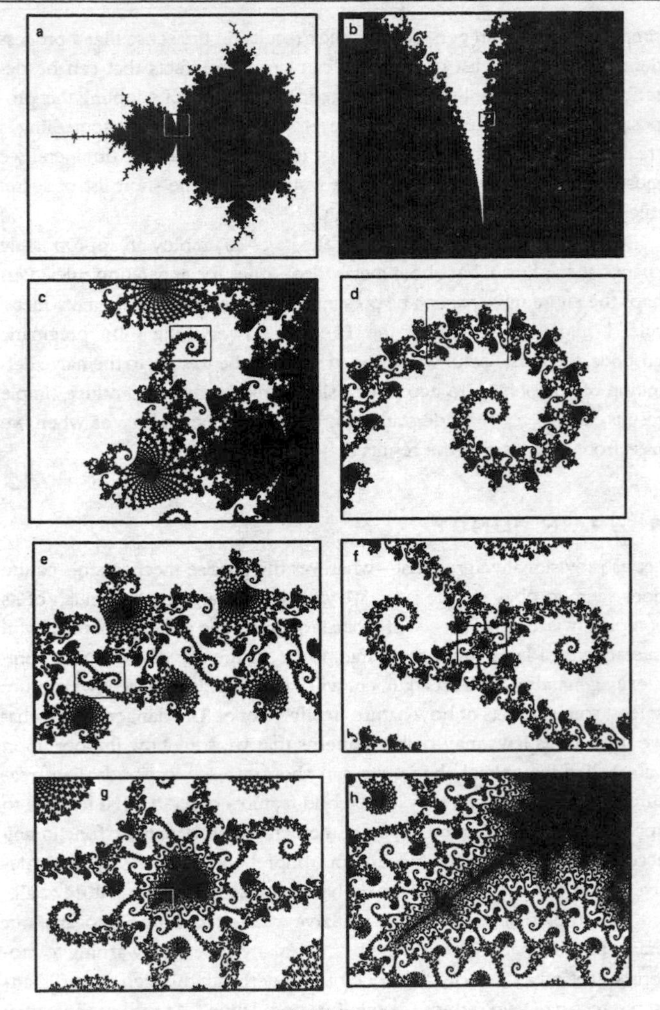

FIGURE 4

The Mandelbrot set (a). If any region of the set is magnified, new and intricate detail appears, as in the sequence of blow-ups, b–h. (Peitgen and Saupe, The Science of Fractal Images*)* simple processes can generate complex results, in the sense that a process defined by a small list of numbers can produce effects that can be described only by huge lists. One interesting side effect of adopting this proposal is that it then becomes very hard to tell whether or not something is "really" simple. We can no longer just look at it and count numbers; we must ask ourselves whether it might result from some short list of secret rules. Which rules? The sky's the limit.

Now Pandora's cat is really out of the bag—to employ an appropriately chaotic metaphor. What about meta-rules—rules for generating rules? Perhaps the entire universe can be prescribed by one simple meta-meta-meta-rule! (This is the dream of the Theory of Everything.) On pragmatic grounds, the most useful way forward seems to be to stick to the naïve definition of complexity, to accept that simple rules may not produce simple results, and to try to understand just how complexity behaves when we pass from a process to the results that it generates.

■ BRAIN PUNS?

Let us provisionally agree that—whatever the precise mechanism—nature does seem to obey simple laws. We've just been arguing that much of its complexity is a consequence of those laws, resulting either from the lawful interactions of large numbers of interactors, or the use of chaos as a complexity generator. Such an argument will collapse unless those laws capture at least some aspects of how nature "really" works. The danger is that what we think of as laws may be just patterns that we somehow impose upon nature, like the animal shapes we can choose to see in clouds. Our treasured fundamental laws may just be odd features of nature that happen to appeal to the human mind. If so, then much of nature may be functioning according to processes that we cannot comprehend, and consequences derived from our imaginary laws may bear no resemblance to nature at all.

The problem is that, because we have evolved to think in slogans, our brains play strange tricks. On the back of many buses is a warning to motorists: "Let buses pull out." On one bus, underneath this sign, a graffiti artist wrote: "and help reduce the minibus population." As well as the verbal pun there is a brain pun: the idea that minibuses are how buses breed. Our mental wires are getting crossed somewhere around the slogan "Babies are small adults."

We perpetually do this kind of thing, and we laugh when we notice what we're doing; but because our brains are attracted to puns, we have to be careful not to read too much into what might be an accidental similarity. Are what we imagine to be laws of Nature just brain puns, partial congruences between what our neurology can imagine and what really happens? For example, we may think we see an analogy between body hair on a dog and forests in Amazonia, but it does not follow that we can usefully compare Brazil to a poodle. Some Freudians see everything as sex, as phallic or pudic symbolism; but the only things that are neither convex nor concave are flat. This is not to deny many Freudian metaphors, but it does suggest they can be overstated. As Freud himself said, "sometimes a cigar is only a cigar."

Moreover, it is undeniable that the patterns we can make explicit are limited by the material available to our imaginations. In 1963 Benoit Mandelbrot introduced the new concept of a "fractal," a geometric form with fine structure on all scales of magnification. That concept has since become a remarkably pervasive influence in scientific thought. Before 1963, one of the simplicities that pervades the 1990s picture of the world was—missing. Just as the concept "sphere" unites raindrops, planets, and suns, so we now perceive a unity between such diverse objects as trees, clouds, and coastlines. They are irregular, but with the same kind of irregularities. Before the simplicity "fractal" was introduced, it was not only impossible to express this unity, it was pretty much impossible to notice it.

This same example, however, does make us ask whether our mental patterns are genuine reflections of reality. No matter that trees, clouds, and coastlines existed before Mandelbrot captured their geometry in felicitous phraseology; so, too, did planets follow elliptical orbits before Kepler brought this aspect of their motion to human attention. Nature can use a pattern before humans have learned to perceive it. Does a tree "know" it is a fractal? In at least some respects it seems to. The form of its branches is a consequence of the genetics and chemistry of the plant. The evidence before our eyes is that this consequence can be expressed as fractal structure. Mathematical gadgets called L-systems successfully mimic a huge variety of plants and flowers. It seems unlikely that this is total coincidence, mere geometric pun; more likely, the combinatorial rules of L-systems capture, albeit in idealized form, some deep mathematical consequence of the workings of plant genetics and chemistry. We're happy that Kepler's elliptical orbits capture a high-level mathematical consequence of the workings of gravity; why can't something similar govern the shape of trees? They must derive their characteristic shapes from *some* general principles.

Assuming, of course, that there really is such a thing as a tree; that is, that there is a general, global concept that matches how nature "sees" trees. Human beings don't understand trees as well as they think they do. We'll prove it to you. There are at least two radically different kinds of tree, trees whose lifestyles are as different as those of a rock star and a monk. No, we don't mean the standard textbook deciduous/coniferous distinction, but something far more dramatic. It is this: some trees have mycorrhiza, some don't. Mycorrhiza are fungi that are symbiotic with the tree and convey substances to its roots. Trees with mycorrhiza and trees without them look equally fractal, but belowground they differ, not so much in shape as in function. The presence or absence of mycorrhiza makes a huge difference to how the tree works. For most of its lifetime humanity has got most trees completely wrong, so obsessed by what it sees aboveground that it never thought to look carefully below. And this despite a lot of signs that now seem "obvious," such as the rings of toadstools that appear around the stump after a tree has been felled. Oh, so *that's* why . . . Quite.

We want to raise a question now, but not answer it until chapter 11. There's a danger of circular logic: The sole source of the alleged patterns "fractal" or "mycorrhiza" might be our minds. They may be just brain puns waiting to inflict themselves upon the universe. Whenever something in the world fits that pattern, we label it as treelike. Is it, then, such a surprise that treelike things (such as those plants that we have labeled trees) resemble fractals? And even if fractal structure is "really" there in trees (because of genetics and chemistry), in coastlines (ditto geology), and in clouds (ditto meteorology), does a tree "know" it is a fractal in the same sense that a cloud "knows" it? Did it "know" it was a fractal before Mandelbrot invented the concept? The same goes for mycorrhiza: Is everything that looks like a colony of fungi really like every other such colony? Does a tree "know" it has mycorrhiza? Are fractals and mycorrhiza not puns but *meta*-puns, valid individually but meaningless as a generalization? And even if *we* find them convenient descriptors of reality—which increasingly seems to be the case—do they have the same intimate connections with it as, say, Newton's laws do with the orbital motion of Mars?

GAMES MATHEMATICIANS PLAY

Brain puns, if they are no more than that, are just crossed lines in the brain's telephone network—incongruent with reality, and useless for understanding it. But some things that look like brain puns can be useful. These are metaphors, and they are useful if their congruence with reality goes deeper than mere surface form. Forms are the result of processes, and congruences of processes are metaphors with genuinely useful content. Because we cannot experience the universe objectively, we sometimes see patterns that do not exist. About two thousand years ago, one of the strongest pieces of evidence for the existence of a geometer god was the Ptolemaic theory of epicycles. The motion of every planet in the solar system was held to be built up from an intricate system of revolving circles (Fig. 5). How much more mathematical can you get? But appearances are deceptive, and today this system appears neither mathematical nor sensible. The process that it describes is not congruent to reality. Indeed, the epicycle process can be adjusted to model any kind of orbit—even a square one, if you let the circle rotate at varying speeds. What explains too much, explains nothing. The circles existed only within Ptolemy's head.

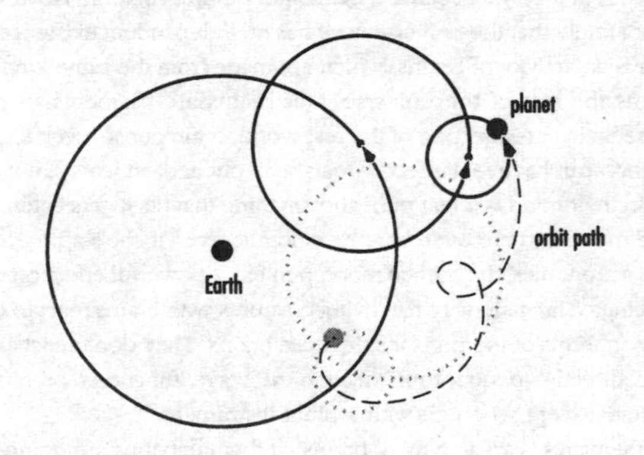

FIGURE 5

Ptolemaic epicycles. A planet moves in a circle whose center moves in another circle whose center . . . etc.

To err is human. Are those things that we so hopefully call laws of nature brain puns, metaphors, or "real" laws? Reality has a habit of throwing up phenomena that don't fit our metaphors. At one time, if you put a silver threepenny piece into a London Underground sixpenny ticket machine, it spewed forth its entire roll of tickets. That wasn't the metaphor the engineers who designed it had had in mind. But it was they who'd made a mistake: the ticket machine was obeying a deeper metaphor: nature's.

So are the patterns that we profess to detect in nature brain puns or genuine laws? The verdict is not yet in, but they *could* be puns. In recent years a fecund mathematics has generated innumerable "new" mental images, such as catastrophes, chaos, fractals, that might be advance warning of new simplicities in the world. Each extends the list of patterns that we can name, recognize, and manipulate. It is not clear that all such patterns must necessarily prove operationally congruent to reality. They may describe games that mathematicians play, but that have nothing to do with the world outside human brains.

Up till now we've talked about what happens in brains as if it is entirely divorced from the physical world. Brains are intricate organs that can build their own models of reality, tiny mental universes. The universe that we experience is in a very real sense a figment of our imagination. However, this does not imply that the universe itself has no independent existence. Imagination is an activity of brains, which are made from the same kind of materials as the rest of the universe. Our brains are "figments of reality." Because brains are also part of the real world, brain puns—even if they *are* just puns—can have real effects. Goats seek out and eat leaves, not chlorophyll. In their mind is a leaf pun, and anything that fits it gets eaten. Lovers see the moon and are overcome by emotion, even if she's a physicist and he's an astronomer; the mental moon pun has a powerful effect on real human beings. The main way that living creatures with brains react to the universe is in terms of the puns inside those brains. They don't react anything like as directly to quantum mechanical wave functions or molecular structure—except when hit with a blunt instrument.

Mathematics is an activity of brains, and while brains are doing mathematics then in some sense they must function according to mathematical laws. And if brains really can do that, why not goats, leaves, and chlorophyll molecules too? Our minds may indeed be just swirls of electrons in nerve cells; but those cells are part of the universe, and they evolved within it. The swirls of electrons have survived millions of years of natural selection for congruence with reality. What better way to build simplified models of the world than to exploit simplicities that are actually there? Brain puns that get too far removed from reality are not useful for survival.

Intellectual constructs, like epicycles or laws of motion, may either be deep truths about nature, or clever delusions. The task of science is to provide a selection process for ideas that is just as stringent as that employed by evolution to weed out the unfit. Mathematics is one of its chief tools, because mathematics mimics the pictures in our heads that let us simplify the universe. But unlike those pictures, mathematical models can be transferred from one brain to another. Mathematics has thus become a crucial point of contact between different human minds; with its aid, science has come down in favor of Newton and against Ptolemy. Even though Newton's laws may be delusions, they are much better delusions than Ptolemy's. Perhaps we have created a geometer god in our own image; but if so we have done it by using the basic simplicities that nature supplied when our brains were evolving. Only a mathematical universe can develop brains that do mathematics. Only a geometer god can create a mind that can delude itself that a geometer god exists.

■ THE *GOON SHOW* OF SCIENCE

That's a rather subtle point: We can be more positive. There is abundant evidence that the simple mathematical laws discovered by physicists really do exist; they are real, not just artifacts of human imagination. We've probably got them wrong, but we've come fairly close. The evidence is the success of technology. As we hinted earlier, jumbo jets would fly badly on imagination alone. The simplicities of physical laws must in some sense be genuine simplicities, buried in what at first sight appears a complex world, because they work; they cannot be just oversimplifications made by an inadequate intelligence. If only for this reason, physics has been spectacularly successful, either in explaining things, or at least in persuading us that it has explained them. The patterns exploited by physics may be imperfect, mere approximations; but they are good approximations, and that cannot be just coincidence. If we *just* filter the universe through mental templates and discard anything that doesn't fit our prejudices, then the bits we discard will sneak up on us in nasty ways.

Biologists are not quite as fortunate as physicists, but at least a biologist can give us a glimpse through a microscope and show us creatures and structures beyond our wildest imaginings. The generalities observed by biologists, such as the ubiquitous role of DNA in reproduction, are also real. Economists, too, see patterns—microeconomic ones, such as cash flow; macroeconomic ones, such as inflation, unemployment, the balance of trade, the Dow Jones index. They demonstrate their mastery of these patterns by explaining in enormous and erudite detail precisely why we are in the middle of a slump. Control is not as straightforward as description: They often omit to mention that the slump was intended to be a boom.

Mathematicians are less fortunate still. Their simplicities are by definition imaginary—in the sense "of the imagination" rather than "chimerical." This is why it is impossible to televise mathematics: Mention a seven-dimensional hypersphere and the producer will want to show the audience one. Mathematics, like the revered *Goon Show*, works far better on radio, with an audience that is by habit imaginative and reflective. In many respects mathematics is the *Goon Show* of science: Compare the surreal quality of a typical piece of Goonery, "Shut the door and bring it in after you," with the mathematician's "A Klein bottle is a surface with only one side." To those who wish to use mathematical concepts to determine the System of the World, the proof that math works is the congruence between the patterns of mathematical imagination and the regularities of the universe. This can't be just coincidence, a result of mental filtering that selects only successes and conveniently forgets failures. The mathematical picture of nature works too well and too often. God may not be a mathematician, but he can do sums with the best of them.

■ ONE FINAL QUESTION . . .

The history of science, broadly speaking, is the tale of a lengthy battle to dig out the secret simplicities of a complicated world. It is an astonishing story of insignificant humanity's triumph over huge mysteries. Those unusual individuals who turn up major new simplicities are revered for generations, if not millennia. Mathematicians still speak in awe of Archimedes, who dates from the third century B.C. Cosmologists no longer accept the epicyclic theories of Ptolemy—the motion of the planets as wheels within wheels within wheels—but they remember his name.

The central aim of science is to render the complexities of the universe transparent, so that we can see through them to the simplicities beneath. We've dug down through layer upon layer of ever more simple and ever more general rules; we've explained how simple laws can give rise to complexity. But at the bottom of the funnels that lead downward from the complexity of our daily lives, we find two apparently contradictory things: order and chaos (Fig. 6). And chaos, whether it has cryptic order or not, looks too disorganized to explain the complexities of life, the universe, and everything.

This raises a question that will occupy our attention throughout the whole of this book. How can simple laws, even simple laws with built-in chaos, explain *organized* complexity?

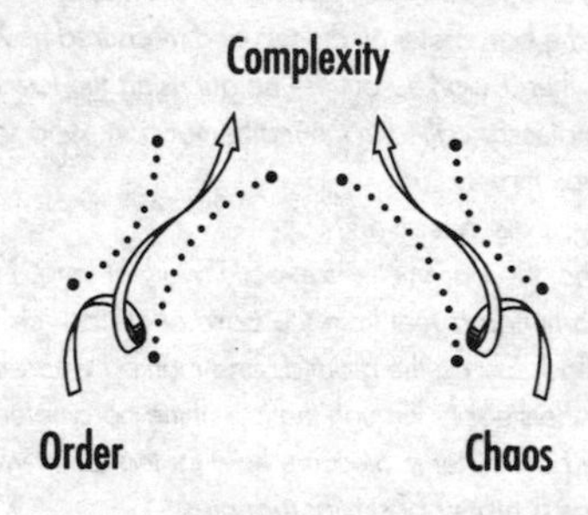

FIGURE 6

Complexity's at the top, but what's at the bottom of the funnel?

2 THE LAWS OF NATURE

An engineering student, a mathematics student, and a business student were on a field trip. Their task was to find the height of a church steeple using a barometer.

When they returned, they reported to their professor. The engineer said, "It's about 150 feet high."

"How did you discover that?" asked the professor.

"I dropped the barometer off the top and measured how long it took to hit the ground; then I worked out the height using the laws of motion."

Next, the professor called in the mathematician, who said, "It's 151½ feet high, approximately."

"How did you discover that?"

"I noticed that the barometer is exactly two feet long. I measured off a distance of two hundred feet from the base of the church tower, held the barometer on end, lay on the ground, and marked where the line of sight from the top of the steeple through the top of the barometer hit the ground. Then I used the barometer to measure how far that point was from its foot. After that, it was a matter of similar triangles."

Finally the professor called in the business student, who said "It's 154 feet 7 inches."

"How did you discover that?"

"I was walking round the churchyard, totally baffled, when I noticed the verger tidying leaves. I approached him and said 'Verger, I have here a fine new barometer. If you you tell me how high the steeple is, I will give you this barometer.' And he said it was 154 feet 7 inches."

The original stimulus for the scientific approach to the universe was the attempt to understand things that manifest themselves on a human scale—lights in the sky, and later, trees, tigers, and tornadoes. Today the search for the simplicities that underlie the apparent complexity of the world goes below the atomic level and out to the farthest reaches of the universe. The strategy is to reduce complexities on any given level to simplicities a level lower down. Thus the earth's ecology results from the interaction of organisms; organisms have evolved from much simpler ones; the development of a given organism is specified by the genetic code in its DNA; DNA chemistry results from simple laws about atoms; atoms may be understood as collections of subatomic particles; subatomic particles are explained by quantum mechanics. Space, time, and matter came into being in the Big Bang, and are thought to be governed by a unified Theory of Everything.

That's the general route whereby scientists worked their way from quadrupeds to quarks and quasars. For the moment we shall concentrate on chemistry and physics, the deeper levels of this vision of the universe. Let's just think it through. What's so complicated about chemicals? Basically, there are an awful lot of different ones. We use them for almost everything. A lot of our clothing is woven from man-made fibers—nylon, for example. We build buses from metal, alloys, plastics—all chemicals. Beer is several chemicals mixed up, mostly water and alcohol. Food, like beer, is made of chemicals. Just about everything is made of chemicals: planets, plants, people. One of the chemicals in plants is chlorophyll. One of the chemicals in humans in hemoglobin. Without any of a huge number of chemicals, we can't survive. At the bottom of the funnel leading downward from life you find chemistry.

■ CHEMISTRY IS LEGO

What makes chemistry important to the human race is that substances can change. If you rub two sticks together and set them alight, you get flames, heat, and a heap of ash. If you hold a leg of lamb near the resulting fire, it cooks, becoming tastier and easier to eat—and also less disease-ridden. If you mash up vegetable matter with water and leave it in a bowl for a few weeks, you get a brew that in moderation makes people more sociable and in excess makes them less so. If you mix the right kind of rock with wood or charcoal and heat it to a sufficiently high temperature you get metal for making swords or fishhooks. If you make extracts from suitable plants you can dye cloth to get interesting colors. Chemistry leads to cooking, warfare, and fashion.

The ancient Greeks contended that all matter is composed of four elements: earth, air, fire, and water. For example, when you heat wood, you see flames coming off, and smoke, and what's left is crumbly gray ash; this suggests that wood is made from a bit of fire, a bit of air, and quite a lot of earth. Some of the Greeks also suggested—on the basis of philosophical principles rather than experiments—that all matter is composed of tiny indivisible particles, known as atoms. The theory of elements is so malleable that it can be made to fit almost any observed phenomenon; nonetheless, it contained an important idea: that the substances that we find in the world around us are all made from a small number of simpler ones. The theory of atoms was until very recently out of experimental reach and—again until recently—largely unhelpful even if it turned out to be true. One of its modern consequences, though, is an effective theory of elements.

Chemistry went through a lengthy period of empirical development—unstructured messing around. It began as a less respectable endeavor, alchemy, which tackled big problems such as finding a "philosopher's stone" that would turn lead into gold, but didn't solve them. What it did do was get people interested in the curious changes that occur when different substances are mixed together and heated, or otherwise placed under some form of stress. It also inspired the development of a range of laboratory equipment, such as retorts and crucibles, to hold the substances being stressed, and a range of important techniques such as crystallization, filtration, and distillation. By the time of Chaucer, alchemy had become a fertile ground for frauds and tricksters, whose faked demonstrations of the creation of gold were intended to extract money from wealthy patrons, in order to carry out further experiments. (Today this technique is called writing a grant proposal.)

The Greek picture was wrong in detail, but it contained the germ of a good idea. Most substances can be broken down into simpler ones, but the simplest substances, the elements, cannot be. Why not take this as the defining property for elements, and see what you get? By this definition, elements are both the simplest substances that exist, and the basic building blocks for all the others. They are the "prime numbers" of chemistry. However, it is now a matter of experiment, rather than of theory, to find out which substances are elements. It turns out that there are more than four; the correct number is roughly a hundred. Among them are hydrogen, oxygen, sulfur, carbon, iron, and chlorine. Salt is not an element: it is composed of sodium and chlorine. Sugar is made from carbon, hydrogen, and oxygen. Sulfuric acid can be broken down into sulfur, hydrogen, and oxygen. Hemoglobin contains, among other things, iron. Chemically pure substances that are composed of distinct elements are called compounds.

None of the classic Greek elements is actually an element. Water is a compound; air and earth are mixtures. Fire gave rise to one of the most famous blunders in science. Around the mid-1700s, it was believed that when a chemical burns, it gives off a substance known as phlogiston. This theory explained such diverse observations as the flames emitted by burning wood (signs of the emerging phlogiston) and the powdery nature of ash (a solid crumbles when it loses its phlogiston). The idea of phlogiston explained how smelting produces pure metal. Take a quantity of dephlogisticated lead (ore) and heat it with wood, which contains large quantities of phlogiston. The ore absorbs phlogiston from the wood, becoming pure metal; meanwhile the wood turns to ash as it loses its phlogiston. In retrospect, we can see where the phlogistonists went wrong. Combustion is a gain of oxygen rather than a loss of phlogiston; effectively phlogiston is "negative oxygen." Any chemical reaction that involves the transfer of oxygen can equally be viewed as a transfer of phlogiston in the opposite direction. The two theories are, in a sense, logically equivalent. However, once chemists began to make their subject quantitative, by weighing the chemicals that were reacting, it became clear that phlogiston must have negative mass. "Dephlogisticated lead"—now known as lead oxide—weighs more than the lead produced by the smelting process. The phlogiston theory was abandoned.

The classification of substances into elements and compounds was the first big success of the philosophy of reductionism, a technique still firmly at the forefront of scientific methodology. In a reductionist approach, complicated systems are analyzed into simpler constituents, linked together by relatively simple rules—laws of nature. The number of chemical elements is relatively small; but despite this, the complexity of compounds is perfectly natural, because even a small number of building blocks can be combined in an incredibly large number of ways. One Lego set can make an

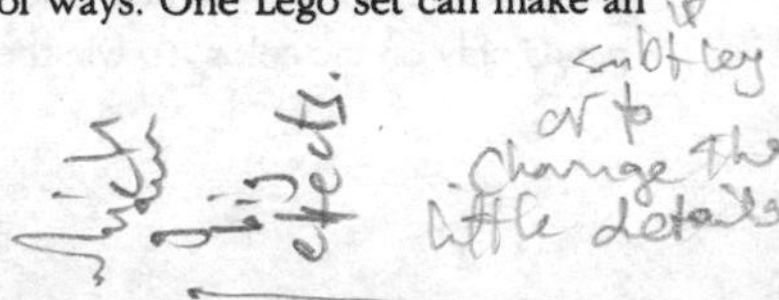

awful lot of toys, even though the number of different types of pieces is small; you just have to join the bits together in different ways. Chemistry is Lego with elements as pieces.

PLACE SETTINGS

The whole numbers 1, 2, 3, 4 . . . can all be obtained from the single "element" 1 by repeated addition, for instance $7 = 1 + 1 + 1 + 1 + 1 + 1 + 1$. Does this mean that 1 is the only number that matters in arithmetic? Of course not. The operation "+" is also important; indeed, thanks to + arithmetic is extremely complex, because + leads from 1 to all the other numbers. A reductionist approach to a phenomenon does not simply refer all questions to properties of its constituents as individuals. We must also understand the rules for putting those components together.

Such considerations lie at the heart of chemistry. For instance, if you mix up hydrogen and oxygen in a flask, they form a gas. However, a little heat or a spark causes the mixture to explode, and then water condenses on the inner surface of the flask. So water is some kind of combination of hydrogen and oxygen. Just what kind of combination is explained by the second Greek theory, that of atoms. Over a century and a half ago, every respectable chemist believed it. What evidence led them to place such confidence in particles so tiny that there was no conceivable way to detect them? If you weigh the constituents you find that water is always made from specific and invariable proportions of hydrogen and oxygen. In fact, all compounds are made up from specific and invariable proportions of their constituent elements. Chemists established huge lists of these characteristic proportions, and asked how the proportion of any particular element varies from one compound to the next. It soon became clear that for any chosen element these characteristic proportions are always whole-number multiples of some common quantity—no matter what compound the element occurs in. In different hydrogen compounds you find twice the amount of hydrogen, or three times, but never 3.74563 times as much. It's a striking and surprising fact, and it demands an explanation.

Here's an analogy. Pestco Supermarket is running a promotion for a mystery product. It gives away free boxes of it: a red box, a blue box, a green box, and a yellow box. The number of objects inside each box depends only on the color. To win the grand prize of a week in the Bahamas you must guess how many objects there are in each box. Of course, you're not allowed to open them, but nobody says you can't weigh them. You find that the weights are always 17 ounces, 27 ounces, 37 ounces, or 47 ounces. It would be hard not to conclude that the box weighs 7 ounces, and contains either 1, 2, 3, or 4 identical objects each weighing 10 ounces.

The chemical conclusion seems equally unavoidable. Compounds are not just made up from elements, but from whole-number multiples of specific quantities of elements. The chemical universe comes in quantitative lumps. Moreover, chemical reactions just redistribute the existing lumps, with every single lump neatly accounted for. Water (two lumps of hydrogen, one of oxygen) plus sulfur trioxide (one of sulfur, three of oxygen) produces sulfuric acid (two lumps of hydrogen, one of sulfur, and four of oxygen). The numerology of compounds suggests that they are made up of whole-number multiples of the smallest possible quantities in which their constituent element can occur. What *is* the smallest possible quantity—and why should there be such a thing? Atoms! Conclusion: Compounds are made up by combining whole numbers of atoms; moreover, all atoms of any particular element appear to be identical.

One further consequence seems equally inescapable. The business with weights implies that a quantity of pure sulfuric acid containing, say, 101 sulfur atoms, must contain precisely 202 hydrogen atoms and 404 oxygen atoms. How is this agreement ensured? Do the gases hydrogen and oxygen count the number of sulfur atoms, multiply respectively by two and four, and then hand over the correct number for combination, like a parsimonious but honest accountant? It hardly seems likely. There must be a simpler explanation. Suppose a catering company is asked to supply 101 knives, 202 forks, and 404 wineglasses for a dinner dance. It is certainly possible that there are two guests, one requiring 100 knives, 2 forks, and 202 glasses, the other 1 knife, 200 forks, and 202 glasses—but it's not likely. It is far more plausible that there are 101 guests, and each has a place setting of 1 knife, 2 forks, and 4 glasses. Slightly strange numbers, but clearly a terrific party.

In other words, sulfuric acid is not just made from atoms of sulfur, hydrogen, and oxygen mixed up at random in the proportions 1:2:4. It is made from "place settings" that consist of one sulfur atom, two hydrogen atoms, and four oxygen atoms. These place settings are called molecules. So pure substances are made from large numbers of identical molecules, and each molecule is a specific collection of atoms of various kinds. The concepts "atom" and "molecule" are quantitative, whereas the concept "element" is qualitative. Each type of atom determines a unique element, made from that type of atom and that type alone; but in bulk quantities of elements, those atoms are arranged in molecules. A molecule of oxygen contains two oxygen atoms; a molecule of nitrogen contains two nitrogen atoms. Bulk carbon employs many different arrangements of atoms, such as those in graphite or diamond. A molecule of helium contains just one atom, so there the distinction between atoms and molecules becomes one of context.

"Theories destroy facts." The simple assumption that everything is made up of whole numbers of atoms removes the need for huge lists of percentages of constituent elements. It is replaced by a short list of (proportionate) weights of the hundred or so types of atom, together with a rather longer list of chemical formulas.

It pays to take a closer look at these weights. The hydrogen atom turns out to be the lightest, so it makes sense to choose units in which the hydrogen atom has weight 1. In fact—for reasons to be described later—the modern choice is 1.008 for hydrogen. The resulting numbers are called atomic weights. The atomic weights of the first few elements are 1.008 for hydrogen, 4.00 for helium, 6.94 for lithium, 9.01 for beryllium. With very few exceptions—the worst being chlorine, with atomic weight 35.46—atomic weights are very close to whole numbers. Is everything made up from hydrogen? The numerology is suggestive, but no known *chemical* process can turn hydrogen into any other element.

Just as the important connection between the numbers 7 and 1 is the operation +, so the important connection between elements and compounds is the manner in which the elements combine, about which we have as yet said absolutely nothing. In any case, the whole thing may be an illusion, a coincidence: After all, 35.46 is about as far away from a whole number as you can get.

■ ATOMS: THE INSIDE STORY

Let's leave that one on the back burner and ask a different question. When two gases, hydrogen and oxygen, combine to form water, why is the result a liquid? Why do a yellow powder (sulfur), a reddish metal (copper), and a colorless gas (oxygen) combine to produce a blue solid (copper sulfate)? Those are ambitious questions, which we won't answer in detail here, but we really must start to make some kind of sense of the physical properties of chemical elements. Presumably they reflect features of their component atoms. Which features?

The elements all fall into broad classes. There are metals, such as copper, iron, sodium, potassium; gases, such as hydrogen, helium, nitrogen, oxygen; nonmetallic solids, such as boron, carbon, sulfur. Sodium and potassium have a strong family resemblance, and so do many other pairs. A number of scientists came up with various "numerological" patterns, but by far the most successful scheme was introduced by Dmitri Mendeleev in 1869. He developed a "periodic table," which exhibited the family resemblances between elements. In a fairly modern form, it lists the elements in order of atomic weight and arranges them into groups. (It so happens that the sizes of the groups are 2, 8, 8, 18, 18, and 32, a rather curious set of numbers to which we shall return.) The table's structure wasn't perfect, but it was good enough for Mendeleev to predict the occurrence of ten "missing" elements. Eight of these are now known to exist. He also predicted their properties, by comparing the missing elements with the corresponding ones in other groups.

The explanation of Mendeleev's table requires yet another reductive step, from atoms to their internal constituents, and goes back to the physicist Niels Bohr. According to his theory, atoms are not indivisible, but are made up of three kinds of particle: protons, which carry a positive electric charge; neutrons, which have no charge; and electrons, which carry a negative charge exactly opposite to that of the proton. Protons and neutrons are close together near the atom's center, forming its nucleus; electrons orbit the nucleus at various distances. Neutrons and protons have almost equal masses; electron masses are far smaller, almost negligible by comparison. The hydrogen atom is made up of one proton plus one electron, so the mass of the proton (and hence also of the neutron) is approximately that of an atom of hydrogen: atomic weight 1. Helium is two protons, two neutrons, and two electrons: atomic weight roughly 2 + 2 + a tiny bit = 4. Carbon has six of everything: atomic weight roughly 6 + 6 + a tiny bit = 12. Chlorine has 17 protons, 18 neutrons, and 17 electrons: atomic weight roughly 17 + 18 + a tiny bit = 35.

We now know that the curious numbers in Mendeleev's periodic table reflect the way in which electrons are arranged. They come in concentric "shells" surrounding the nucleus, which contain 2, 8, 8, 18 . . . electrons for arcane reasons of quantum theory. An element's chemical properties—its interactions with other elements—are determined by its electrons. Those in the outermost shell are the most important for chemistry, because the outer electrons interact most easily with the electrons of other elements. In fact, the broad chemical properties of an element depend only on the number of electrons in its outermost shell. Sodium, for example, has two, eight, and one electrons in successive shells, whereas its cousin potassium has two, eight, eight, and one. The outermost shell has one electron in both cases, which explains the chemical similarities between the two elements. However, there's an additional complication. Although the larger shells can contain up to eighteen electrons, they don't "like" to do this when they're on the outside. Once an outer shell contains eight electrons, it "fills up"; if you try to add more, they start to form a new shell farther out. When that shell gets big enough, extra electrons can be added to the original, apparently full shell—which is now one layer farther in. This may sound weird, but it's essentially a problem of stability: Any electrons over the eight get squeezed out unless there's something outside to help keep them in. Atoms with a full outer shell—two electrons if there's only one shell, eight otherwise—are unusually stable and chemically inert. These are the "noble gases," helium, neon, argon, krypton, xenon, and radon. They are not so much true nobility as snobs, refusing to combine with any other atoms, themselves included.

Bohr's theory represents a second triumph of reductionist thinking. Its modern successors are more elaborate, and replace the picture of particles in orbit with more abstract mathematical schemes, but they explain the same observational facts in the same general fashion. The picture of atoms as combinations of protons, neutrons, and electrons also tidies up another difficulty in the list of atomic weights: the exceptional cases such as chlorine. If most elements have whole-number atomic weights, but a few do not, it makes sense to view those exceptions with some suspicion. A plausible explanation requires little thought: Maybe what we think of as chlorine is actually a mixture of two different elements, just as air is a mixture of several kinds of gases. The twist, for chlorine, is that there seems to be no *chemical* way to distinguish the two components of the alleged mixture, so as far as chemistry goes they appear to be the same element.

Here's how it goes. Take an atom of chlorine, of atomic weight 35, and henceforth call it chlorine-35. Add two neutrons to its nucleus, increasing the atomic weight to 37 and yielding a new element—call it chlorine-37. Why do we keep the name "chlorine"? Remember that an element's chemical properties are determined by its electrons, especially the number of electrons in the outermost shell. Not only does the outermost shell of chlorine-37 have the same number of electrons as chlorine-35; *all* shells of electrons in chlorine-37 are identical to those in chlorine-35. Only the nuclei are different. No wonder no chemical properties can distinguish the two! Such chemically identical but subatomically distinct elements are called isotopes. The observed atomic weight of chlorine, 35.46, arises because it is a mixture of chlorine-35 and chlorine-37 in proportions of roughly three to one.

Notice that chemical properties are determined by an element's complement of electrons, but atomic weights are determined by its protons and neutrons. It is this difference in the properties of subatomic particles that accounts for the imperfect numerology of chlorine. Pursuing this line of thought leads relentlessly deeper into the atom. But before we head in the same direction, let's give chemistry a human face.

CARBON SCHIZOPHRENIA

Although Lego is very flexible, you can't fit pieces together in a totally arbitrary manner. The little pimples on one block have to push into the sockets in another. Chemistry also has assembly rules. For example, it takes two atoms of hydrogen and one of oxygen to make one molecule of water. Why not one of each? We've said that the chemical properties of atoms are mostly determined by their outermost layer of electrons. A full outer shell has eight electrons—except when the outer shell is the first shell, in which case there are only two—and atoms with full outer shells are chemically inert, the noble gases. Full shells are unusually stable. Let's count electrons. An oxygen atom has six electrons in its outer shell, but needs eight to be really stable, so two are "missing." One hydrogen atom can supply one electron. Deduction: You need two hydrogen atoms to fill up the outer shell of one oxygen atom. It's pimples fitting into sockets, just like Lego. Oxygen has two sockets, hydrogen has one pimple. Two hydrogens can plug the sockets in one oxygen.

Carbon has four outer electrons, but since it needs eight for a complete shell, it is also missing four electrons. It looks as if it has four pimples, but equally it can be seen as having four sockets. The electronic pimples of four hydrogen atoms will fit neatly into the sockets of one carbon atom, making the molecule methane, the natural gas that you burn in your oven. This is the electronic theory of the chemical bond—or, more accurately, of one particular mechanism, known as *covalent bonding*, by which atoms link together to form molecules.

Carbon, indeed, has a far more complicated chemistry than any other element. The schizophrenic nature of its outer electron layer—four pimples or four sockets, four electrons present or four missing—is largely responsible: It permits a huge variety of stable combinations. Carbon, uniquely, can form covalent bonds with itself. If it were not for the versatility of carbon, life would be very different. Indeed, life as we know it would be impossible, because only carbon can form structures complex enough to support life. Virtually all of the chemicals that are found in living creatures involve carbon. For this reason, carbon compounds are said to be organic.

Organic chemistry is crucial to the understanding of life. We don't want to write a textbook on chemistry, but we do need to capture a little of its flavor. The simplest organic compounds are hydrocarbons, composed only of carbon and hydrogen atoms. They tend to fall into families. One starts with methane, which as we've just mentioned has four hydrogen atoms plugged into a central carbon atom. Ethane, its nearest sibling, has two carbon atoms plugged into each other, which leaves room for six hydrogens. Next comes propane, a row of three carbons surrounded by eight hydrogens. Although these molecules are really three-dimensional, they're often drawn as a long line of carbon atoms, like the rowers in a boat, with the hydrogen sticking out like oars: two per rower arranged one on each side, just like a real boat, plus one at the bow pointing forward and one hanging out of the stern. Twice the number of rowers, plus two. It looks as if this pattern could continue indefinitely, building up long chains of carbon atoms surrounded by hydrogen. Indeed, it can: Chemists have synthesized chains containing up to 100 consecutive atoms of carbon, which perforce must have 202 atoms of hydrogen.

One of the most important units in organic chemistry first occurs in the compound benzene, which has six carbon atoms and six hydrogen atoms.

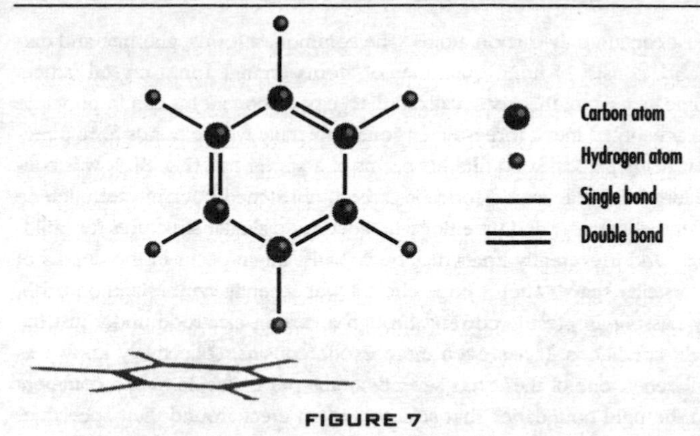

FIGURE 7

The benzene ring

At first sight this has far too few hydrogen atoms; a "rowboat" pattern like that described above would require fourteen hydrogen atoms to plug all the sockets in the six carbons. Friedrich Kekulé—allegedly through a dream of a snake eating its own tail—realized that a closed ring of six carbon atoms would work (Fig. 7).

This benzene ring occurs in many organic compounds; it helps "rigidify" the molecule by adding "cross struts" to brace it. The shortcomings of the pimple/socket approach are also evident in the benzene ring. We've drawn a single bond (—) to show that one pimple fits into one socket; and a double bond (═) when two pimples fit into two sockets. Single and double bonds are drawn alternately around the ring; this pattern is dictated by the requirement that carbon atoms should possess four sockets but hydrogen atoms only one pimple. However, there is a second way to make the number of bonds fit: On the central ring of carbon atoms, change all single bonds to double and all double bonds to single. Which actually occurs? Real benzene behaves like neither of them, but more like an average of the two. Is it a mixture? No. It's a "resonance structure," which can't adequately be represented by this kind of diagram, in which the six carbon atoms share their electrons symmetrically. The closest that our diagrams could get to the structure would be to make each carbon–carbon bond consist of one and a half pimples.

So, by using hydrogen atoms to plug a few spare sockets, you can assemble six carbon atoms into a ring. Elemental carbon, on the other hand, must contain only carbon atoms. The commonest forms, graphite and diamond, consist of huge quantities of atoms arranged in a crystal lattice. Some years ago, theorists realized that sixty carbon atoms can in principle be assembled into a three-dimensional molecule whose bonds form a regular mathematical solid like the seams of a soccer ball (Fig. 8). It was conjectured that this exotic form of carbon, christened buckminsterfullerene after the engineer and inventor who advocated similar structures for buildings, and irreverently known as buckyballs, might occur in the depths of interstellar space. After a huge effort it was recently synthesized on earth, by passing an electric current through a carbon electrode under just the right conditions. It has even more exotic cousins, collectively known as fullerenes; one of these has seventy atoms per molecule. It is a comment on the rigid boundaries that scientists often erect around their specialties that the experimentalists who first discovered buckminsterfullerene spent a considerable time wondering whether such a structure was feasible mathematically and chemically—unaware that any mathematician would have told them what it was. In the trade the shape is known as a truncated icosahedron, meaning that you start with a twenty-sided solid and shave off its corners. It belongs to a list of standard shapes called polyhedrons, meaning "many-sided."

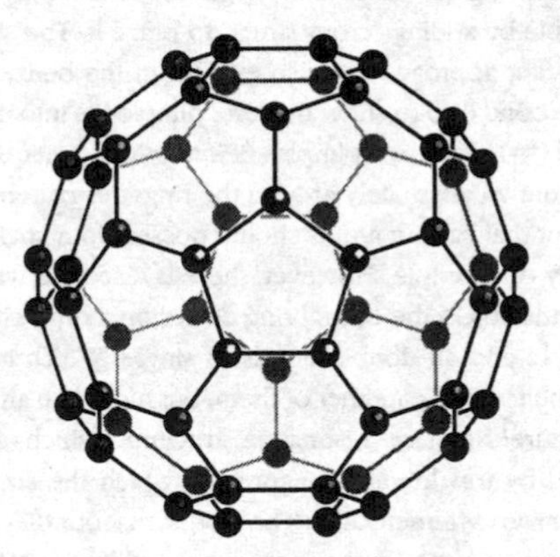

FIGURE 8

The buckminsterfullerene molecule: a truncated icosahedral cage formed from sixty atoms of carbon

Furthermore, theoretical chemists had already publicized dozens of papers predicting what this hypothetical molecule's properties would be, if it existed. Fullerenes are now attracting a lot of attention; in the first place because of their potential for creating an entire new range of organic compounds, whose uses are as yet undreamed of; and in the second place because they're fashionable. We still don't know if they occur in space; but—to cap the tale—natural fullerenes have very recently been discovered lining cracks in rocks. Making fullerenes is rather easy, for nature as well as humans. Nature knew this all along, but we've only just noticed.

QUANTUM UNCERTAINTY

The reduction from chemicals to molecules to atoms bases the laws of chemistry upon those of physics. The nature of chemical bonds and the material properties of chemicals lead deeper still, to subatomic physics. In the simplest picture, that of the Greeks, atoms are indivisible particles, tiny lumps of matter that can't be pulled apart into even tinier lumps. In the picture of classical (meaning "nonquantum") physics, atoms are divisible, but their component electrons, neutrons, and protons are pretty much like Greek atoms. The main difference is that these particles have not just mass, but electric charge. Another attribute must also be introduced: spin. Some aspects of the behavior of electrons make sense, in this classical picture, only if you think of tiny particles spinning on an axis, like a top. Physicists looked inside the atom, and at the bottom of their mental funnel they saw what at first sight seemed to be Newtonian mechanics—yet more confirmation, if any were needed, of the truth of Newton's laws.

Unfortunately, other aspects of the behavior of electrons made no sense at all within the Newtonian frame. In particular, in Newtonian mechanics any body with electric charge that orbits another body—for example, the electron orbiting the atomic nucleus—should radiate its electrical charge away into space. But electrons stubbornly hang onto their charge. Niels Bohr saved things for a while by assuming that electrons obey Newtonian mechanics except when they don't; but as more and more evidence of bizarre behavior among the subatomic particles piled up, it became clear that something more radical was needed at the bottom of the atomic funnel. That something was quantum mechanics. We don't want to get too entangled in this highly complicated area, but we do want to show you just where the reductionist approach led, how far away from ordinary human experience it went as it dug deeper and deeper into the hidden strata of nature's regularities.

The quantum world is a world of uncertainties. There is no way to predict when a radioactive atom will decay. Indeed, there is no difference between an atom that is about to decay and one that isn't. Albert Einstein objected to this picture of "God throwing dice," but the evidence is that the uncertainties really are present. Things that classical physics views as particles—such as electrons—sometimes behave like waves and spread themselves out. Things that classical physics views as waves—such as light—can also behave like particles. In classical mechanics, waves have associated with them a number known as their frequency: how many waves pass a given point in a given time. They also have an energy. The energy of a classical wave can be any number whatsoever and does not depend on its frequency. But in quantum mechanics the energy of light must always be proportional to a whole-number multiple of a particular number, Planck's constant. The proportionality factor is the frequency of the light. Effectively, light of a given frequency comes in tiny packets with a fixed energy. This sounds just like a description of particles. Planck's constant is incredibly tiny, so the range of energies is almost continuous, but physicists can make such accurate measurements that the forbidden gaps show up clearly. Moreover, those gaps actually help us to make sense of other observations.

In quantum mechanics, if you measure something you usually disturb something else. There is an irreducible inability to know everything you'd like to at the same time. For example, imagine measuring the position of a football on a field. In classical mechanics you can shine a very, very weak light onto it, and the reflected light will tell you where the football is. You disturb it a tiny, tiny amount; but not enough to matter. In principle, you could observe the position as accurately as you wished. In the quantum world, however, there is a lower limit to the energy that you impart to the football when you measure it—thanks to Planck. It's like staggering round the field blindfolded, aiming large kicks in random directions. Eventually you make contact with the ball and shout, "I got it!" Then somebody says "Terrific, but where is it *now*?" and you feel rather silly. The actual problem in quantum mechanics is a little more subtle: You can measure where the ball is, but to do so you must lose information about how fast it's moving.

Or you can measure how fast it's moving, but then you don't know where it is. This is the uncertainty principle of Werner Heisenberg, and it exemplifies the weirdness of quantum laws of nature.

Despite its lack of resemblance to our familiar commonsense world, quantum mechanics can be set up as a self-consistent system of mathematical rules; and it works. It works brilliantly. The problems come when we try to interpret what its rules mean. Observable quantities, such as the spin of an electron, just don't behave like classical observables. For example, an electron can have either "up" or "down" spin—imagine a top spinning ("up") and tilt its axis through 180° to get "down." But the spin can also be in a "superposed" state, say 50 percent up and 50 percent down. This is *not* the same as a horizontal spin axis. It is more like a spin that is not actually known, but that has a 50 percent chance of being up and a 50 percent chance of being down. Quantum observables are inherently probabilistic—until, that is, you observe them. Then, suddenly, your measurement seems to force the electron to choose: either up or down. It's as if an electron's spin is a coin, flipping over and over indefinitely, until you observe it by making it land on a flat surface.

Every quantum object has a "wave function" that describes the probability cloud of states in which that object can exist. According to the "Copenhagen interpretation" of the mathematical rules, any measurement forces the wave function to "collapse" to specific certainties rather than indeterminate probabilities. Until you make such an observation, the system will be in a state that superposes all of the possible things that might be happening.

This unsatisfactory state of affairs was dramatized by Erwin Schrödinger, in a thought experiment involving a very unfortunate cat. The cat is imprisoned in an impenetrable box, along with a radioactive atom and a device that detects whether that atom has decayed. If this detector is activated, it triggers a capsule of poison gas that kills the cat. Now, radioactive atoms decay at random, so from outside the box we can pick our moment and then ask whether the atom has decayed yet. Because the detector releases poison gas only if the atom decays, this is the same as asking whether the cat is alive or dead. Now, since everything is inside the impenetrable box, we can't make any measurements. Therefore the atom is not in a pure state; it is neither "decayed" nor "not decayed," but in some probabilistic superposition of the two. This implies that the wave function of the cat is also in some probabilistic superposition of the corresponding states "dead" and "alive," because we know that the state of the cat is determined by that of the atom. But surely a cat is either alive or dead?

According to the Copenhagen interpretation of quantum mechanics, it isn't. Not until you open the box. Then you collapse its wave function, and either it instantly dies, or it leaps thankfully out of the box and vows never again to take part in silly experiments.

There are many problems with the Copenhagen interpretation. If it's right, what would it be like to be Schrödinger's cat *before* anyone opens the box? Isn't the cat an observer anyway? Eugene Wigner worried about this; he invented a hypothetical invisible friend to sit in the box with the cat and keep an eye on it. Then, even if you don't agree that a cat can be an observer, there is a genuine observer in the box—but he can't communicate his observation to the outside world. Wigner's friend sees a collapsed wave function, but Wigner doesn't! We're just as worried about the Copenhagen interpretation as Wigner was, and we'll further consider Schrödinger's cat in chapter 8.

■ BIG BANG

Instead of delving ever deeper into the atom, looking at the internal structure of chemistry, we can look at the chemistry that goes on in the universe around us. We can, for example, measure the proportions of various chemical elements in as much of the universe as we can reasonably observe. Because of quantum mechanics, chemical molecules betray their composition by affecting light that passes through them, the analysis of which is called spectroscopy. When you shine a beam of sunlight through a prism it separates into its constituent colors, forming a multicolored band of light called a spectrum. Each color corresponds to light of some particular wavelength. Within the sun's spectrum are dark lines where molecules have absorbed light of a wavelength that resonates with their characteristic vibrations. These absorption lines are a kind of molecular signature that sends the outside world messages such as "Carbon here." Using spectroscopic techniques, astronomers found that they could learn more about the chemistry of the interior of a star than they could about the chemistry of the interior of the earth. The reason is that light from stars can pass upward from the deeper layers to the outer ones, carrying the spectroscopic signature with it; the earth is less transparent.

Using such methods, it becomes clear that the universe contains an awful lot of hydrogen, a fair amount of helium, and rather arbitrary smaller proportions of other elements. In the spirit of finding mathematical regularities, it would be nice to finish everything off by devising a theory that predicts those proportions. This can be done, but it is dramatically different in flavor and complexity from anything we have previously described. It depends on the past history of the universe—or at least of this particular region of the universe. Elements are mainly formed inside stars as a result of nuclear reactions. The proportions of different isotopes that are produced depend upon the details of these reactions. Morover, some isotopes are radioactive and decay, over various periods of time, into other isotopes. The value 35.46 for chlorine is a consequence of cosmology and history, not chemistry.

Indeed, we must go back not just to the chemistry of stars, but to the manner in which the stars themselves appeared on the scene. That takes us back in time to the instant at which, cosmologists believe, space and time got started in the Big Bang. The Big Bang has the virtue of explaining two major astronomical observations. First, the universe appears to be expanding, as is confirmed by spectroscopy applied to distant bodies—originally galaxies, but nowadays mainly the highly energetic objects known as quasars. The farther away the body is, the further the lines in the chemical signature shift into the red part of the spectrum. This "red shift" is related to velocity, and its existence implies that distant parts of the universe are moving away from us. The universe is getting bigger. This expansion can be viewed as an ongoing consequence of the Big Bang, which gave everything a big push from which it has yet to recover. The second observation is the cosmic background radiation, an almost uniform level of radio noise corresponding to a temperature of 3°K (*minus* 456°F): this is the "echo" of the Big Bang. Very recently the COBE satellite has observed faint ripples in the background radiation, the irregularities that were needed to "seed" the formation of stars.

So the proportions of chemicals in the universe depend on the nuclear reactions inside stars, and the existence of stars follows from the ripples in the Big Bang. When you put all this together, you can calculate the proportions of elements in today's universe, and it comes out pretty much correct.

That's impressive: Even stars are made of chemistry. The mathematical theories that lead to the Big Bang combine two things. One is Einstein's theory of relativity, the idea that gravity is bent space-time. The other is quantum mechanics. As mentioned in chapter 1, we find quantum mechanics down two funnels—one leading from chemistry, the other from cosmology. That certainly makes quantum mechanics look pretty fundamental. It provides a stable intellectual framework for thinking about the universe, both in the small and in the large; and it must therefore be congruent to at least some aspects of how the universe works.

However, particle physics has become a lot more complicated than the old story of electrons, protons, and neutrons. There are a host of other particles, such as quarks and gluons. These have to be included because they make sense of particle-accelerator experiments, but there are big difficulties if you try to make current particle physics mesh really effectively with gravity. A genuine unified theory is still lacking. Many physicists think that they are on the track of a Theory of Everything. The phrase may sound pretentious, but if the same underlying laws can explain matter, gravity, space, and time, then maybe they deserve a little bit of innocent hype. These physicists think that we are close to the bottom of the ultimate funnel, and that absolutely everything leads down to the same underlying laws (Fig. 9).

ANTHROPIC PRINCIPLES

A curious coincidence emerges from the quantum-mechanical link between cosmology and chemistry. Organisms rely heavily on the element carbon, because it alone can form the complicated chemical structures needed for something as esoteric as life. (Maybe silicon could do this too, but certainly nothing else.) Carbon, like all elements heavier than hydrogen, is born in the nuclear furnaces of stars. That is, the whole elaborate mechanism of the Big Bang is needed to explain why carbon forms at all.

Not only that, the formation of carbon in stars is possible only because of an unusual quantum-mechanical coincidence in that element's energy levels. This makes it a lot easier for carbon to form than would otherwise be the case. Now we are led inexorably toward the "anthropic principle," which says that it is really rather silly for a carbon-based life-form to ask why the structure of the universe and the laws of quantum mechanics just happen to permit the formation of carbon. If they didn't permit this, the ar-

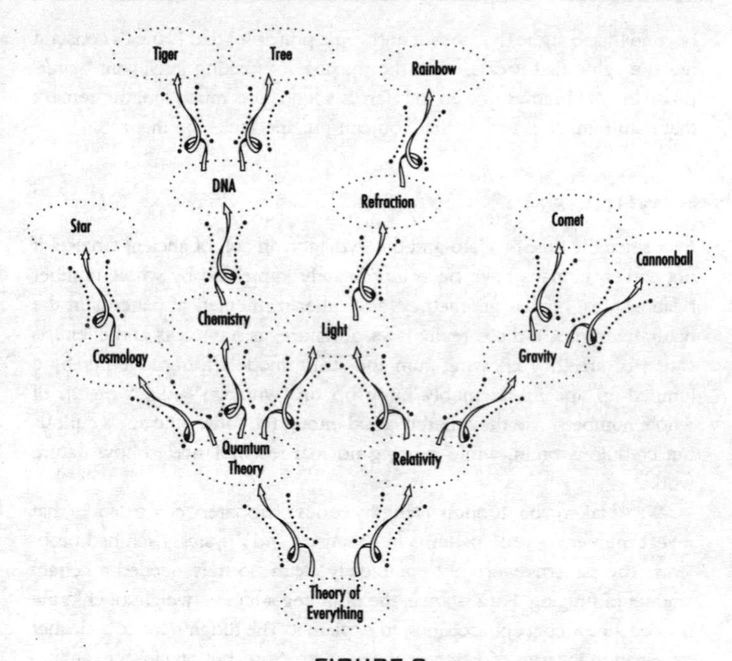

FIGURE 9

Deep down the funnels lies the Theory of Everything.

gument goes, then the carbon-based life-form wouldn't be around to ask the question. It's just like the survivors of a major accident sitting around afterward asking themselves, "Why did *I* survive?"—forgetting that had they not done so, they wouldn't be asking that question.

That's really a very logical point, but it has a sting in its tail. Physicists would like to explain the values of various "fundamental constants" of nature—for example, Planck's constant, whose value is 6.6×10^{-34}. This is extraordinarily small, but even so the number is of vital importance. It turns out that if Planck's constant were changed by just a few percent—say, to 7×10^{-34}—then stars wouldn't be able to form carbon. So we wouldn't be here to ask why Planck's constant has that particular value. We can deduce the value from the fact that we're around to ask what it is. But we still don't understand the physical reasons for that value to occur. Some people go beyond this to argue the "strong anthropic principle" that Planck's constant has the value that it does with the purpose of making intelligent beings possible. We'd rather not go that far: It sounds too much like the remark that nature made noses so that you can put spectacles on them.

WHOLISM

Ever since the time of Plato and the Pythagorean cult of ancient Greece, if not earlier, humans have been inordinately impressed by whole-number relationships, regular geometries, and other mathematical patterns in the natural world. Are these really laws of nature, as physicists and chemists claim, or are they chewing-gum-and-string models flung together by a jumped-up ape irredeemably hung up on "wholism"—the worship of whole numbers? Are they merely good enough to satisfy the ape's curious but credulous brain, while bearing no real resemblance to how nature works?

We've taken you through a lengthy series of apparent coincidences that reveal "numerological" patterns in chemistry and physics. Each had problems: The patterns were not completely clean, so they needed a certain amount of fudging. For instance, the noninteger atomic weight of chlorine needed a new concept, isotopes, to explain it. The fudging led to a cleaner explanation in terms of a "lower" level of structure, but on closer examination this new level was itself not entirely clean either—atoms aren't just made from protons, neutrons, and electrons; these in turn have their own substructure. . . . It's fudges all the way down! The explanations all worked well enough, but with quirks: The characteristic proportions of elements were explained by atomic theory, but with curious and arbitrary values for the weights of individual atoms; then the weights of atoms were explained by Bohr's theory of electron shells, but with curious and arbitrary rules for assembling them, and with a lot of fiddling around with isotopes. Then we used quantum theory to explain the rules of the shells, and invoked cosmology to account for the preponderance of different isotopes—but quantum theory and cosmology themselves have oddities.

It does make one wonder whether the laws of physics and chemistry are real, or brain puns. In order to resolve this question, we shall enlist the help of beings whose thought processes do not necessarily follow ours.

Their role is not so much to answer tendentious questions about the Meaning of Everything as to push our own minds in new directions, to stimulate lateral thinking. Naturally, everything they say must be taken with a pinch of salt.

Or, in this case, a pinch of buckminsterfullerene.

■ FUELS RUSH IN

The spacecraft *Thighbone* is in orbit around the distant world of Zarathustra. It is low on fuel—buckminsterfullerene in a benzene solvent. It has just enough to land safely, but not to take off again. If Zarathustra were a world of the Civilized Combine there would be no problem: drop in, make a quick trip to the fullerene station, and head for home. Unfortunately, it is not; but according to the *Good Galaxy Guide*, Zarathustra is inhabited by intelligent aliens. Its technology is listed as being (a) advanced, and (b) weird. Having assessed the probabilities, the flight computer comes to the only possible decision: Land and hope. It ejects a precisely calculated quantity of reaction mass at an equally precisely calculated velocity. *Thighbone* drops neatly out of orbit, and touches down having consumed the absolute minimum of fuel required for a safe landing. The gauges conform to predicted values to four decimal places.

The next step is to contact the local life-form. This proves easier than expected: No sooner has the dust settled than a shiny purple sluglike beast slithers into view, driving ahead of it a group of eight fluffy yellow creatures, rather like miniature ostriches. The scene is reminiscent of a shepherd with a flock of sheep. The slug turns its flock to head straight for the ship. Soon the ostriches are milling about near the airlock, and the slug is climbing over the cabin, leaving slimy tracks across the glass.

The final step is to communicate with the creatures. This is nowhere near as straightforward as the elegant landing; xenobiology does not obey laws as clean as Newton's. The creatures emit a remarkable range of noises, from deep booms to batlike squeaks, changing pitch with astonishing rapidity. No human voicebox could produce such sounds. However, the appropriate technology is to hand. Captain Arthur orders the translator program to be run, and after a few minutes spent analyzing the bedlam outside to acquire a good linguistic database, the red light goes off.

CAPTAIN ARTHUR [*Clears throat.*]: Good day to you, creature from another world.

NEEPLPHUT: No, *you* are the creature from another world. *We* live here. How strange that you are so like/unlike us (delete whichever is inapplicable).

CAPTAIN ARTHUR: We come in peace. I am Captain Arthur, and this is my colleague Stanley.

NEEPLPHUT: I am Nyplefuffl-pofta-Gostaphut Third Remove, but you can call me Neeplphut. You must be out of fuel, judging by the careful way you were driving.

STANLEY: Captain, that's wonderful! They must understand Newton's laws of motion, otherwise they wouldn't be able to make sense of our trajectory. I bet their physics and chemistry are as good as ours, too.

NEEPLPHUT: I recognized that you were using Pnurflpeef's laws of motion the moment we saw how you were optimizing fuel consumption. Odd that you should use something so outmoded, but it *is* an excellent approximation, so we assume you need to reduce computational overheads. Incidentally, the Regulations say you can/cannot park here (delete whichever is inapplicable).

CAPTAIN ARTHUR: Um . . . can I delete "cannot"?

NEEPLPHUT [*after hurried consultation*]: You must delete "can," according to the Regulations here. [*Slug waves something indistinct but mildly revolting in shape.*]

CAPTAIN ARTHUR: Are you a traffic warden? You look like a slug.

NEEPLPHUT: No, I am one of the eight yellow ostrich things. The purple slug is the Regulations. It is a composite hive-bureaucrat, and I have already told you what it said about not parking here.

CAPTAIN ARTHUR: Sorry. You looked just like a flock of sheep being herded. That made us think the slug was the intelligent one.

NEEPLPHUT: That concept is definitely inapplicable.

CAPTAIN ARTHUR: Eh?

NEEPLPHUT: I do not know what you are talking about. What are a "flockofsheep"?

STANLEY: There's an incongruence, Captain. The translator can't cope with concepts that have no parallel in their experience. They can't be expected to cope with accidents of terrestrial biology, only with universal abstractions.

CAPTAIN ARTHUR: I hope it can cope with "buckminsterfullerene," or we're done for.

NEEPLPHUT: What are "buckminsterfullerene"?

CAPTAIN ARTHUR: Oops. [*In worried tone:*] It's what our fuel is made of.

NEEPLPHUT: I would say you are in trouble. Your translator is suffering a conceptual incongruence.

STANLEY [*under his breath*]: It's also having trouble with singular and plural. It'll take me weeks to straighten out the neural network after this.

CAPTAIN ARTHUR: There must be some way to get the concept across. Are you people any good at chemistry?

NEEPLPHUT: What are "chemistry"?

CAPTAIN ARTHUR: I think we'd better start at the beginning. Um . . . I am holding up *one* finger. Now I am holding up *two* fingers. . . . [*He works his way through arithmetic, atomic weights, and the periodic table. Neeplphut learns human notation for the elements up to carbon, and is shown how to draw structural diagrams of molecules. None of this appears to cause any difficulty. Although little of it is state-of-the-art Zarathustran science, the parallels are unmistakable. He gets unusually excited about the eight electrons in a full shell, for reasons that he tries very hard to explain but is completely unable to.*]. . .Now, what we need is a good quantity of a form of carbon. Like this. [*Draws diagram of buckminsterfullerene molecule and holds breath.*]

NEEPLPHUT: Oh, yes, buckyball. You can find the stuff in any hypermarket. [*In conspiratorial tone:*] My cospouses use it for plumage makeup, très sexy.

STANLEY: Um . . . Captain, we can't just use dry buckminsterfullerene. It needs a benzene solvent.

CAPTAIN ARTHUR: Yes, I keep forgetting that, it always comes out of the pumps as a fluid. Well, if they sell buckminsterfullerene for cosmetics, benzene should be no problem.

NEEPLPHUT: What are "benzene"?

CAPTAIN ARTHUR: Like this. [*Draws diagram of benzene molecule.*]

NEEPLPHUT: Hmmm. *Tri*-cky.

CAPTAIN ARTHUR: What do you mean, tricky?

NEEPLPHUT: Not octimal enough. We never stock it. Mind you, buckyball has sixty atoms, and that is not terribly octimal either, which is strange. [*Reluc-*

tantly:] I suppose we could knock some benzene together for you if you really insist.

STANLEY: Octimal? What do you—

CAPTAIN ARTHUR: Stanley, not right now. (*To Neeplphut:*) What kind of civilization sells buckminsterfullerene in hypermarkets, but doesn't have any stocks of benzene? It's totally outlandish!

NEEPLPHUT: Captain, that is what "alien" means.

ALIEN SCIENCE

There's a serious point concealed in the dialogue. Just how congruent would alien science be to ours? If the "laws of nature" that we think we perceive really are genuine patterns, then they ought to be universal, and Neeplphut should be able to make a sensible translation between human and Zarathustran science. However, if they are puns made by human brains, then Zarathustran brains will make quite different puns, and the dialogue will be meaningless crosstalk between different paradigms.

Could an alien cosmetic fuel a Terran spaceship? Whoever manufactures the buckminsterfullerene, it will have the same chemical properties. We happen to exploit the properties that make it useful as fuel; the Zarathustrans exploit the properties that make it useful as a cosmetic. There is no incongruence in two distinct uses of the same chemical. Is it really conceivable that buckminsterfullerene should be a common substance on Zarathustra, but benzene more exotic? Would it not be clear to Zarathustran chemists that benzene is a simpler molecule of the same general kind—closed rings or cages? Probably, though if their sense of mathematical symmetry is different from ours, they might never have appreciated the connection. Anyway, you can scrape buckminsterfullerene out of cracks in rocks, but benzene evaporates and doesn't get deposited. Clearly what is exotic in some societies may be common in others.

Would the Zarathustrans be able to match their physics and chemistry to ours? Like us, they need them to explain their own biology. Provided they had the same general picture of atoms and molecules, then they might reasonably be expected to have come to similar conclusions about them. Chemistry works—it makes nylon tights and polyethylene sheets, so its patterns cannot be just puns. We could argue, in the same vein, that the Zarathustrans would possess a theory of atomic structure that is also congruent to ours. There would be things they knew that we don't, and conversely; but probably there would be more overlaps than gaps, and we could put their chemistry and ours together to get a consistent structure.

There remains the possibility, of course, that there are aliens so alien that their science is simply "at right angles" to ours. They grew up in the space between the stars, they developed an understanding of gravity but not of optics; they have discovered the hypothetical graviton, the particle that is thought to transmit gravity, but they don't have any idea what an electron is, or an atom, and hence have no concept of chemistry. The problem with this suggestion—so the conventional argument goes—is that the physics of gravitons only makes sense in conjunction with all the other particles such as photons and electrons. Indeed, if a genuine Theory of Everything exists, then the Zarathustrans would have to come up with physics and chemistry that are consistent with it, even if they've not yet got down to that level of generality. Just as we have been forced, for the sake of mathematical consistency, to recognize the existence of exotic particles, so the aliens would be forced to recognize the existence of photons, electrons, and—pursuing that line of thought to its logical conclusion—buckminsterfullerene.

However, maybe there is no Theory of Everything; maybe conventional physics is just a mathematical system that appeals to humans because they like its brain puns. Or maybe there *is* a Theory of Everything, but we've been lured down the wrong set of mental funnels and haven't a hope of finding it. If so, we can't rule out the possibility of truly alien aliens whose science is at right angles to ours—or, worse, whose mathematics is. The aliens would have mapped the same territory as us, but while we were making a road map, they were making a geological one. Communication with such creatures would probably be impossible. But we have no idea what an effective alternative to physics, as it is currently understood, would look like. So, for the purposes of argument, we will assume that intelligent aliens would be able to talk physics and chemistry to us.

Biology, however, could be a very different matter.

3 THE ORGANIZATION OF DEVELOPMENT

A man lies on his deathbed, surrounded by his family: a weeping wife and four children. Three of the children are tall, good-looking, and athletic; but the fourth and youngest is an ugly runt.

"Darling wife," the husband whispers, "assure me that the youngest child really is mine. I want to know the truth before I die, I will forgive you if—"

The wife gently interrupts him. "Yes, my dearest, absolutely, no question, I swear on my mother's grave that you are his father."

The man then dies, happy.

The wife mutters under her breath: "Thank God he didn't ask about the other three."

■ ■ ■

At a dinner party, the hostess introduces the Great Man to her guests. "This is Professor Hackensplacken. He's an authority on crocodiles."

The professor smiles modestly. "My dear lady," he says, "you do me too much honor. It is on the crocodile's eyelids that I am an expert."

A bee buzzing around a flower is such a familiar sight that we seldom think how remarkable it is. A bee is a marvel of microengineering that can hover, recognize the right kind of flower, milk it of its nectar, and fly back to the hive with the booty. We may be able to cross the Atlantic at twice the speed of sound, but the combined expertise of the entire human race can't make a bee. It can't make anything as compact as a bee that can fly like a bee—or fly at all. Anything we made would have a hard time recognizing flowers, let alone distinguishing honeysuckle from daffodil. A bee is a complex creature indeed.

In some ways the complexities of biology are similar to those of chemistry. Instead of millions of compounds, there are millions of different kinds of animals and plants; but just as chemical compounds are made from a few kinds of atom, so living organisms are made from a few kinds of cell—nerve cells, muscle cells, blood cells . . . It is true that cells have a complicated internal structure, but so do atoms. We might even hope to find something like a periodic table of living cells, some simple key to an underlying mathematical regularity.

On the other hand, life is very strange. Even though we are surrounded by living creatures from the moment of our birth, almost all of us find the whole idea of life rather awesome. Many people object violently to the idea that a living creature is really just a rather complicated machine; they find the comparison demeaning. To such people, a mathematical classification of life would presumably appear even more demeaning.

There are at least two distinct kinds of complexity in a bee. One is the complexity of structure of a single (adult) organism. The other is the complexity of its development. A bee begins life as a single cell, a fertilized egg no more than a tenth of a millimeter across. How can all the complexity of the adult insects be compressed inside so tiny an egg? And if it *can't* be so compressed, where the devil does it come from? These are difficult questions, and we must think about them on several levels. Accordingly this chapter, alone in the entire book, opens with *two* jokes. One is an ordinary joke, the other a meta-joke.

For the moment, we shall concentrate on complexity of structure, but it will soon become clear that it is inseparable from the complexity of development. This is not really surprising. A car is quite a complicated piece of machinery in its own right; but cars don't just appear in the world, ready for the road, tanked up with gas, tires inflated, with a polka-dotted dog nodding inanely in the rear window. Cars have to be made, and it is the process of manufacture that introduces the complexity. Organisms have to be made, too; but they are far more complicated than cars. Cars don't grow from eggs, but most organisms are also organism factories, capable of making new organisms of the same kind. We should be thinking about car factories that make cars that grow into new car factories. *That's* complexity for you.

HOW TO MAKE A VIRUS

We've seen that chemistry is really just complicated physics, with Lego-block variety as the source of complexity. In the same way, we can ask whether biology is just very complicated chemistry. Over the centuries, we humans have become very competent chemists. We started out trying to turn lead into gold and ended up turning oil into nylons. Today we exploit our chemical expertise to clothe ourselves, to kill pests, to grow more crops, and even to supply vital chemicals when our own bodies go wrong. We can synthesize many of these crucial molecules, that is, build them ourselves. Could we also synthesize the chemicals that come together to make a simple creature, put them together as nature does, and make the resulting artificial life-form work?

In principle, yes. In practice, it may be a lot closer than you think, but the gap may also be wider than many scientists imagine.

The simplest living things are generally considered to be viruses, which consist of two important types of chemical: proteins and nucleic acids. Neither type of chemical is simple. Proteins are huge organic molecules—long, crumpled chains whose individual links are selected from a relatively small range of much simpler chemicals known as amino acids. They are the commonest chemical constituents of all living creatures; excluding water, most creatures are at least half protein. You can specify a protein if you know the sequence of amino acids and the shape into which it must be made to fold, just as you can specify a passage from Shakespeare if you can list the appropriate letters, spaces, and punctuation marks in the correct order. With Shakespeare there isn't the secondary problem of crumpling it up correctly, but for a long time biologists wondered what tells a protein how to crumple. It turns out that to some extent the chemistry—indeed, the physics—of chemical bonds crumples everything up the right way automatically. Usually it is no more necessary to tell a protein how to fold than it is to tell salt how to crystallize into cubes. Sometimes, however, the protein has a genuine choice, and then other proteins known as chaperonins are used to give it a nudge in the desired direction.

Nucleic acids are also made by stringing simpler components together in order. They provide chemical instructions on how to build a virus; that is, they control the reproduction of viruses. They fulfill the same function in all living creatures, and we'll return to these remarkable chemicals later.

A simple virus has a central core of nucleic acid, surrounded by a protective envelope of proteins. Usually the structure consists of repetitive protein units, arranged in elegant mathematical forms such as helices or dodecahedrons. Viruses are chemical machines and cannot reproduce on their own; for that, they rely upon more complex life-forms. Bacteria are tiny chemical workshops, capable of building lots of different chemical machines. Many viruses are parasites on bacteria: They invade them and subvert their chemical workshops so that they make not new bacteria, but new viruses. A virus is a cuckoo in a bacterial nest, and it makes its host do even more of the work than a cuckoo does.

Most viruses, and *all* other organisms, use one particular form of nucleic acid to carry information across the generations. It is known as deoxyribose nucleic acid, mercifully abbreviated to DNA. When the DNA molecule is broken down it yields a number of components: a sugar (deoxyribose); simple compounds of phosphorus known as phosphate groups; and four more innocent-looking chemicals called adenine, thymine, guanine, and cytosine. (As we shall see later in the chapter, these four components turned out to be crucial to modern biology, but you'd never guess it from their chemistry.) A DNA molecule is threadlike, and truly enormous, so it must be packed into the cell's nucleus in a highly compressed way. The DNA in a single human nucleus would be five feet long if it were pulled out straight—a mile of wire in an aspirin. We can synthesize DNA without the assistance of living organisms, but only in short segments up to a hundred components long. By borrowing chemical assistants known as enzymes from living cells, we can take a leaf from the virus's book and subvert the chemical workshop, persuading bacteria to make designer DNA thousands of units long. There is no obvious theoretical limit to the size of the molecule that could be produced in this way.

Viruses are not especially interesting if you want to think of them as life-forms, because they depend for most of their vital functions on their bacterial host. The bacterium, a far more complex system, does all the complicated chemistry for them, and releases a swarm of copycat viruses, which can go on to invade other bacteria. We can understand viruses in purely chemical terms: complicated chemistry, to be sure, but not chemistry so complicated that the effects it produces are best understood from a totally different perspective. However, when we ask whether living things are complicated chemistry, we're talking about *independent* living things.

After all, written documents are even simpler chemistry than a virus, and they can reproduce very effectively—by parasitizing a human being with a photocopier. That doesn't count as independent life, and neither does a virus.

CHEMICAL WORKSHOPS

The simplest independent living creatures are the bacteria—"germs"—upon which the viruses hitch their ride to reproduction city. Bacteria don't have a central core, but they do contain that all-important DNA, a loop of which is attached to the bacterial wall at one point. They also have an outer coat containing a lot of complicated sugars, some fats, and some proteins—chemical molecules, every one. The coat's job is to keep the outside world where it ought to be—outside—and to keep the inside of the cell where *it* ought to be. (You can't survive as an independent if you get too mixed up with your surroundings.) Inside the bacterium is a kit of specialized chemical tools—assemblages of nucleic acids and proteins whose job is to make hundreds of other kinds of protein. These in turn do most of the chemical work that keeps the bacterium alive. There are proteins that latch onto passing food molecules and drag them inside, and others that pump water or other chemicals in and out. (When the mood is right, bacteria quite enjoy getting mixed up with bits of their environment, as do we in restaurants.) There are proteins that make chemical energy available for other processes. This is manipulated as high-energy chemical bonds, especially those associated with bases and phosphates; and it is often stored in the form of sugars and extracted by turning them into alcohols. Some bacteria can use the energy in sunlight, or extract it from rusting iron.

Bacteria, along with other living things that organize themselves in this "chemical workshop" manner, are called prokaryotes. Although they are simpler than many creatures, don't imagine that they are therefore very simple indeed. The metaphor of a chemical workshop is well chosen; if it is inappropriate, that is only because a workshop—think of a really modern one with hundreds of computer-controlled machine tools—is much simpler than a bacterium. We know how to build a workshop, but we can't yet build a bacterium. Prokaryotes have been around for a very long time. Fossils, and some fascinating fossil chemistry, reveal their presence in some of the very oldest rocks, dating from only five hundred million years after the earth formed. They were the dominant life-forms for the next three billion years. Without their interminable labors, the chemical constitution of our planet's surface would have been very different. For example, many beds of iron ore were laid down by bacteria that lived in the ancient oceans and contained their own tiny magnets to tell them which way was—no, not north: *down*. They can do this because the Earth's magnetic field is tilted as well as directional.

The biggest change that those ancient prokaryotes made, however, was far more important than producing iron ore. It was producing oxygen. They didn't even want it for their own use; it was a by-product of their solar-power processes, dumped into the environment by innumerable chemical workshops. The oxygen built up in the atmosphere. This was bad news for those bacteria that lived near the surface of the ocean or upon its shores, and it forced them to change their chemistry so that they could survive in an environment permeated with such a corrosive gas. We oxygen breathers tend to have a rather relaxed attitude to oxygen, but it's a violently reactive chemical, and even we spend a great deal of time painting metal surfaces to keep them from rusting away and hanging up bottles of oxygen-proof material, which we refer to as fire extinguishers. Without oxygen, none of this would be necessary; but then, without oxygen, we wouldn't be here.

THE MAGIC OF DEVELOPMENT

About a billion years ago, when substantial quantities of oxygen had built up in the atmosphere, some of the other organisms used it to invent a new and much more efficient chemical process to extract energy from sugar: They burned it. Not by piling it up in a bonfire, you understand, but by making it combine with oxygen, taking it beyond alcohols, all the way to carbon dioxide. This trick opened the way to much more complicated organisms, called eukaryotes, whose story we take up in the next chapter. These organisms didn't just live, grow, divide in two, grow, eat each other, die or grow, as prokaryotes had done for three billion years. They developed. They began as one kind of complicated chemical system—most familiar to us as the single cell of an egg—and grew from that, in an inordinately complicated series of changes, into creatures that adopted a very different approach to the business of feeding, growing, and, most important, reproducing. Some remained single cells, but even a single-celled eukaryote, such as an amoeba, has far more structure than any prokaryote; it is a chemical factory, not just a workshop. Unlike, say, a bacterium, it confines its DNA within a central core, the nucleus.

The rest of the eukaryotes became multicellular creatures. If viruses are chemical machines, and bacteria chemical workshops, then we can liken such eukaryotes to industrial complexes. Worms, fish, people, they are composed of small or large numbers of cooperating cells, each with its own nucleus. At no stage in its life history does a bacterium ever have to increase its size by more than double, and it keeps pretty much the same structure; then it divides in two and starts the process again. It is easy to believe that this is just chemistry, because simple structures like crystals can do similar things. (So, indeed, can flames, which reproduce rapidly by purely chemical means in any favorable environment.) But human beings, in contrast, have to grow millions of times larger than the initial egg, and millions of times more complicated. Only after this explosion of complexity are they able to produce a new egg, to keep the process rolling.

This is the real magic trick that complicated living things, "higher" animals and plants, reveal to us. This is what cannot be imagined as "just" chemistry, as a complicated cousin of the growth of crystals. A tiny human egg, a sphere about a tenth of a millimeter across containing a few thousand kinds of chemicals and chemical construction kits, doesn't just copy itself, make an almost exact duplicate. Not at all. Instead, a human egg takes in all kinds of other chemicals from its environment and uses them to build something quite different from an egg: an embryo. The embryo is much better than the egg is at extracting food and energy from its surroundings—in this case the lining of its mother's womb—as a result of which it, too, develops into a different, much more complicated system. There's a lot of chemistry involved; but it's so many orders of magnitude more complicated than ordinary chemistry that to call it "just" chemistry is to beg all of the interesting questions.

Soon the developing organism has reached the stage at which it can emerge into the outside world as a baby. The baby turns into an infant, a child, an adolescent; and *only then* does the apparatus for repeating the cycle with a new egg become mature. Even simple animals, like worms and insects, develop. Their simple eggs, which we can easily believe are made "only" from chemical systems, also follow immensely complicated routes.

Insects pass through a larval stage—for example, before you get a butterfly you have to have a caterpillar—and the shape of a larva is totally different from the shape of the eventual insect. Some animals must become four or five different creatures in turn before they get to the point where the egg can reappear. And the same egg can make different kinds of creature, such as a queen bee or the worker bee that started this chapter.

This system is so versatile and adaptable that it can truly perform magic. For example, if an egg is divided into two, either naturally or artificially, then it can produce two babies. Twins. If two eggs of the same kind are fused into one, the usual result is one normal-sized baby. In egg mathematics, $\frac{1}{2} + \frac{1}{2} = 2$ and $1 + 1 = 1$. How can these facts be explained in purely chemical terms?

MONKISH HABITS

People can have blue eyes or brown; pink skin or brown; and red, black, brown, or blond hair. These features are called characters. Their inheritance, from one generation to the next, is a rather chancy thing: for example, two black-haired parents can produce a red-haired child. Nevertheless, we tend to talk as if children are assembled from pieces of their parents: "She's got her father's nose." But, if so, where did her mother's nose go?

It used to be thought that the characters of the child were a blend of the father's and the mother's. According to this theory a short father and a tall mother should have medium-sized children. However, by the same token, a redhead and a platinum blond should have children with pink hair. Punk hairdos aside, pink-haired children are rather thin on the ground. So whatever mechanism it is that determines hair color, it can't be blending.

The correct mechanism was discovered in the 1860s, by a monk called Gregor Mendel. He bred peas with various characters, crossed them (that is, used pollen from one to fertilize the others, or itself), and observed that the proportion of offspring with particular characters obeyed definite mathematical rules. If you cross a plant that produces yellow peas with another similar plant you might expect always to get offspring that produce yellow peas. For some pairs of plants, this is the case. But for other pairs, three quarters produce yellow peas and one quarter green. At the heart of heredity, Mendel found a simple numerical ratio of 3:1. Then he looked at combinations of two different characters—smoothness or wrinkliness as well as yellowness or greenness. Now there are four possibilities—smooth yellow, smooth green, wrinkly yellow, wrinkly green—and Mendel found equally simple proportions: 9:3:3:1. The mathematical simplicity of those numbers was impressive. And Mendel could even explain these simple proportions. Give "yellow" and "smooth" the value 3, and "green" and "wrinkly" the value 1; these are the proportions that you find for those characters when you look at them alone. Then multiply the values together when characters occur in combinations. For instance, "yellow wrinkly" gets the value "yellow" × "wrinkly" = 3 × 1, which is 3. The numbers you get for the pairs are 9, 3, 3, and 1; so the magic 9:3:3:1 proportions are governed by a kind of hereditary arithmetic. Might there be laws of heredity, as there are laws of motion and laws of chemistry?

Later generations, armed with the tools of statistics, found that Mendel's numerical results were a little too tidy: The good monk had wholistic tendencies, like most of us, and might unconsciously have chosen what to do with borderline cases to get the results he hoped to obtain. But by then his central idea had been demonstrated over and over again in less subjective experiments. Mendel realized that there must be factors that determine the color of the peas, and that these factors must be inherited from both parents. Nowadays these factors are called alleles. Alleles are related to, but distinct from, genes, a word that has made its way into everyday language. The story is complicated because for a long time the physical basis of heredity was unknown; geneticists observed what genes did, but had no idea what they were. Alleles are relatively straightforward: They are the differences between genes—the different forms that a gene can take. In any pea plant there is a gene for the color of the peas that it produces; that color may be either yellow or green; therefore the gene "pea color" has (at least) two alleles, yellow and green. Genetic experiments are really carried out on alleles, because you can observe the effects of alleles even if you don't know what a gene really is. Differences in characters are signs of an underlying difference in genes, so alleles match up with character differences.

For simplicity, let's go back to Mendel's terminology and call the things factors. The simplest theory is that each plant possesses a single factor that determines its pea color. The offspring of a yellow-pea parent and a green-pea parent thus inherits two factors. But it can't keep them both; like its parents, it must possess a single factor for color. Does it perhaps select at random between its parents' factors? If so, on average, half the time the offspring will be yellow-pea, half green-pea. But that's not what happens; the ratios are 3:1, not 1:1.

Another possibility is that the offspring inherits two factors, one from each parent. This is what actually happens. But then the parents, being of the same species, must each have had two factors as well. The child inherits one from each parent, selected at random. Call the factors Y and G, so that the possible pairs are YY, YG, GY, and GG. If a pea plant has factors YY or GG, then it is clear what color its peas should be; but what about YG, where the two conflict? Mendel's answer is that in such cases a very simple rule applies: The same factor always wins. Such a factor is said to be dominant; the other is recessive. In peas, Y is dominant. Black hair is dominant in humans, red recessive. So two redheaded parents can't produce black-haired children, but two black-haired parents can produce red-haired children, provided each has a recessive factor for red hair and a dominant one for black. At any rate, each of the pairs YY, YG, and GY leads to offspring with yellow peas; only GG leads to green-pea offspring. Note the numbers: three pairs yellow, one green—the magic 3:1 ratio.

■ GENES AND CHINES

Fine, but what *is* a gene? Consider any character of the developed organism—for example, the color of peas in a pea plant. We know that the color can differ; moreover, we know that such differences can be transmitted by crossbreeding. There exists some kind of genetic variation whose effect, all else being equal, is to vary the color of the seeds. As a convenient shorthand for this convoluted statement, we say that the plant possesses a gene for pea color. We do not mean by this that somewhere within the plant is something whose sole function is to decide what color the peas are. In a complicated, interacting system, cause-and-effect relationships are never that straightforward. What we mean is that we can find something in a plant such that, if it is removed or changed, the color of its peas changes.

As an analogy, imagine trying to understand the workings of a car by removing components one at a time, to test the theory that cars are governed by chines (pronounced "sheens" and named by analogy with "gene" and back-derivation from "machine"). If the wheels are removed, the car fails to move: It possesses "a chine for motion." It also possesses chines for wiping the windshield, annoying pedestrians, vibrating at sixty-two miles per hour, or skidding. Although removal of the wheels causes the car to remain motionless, it does not follow that the wheels cause motion. Removal of the engine also causes the car to fail to move; and the same goes for the gas tank, the transmission, even the nodding dog if the driver is superstitious and won't leave home without it. So the chine for motion is not the wheel alone; chines are not the same as components. Chines for skidding, or vibrating at sixty-two miles per hour, are very subtle indeed. Chines can't be observed directly; their presence is inferred from "mechanical alleles" such as move/stay or vibrate/don't.

In an imaginary population of cars that could breed, swapping mechanical components, we could study the effects of various chines, without ever knowing what a chine really was. We could ask questions like "is the chine for motion inherited?" and answer it by studying cars that lack the motion chine. Do their babies also lack the motion chine? It's just another way of asking, "Do cars that don't move have offspring that don't move?" It doesn't involve dissecting out a chine *as such*.

From this viewpoint, there is a chine for every possible character of a car. That doesn't meant that there's a single mechanical component for every character: For example, speed is a character, but cars do not contain any component, or complex of components, whose sole effect is to make the car go faster. On the other hand, many chines are closely related to single components. Arguably, there is a component that corresponds uniquely to the chine "ability to slow down"—namely, the brake. But the character "ability to slow down" is very closely defined, in terms of our understanding of car mechanics. And even a brake has other functions, such as "park the car on a slope." And you can use the gears to slow down if you want to.

Genes are very like chines: Geneticists tended to think of them as corresponding to bits of an organism's hereditary material, whatever that may be, but—certainly until recently—they studied them indirectly, in terms of alleles. Nobody had seen a factor. It's as if one had to study cars in terms of the problems they develop, but without being able to open the hood and see the bits and pieces.

Eventually Mendel's factors were traced to the chromosomes, tiny bodies within the nucleus of the cell that show up in color when stained. The name chromosome means "colored body" in Greek, and like much scientific jargon is just a fancy way of stating a fact that is obvious and probably irrelevant.

The important thing about chromosomes is that sperm cells and eggs each contain half-sets of chromosomes, rejoined by the act of fertilization. This is done in a surprisingly complicated way. Let's consider people (Fig. 10). Mom and Dad both have forty-six chromosomes, arranged in twenty-three pairs (with the sex chromosomes being a little odd, as are most things about sex). Each of Dad's sperm contain one set of twenty-three chromosomes. These are obtained by taking one of his chromosome pairs, causing it to "cross over" at random so that corresponding genes may be swapped between one member of the pair and the other, and randomly choosing one of the resulting two "shuffled" chromosomes. Similarly for Mom's eggs. The act of fertilization brings together Dad's twenty-three shuffled chromosomes and Mom's twenty-three shuffled chromosomes to make a full complement of forty-six—again, twenty-three matching pairs. Notice that at this stage all the genes on one chromosome in each pair have come from Dad, and all those on the other have come from Mom. Mom's and Dad's genes don't get mixed together on the same chromosome until one generation further along still.

However, whether genes are on the same or different chromosomes in a given pair doesn't affect the proportions in which they affect various characters. The numbers fit Mendel's observations quite well, even if it takes two generations to jumble the genes thoroughly. The probabilities of forming various pairs, bringing together genes from Dad and corresponding genes from Mom, make sense of all of Mendel's patterns. The explanation of his simple mathematical ratios does not require a detailed knowledge of the chemical makeup of chromosomes; in a sense, it doesn't actually matter what chromosomes are made of, as long as they do the things that chromosomes do. Only when we dig deeper, and start asking how they do those things—or when we notice that being fully dominant or fully recessive is actually quite rare—does the chemistry become important. It turns out that the main constituent of chromosomes is one particular chemical, DNA.

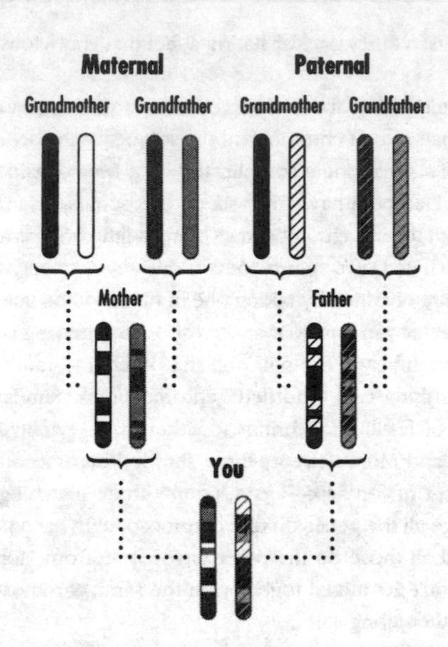

FIGURE 10

Three generations of human genes. Only one chromosome pair is shown, very schematically. There are twenty-three pairs altogether.

■ THE STAIRWAY TO PARADISE

In the 1930s Joseph Needham produced a fine, scholarly book called *Chemical Embryology*. It directed embryologists' attention to the chemical processes involved in—indeed, controlling—development. Needham exhorted embryologists to use the successful techniques of chemistry, instead of just describing the successive changes in form that occur as embryos develop into animals or people. Before Needham, embryology was just developmental anatomy. But he failed to ignite the torch of enlightenment. A quarter of a century later, James Watson and Francis Crick wrote a letter to the journal *Nature* that lit the funeral pyre of any explanation of biological development *not* founded in chemistry. Their bombshell was the chemical structure of DNA, the celebrated double helix—or, as Crick with characteristic immodesty described it in the local pub, the secret of life. Several other scientists played key roles in the discovery, among them a woman scientist called Rosalind Franklin; but Crick and Watson put it all together.

The DNA molecule is long and thin, woven from two strands, and twisted like a screw thread, or helix. Spiral staircases are this shape, although the commonest ones have only a single "strand." DNA has two strands. Think of an ordinary T-shaped corkscrew being screwed into a cork. Each end of the handle traces out a helical path in space, and these two ghostly helices are intertwined. (Don't confuse this description with the helix on the business end of the corkscrew: that's single.) Fig. 11 below shows the general shape.

The central staircase of the Château de Chambord in the Loire Valley has the same double-helix structure. The staircase and its central pillar have open sides, and you can watch people going up and down. Visitors to Chambord are often surprised to find that what looks like a single staircase is actually two. You can go up one while somebody else goes down the other; you will pass on opposite sides of the central pillar, but not meet. It's not hard to see that this must be the case: One set of steps is attached to one of the intertwined strands, the other is attached to the second strand. Because the strands never meet, neither do the two sets of steps. Indeed, one set of stairs at Chambord was used by the nobility, the other by their servants. Transfer between them is possible only by leaving the stairs at one of the floor levels.

The spiral strands in DNA are made from sugar and phosphate groups. DNA also has "steps" in its double spiral staircase, and they are made from those four crucial chemicals that we mentioned earlier: adenine, thymine, guanine, and cytosine. At any level of the Chambord double staircase there are two steps, one on each separate staircase, and they meet at the central pillar. In DNA, there is no central pillar: the steps of the two stairs are joined together by hydrogen bonds (Fig. 11). The four molecules that form the steps are different shapes and sizes, and it turns out that in order to fit together in the space available, adenine can be paired only with thymine, and guanine with cytosine. If we use the initials A T G C for the four compounds, then A pairs with T, and C with G. Each staircase can use all four, but whenever one has A on one step, then the other must have T on the corresponding step, and so on. For instance, if the first few steps on one staircase read AACGTTTC, then the corresponding steps on the sec-

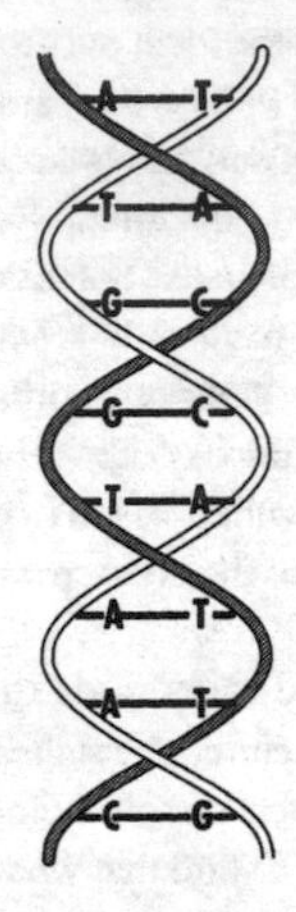

FIGURE 11

The spiral structure of a DNA molecule

ond, complementary staircase, read TTGCAAAG. Just pair A with T, T with A, G with C, and C with G. These four component steps in the DNA staircase are called bases.

■ GENETIC BLUEPRINTS

The bases A, T, G, C are like the letters of an alphabet. In DNA these letters are strung together into long sentences, paragraphs, chapters. . . . The Book of Life? Perhaps the sequence of bases is a coded specification of the organism that eventually develops, a kind of molecular blueprint for the animal or plant to which it belongs. This would make DNA very flexible as genetic material, because you could "write" virtually any description you wanted if you knew the code. Similarly, computers use a binary code, composed of the symbols 0 and 1, but sequences of those symbols can represent words, pictures, *anything*. The DNA molecule is so huge because animals are very complicated things to describe. We really do mean "very." Remember, the DNA in each of your nuclei is like a mile of wire in an aspirin.

If cells carry around these coded descriptions, then every time a cell divides, the description must be copied, so that there are enough descriptions to go around. The fixed pairing of A with T and G with C makes it easy to see a possible mechanism for copying the cell's DNA blueprint. Suppose the double staircase comes apart, unzipping into its two component strands. One strand begins AACGTTTC; the other has the matching pairs TTGCAAAG. But the two strands are no longer joined by hydrogen bonds. There is now room for new molecules of A, T, G, and C to attach themselves to the ends where the hydrogen bonds have been broken (Fig. 12). Because of the matching rules, though, they can't do this at random. The only thing that can attach to the free ends of strand AACGTTTC is the complementary sequence TTGCAAAG—not the old one, for that has split off, but a new copy, assembled letter by letter. And by the same token the only thing that can attach to the free ends of the old strand TTGCAAAG is a new complementary sequence AACGTTTC. We end up with two perfect copies of the original double helix, each containing one strand from the original DNA double spiral.

That particular view of replication created a new set of mental images, widely disseminated in the media. One of the most successful is that of a

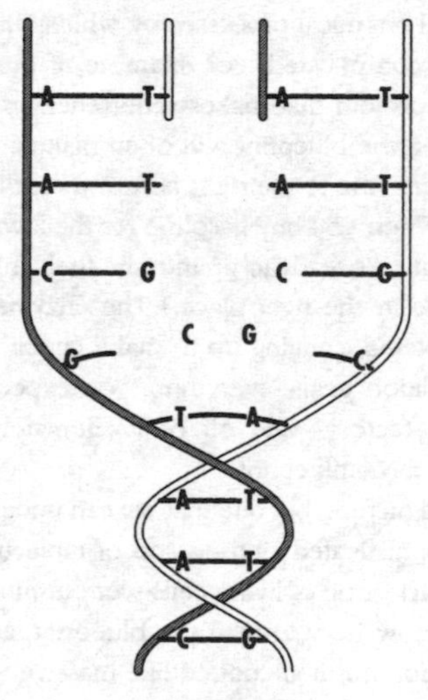

FIGURE 12

Attaching new bases to an unzipped strand of DNA

DNA string as a "blueprint" for development. Just as the blueprint for a car or a personal stereo contains the information that tells you how to build it, so the sequence of bases was viewed as containing the information that directs the construction of the organism. Animal development was simply a matter of following the instructions listed in the DNA that forms its genes. In essence the genetic DNA specifies various proteins, which in turn are used by the organisms for functions determined by their chemistry.

Crick and Watson's discovery was seized on as an explanation of what genes actually are. In this neo-Darwinist viewpoint a gene is a subsequence of DNA. At first it was assumed that each gene identified by classical geneticists would correspond to some connected stretch of DNA, on the apparently reasonable grounds that it didn't need to be more complicated than that. In fact, the DNA sequences that match up with genes are often surrounded by large quantities of junk DNA with no obvious or known function; they may contain apparently useless sequences (called introns), and the sequence corresponding to one gene can even be in the middle of the sequence corresponding to another gene. Nature, it turned out, was not obliged to implement classical genetics in terms of DNA in the obvious way.

In fact, the actual chemical processes by which DNA code leads to an organism are quite complicated. For example, if the DNA specifies an enzyme—a type of protein that makes other chemicals react together in particular ways—then the blueprint will often make a longer protein. The extra bits carry the enzyme to the right part of the cell keeping it inactive until it gets there. (When you buy gasoline for the lawn mower, you carry it around in a can until it's ready to go into the fuel tank: You don't want it to catch fire until it's in the right place.) The enzymes then set to work modifying other proteins, ganging up to make chemical pumps and construction and demolition gear—everything you expect to find in a well-equipped chemical factory—and all of it ultimately comes from the specifications in the DNA blueprint.

It's a complicated picture, but one that we can understand *as* chemistry, albeit of a highly sophisticated kind—a sort of molecular engineering. All the features so characteristic of living cells were turning out to be chemistry. We didn't yet know how to read the blueprint, and we had no idea what a lot of it was for, but it all looked like massive variations on a basic chemical theme.

It turned out that it is important to carry out the instructions in the blueprint in the right order, especially in the developing embryo. The same happens with a cake recipe: If you beat the mix before you've added the eggs and sugar, you get strange results. The chemical factory arranges this by using chemical messengers produced by earlier genes to turn subsequent genes on and off as required. These genetic switches are lengths of DNA on either side of the segments that specify proteins. In fact, in complicated animals and plants, only five percent or less of the DNA specifies how to make proteins; much of the rest is control sequences, which organize the procedures. A lot of the blueprint, in fact, seems to be instructions on how to use the blueprint.

■ GENETIC TIME BOMB

The chemical view of biological development explained more and more of its puzzles, its previously inexplicable "magic." For example, it has been known since the earliest days of genetics that if a change is made in a gene, it need not affect the organism that carries the gene. Instead, it may affect its progeny. A classic case is the difference between left-handed and right-handed snails. Snail shells are spirals, like helices coiled around imaginary cones, and if you look down on them from the tip of that cone they may coil either clockwise or counterclockwise—right or left. It's tempting to think that the direction of coiling must be specified by some gene—after all, the "blueprint" image assumes that every aspect of an organism's structure must be coded in its DNA—but how? Breeding experiments like Mendel's revealed something exceedingly strange. What you'd expect is that there is a gene that specifies, "this organism will develop a left-hand thread," and another gene specifying a right-hand thread, and that the direction in which a snail's shell coils depends on which gene it inherits from its parents. (And then, of course, one might be dominant and one recessive, and you'd expect various proportions of lefts and rights in succeeding generations.)

Not so.

Instead, there is a gene that specifies, "This organism's offspring will develop a left-hand thread." And a corresponding gene for right-hand threads, of course. The effect of the gene is delayed for one generation, like a genetic time bomb.

On the old view of genetics as determining an organism's own characters, this effect is pretty subtle. You have to see "tendency to bear left-threaded children" as a character of the parent, rather than "left-threaded" as a character of the offspring, and once you allow that, what else might you contemplate? But in the new view of DNA as a blueprint, there is a relatively straightforward answer. It lies in the mechanism by which the DNA specifications are turned into proteins. We'll describe this first, and return to the snails a few pages further on.

CODING FOR PROTEINS

Remember that proteins are complicated chemicals formed by stringing together simpler chemicals called amino acids. There are twenty-two of them, some rarer than others, with names like glycine, valine, glutamine, and so on. The DNA blueprint specifies the order in which these are to be assembled— Hey, wait a minute. In DNA there are only four different bases (A,T,C,G). How can they specify so many amino acids?

Here's an analogy. How can a mere twenty-six letters specify all the words in Webster's Dictionary? Answer: Each word is a string of letters. The same goes for amino acids, and we can even make a plausible guess at how big the "code words" are. You can make precisely sixteen different strings from two DNA bases: AA, AT, and so on up to GG. That's not enough for twenty-two amino acids, so we try strings of three bases, of which there are sixty-four—more than enough. It turns out that nature does indeed use strings of three DNA bases, or triplets, to specify proteins; this is the celebrated "genetic code." However, in general several different triplets may correspond to one amino acid. For example, CAA, CAG, CAT, and CAC all specify the amino acid valine, whereas GTT and GTC both mean glutamine. Moreover, three triplets (ATT, ATC, and ACT) don't correspond to a protein at all; they mean "Stop."

How does the chemistry translate this code into proteins? Proteins are not made in the nucleus, where the DNA is, but by bits of the cell known as ribosomes. In principle the DNA could be transported to the ribosomes, but what would happen if part of it got hijacked en route? Instead, the main office prudently keeps the master blueprint and sends a photocopy to the protein works by courier. This copy is in the form of a second type of nucleic acid, ribonucleic acid (RNA). It differs from DNA in interesting and remarkable ways, but all we need to know about it here is that it, too, has a sequence of bases of four types, which are almost the same as those for DNA. (One DNA base, thymine, is replaced in RNA by uracil, a slightly different chemical. It's like writing one letter in a different font, CA**T** rather than CAT; it affects the chemical mechanisms, but not the meaning of the code. Don't take this image literally: In practice, uracil is denoted by the symbol "U.") The cell makes an RNA copy of the DNA sequence in the nucleus and sends it over to the ribosomes. This "photocopy" is known as messenger RNA. Actually the sequence on any chromosome is broken down into pieces, so thousands of short photocopies are sent rather than one long one. And if any gets lost en route, there's another copy on the way.

Yes, but who builds the proteins? Over at the protein works is a team of very specialized, short segments of RNA called transfer RNA. Each recognizes precisely one DNA triplet, catches a molecule of the corresponding amino acid, and glues it in place on the growing protein molecule. Like any factory, the cell has its own office staff and its own specialist technicians. You may think this is a complicated way to go about making proteins, but if you think about it, nothing much simpler would be reliable enough to hold together for billions of years.

A lot is known about how genes produce proteins, but there's much more to making an organism than just throwing together a heap of different proteins. In particular, they have to get to the right place at the right time. The developmental biologist Lewis Wolpert has developed a theory of "positional information" to bridge the gap between chemistry and form. The idea is that certain key chemicals can be distributed in the developing organism so that, for instance, the concentration is higher at one end than at the other. Particular cells in the organism can then use that chemical to work out where they are, and then look up their DNA instructions to find out what to do. In this way the same primal cells can develop into different kinds—muscle, kidney, nerve. It is as if the organism provides its cells with a map (the differences in chemical concentration) and a book (DNA). You look at the map to find out where you are, and then at the book to find out what to do. This is a very flexible system, in principle similar to the way audiences in big sports stadiums can create pictures by each holding up a colored card. To get it right, you have to know where you are sitting and what card to hold up. The main problem with Wolpert's theory is that in some respects it is too flexible; it allows more variation than actually occurs. You can make any picture out of colored cards.

■ SELFISH GENES

Not all DNA codes for proteins. We've already mentioned that there is a high proportion of "junk DNA" too—genes that perhaps were useful in the past, but have degenerated or had their context changed; odd bits of viruses left around but not doing any harm; endless repetitions of short sequences of bases, situated between genes; longish sequences repeated all over the chromosomes, inside or outside genes; introns inside the protein-making bits which get spliced out of the RNA copy to make messenger RNA; and other odds and ends that just seem to be along for the ride. The junk DNA replicates happily along with all the rest. Much of it probably has no function whatsoever *for the organism,* though some may have an important function in the bookkeeping of DNA chemistry. Richard Dawkins introduced the concept of the "selfish gene" (in a book of the same name); basically, the idea is that organisms are just the gene's way of making copies of itself. As Herbert Spencer, Charles Darwin's interpreter to the general public, put it, a chicken is just the egg's way of making another egg. In Dawkins's picture, both chickens and eggs exist only because chicken genes use them for replicational purposes. From this point of view, the sole criterion for a gene's success is its ability to get replicated. Any DNA that can arrange to be replicated into the future has a future. Recall that viral DNA can replicate only by courtesy of a host bacterium. These short, repetitive sequences of bases of parasitic DNA are even more successful: They don't have to go to the trouble of getting protein coats of their own, they don't have to go out into the big wide world looking for a bacterium to sponge off. Like cats in a human household, their reproduction and maintenance is taken care of, even if they do nothing for their host.

DNA, said Dawkins, doesn't "care" what meaning—if any—it has for the organism. If some sequence of bases is involved in anything that promotes its own replication, then you will get more and more copies of that sequence. Conversely, a sequence that has trouble getting itself replicated eventually becomes extinct. Naturally! What else do you expect? Success breeds success. Things that are good at getting copied will soon exist in large numbers; things that don't, won't.

The DNA world is, and always has been, highly competitive. So some sequences do well by evolving into enormously complicated blueprints and maintaining their existence by making bacteria, amoebas, monkeys, and people to look after their continued replication. Other DNA sequences live in the environment created by these blueprints, even live right alongside them in the same cellular factories, and get replicated by the same machinery, without blueprinting anything at all.

DESIGNER GENES

Back to the snails.

While a mother snail's eggs are still in her ovary, before they have been fertilized, they contain a quantity of prepacked messenger RNA, provided by the mother along with everything else. The protein works receives a stack of maternal photocopies, wrapped in packages labeled "Open Now," "Do Not Open Until the First Stage of Cell Division," "Do Not Open Unless Temperature Drops Below 50°F," and so on. Those early stages of development determine, among other things, the direction of coiling of the shell. So the way the shell coils is determined by the mother's genes, not by those of the developing snail itself. Not long after, the snail's own DNA starts making messenger RNA and interpreting the instructions in its own genes; but the mother gives the process a jump start.

Another maternal-effect gene with more dramatic results occurs in the fruit fly *Drosophila*. Geneticists are fond of fruit flies because they breed very rapidly and display a wide variety of easily recognized characters. One *Drosophila* gene leads to offspring that appear to be perfectly normal, but are infertile. Appropriately, the gene is called "grandchildless." How this infertility comes about is rather different from how Mother Snail affects Baby's coils. It involves symbiosis, cooperation by separate organisms in some process for mutual benefit. (Symbiosis is not at all unusual. Inside the stomach[s] of a cow are symbiotic bacteria that help it digest cellulose. Worms and termites have similar symbiotic bacteria for the same purpose.) Inside a developing egg is material which will eventually become the developed organism's own eggs (or sperm); this material is known as germ plasm. In unfertilized insect eggs, there is a population of symbiotic bacteria that are needed to label the germ plasm. The mother's ovary, programmed by *her* DNA, must act to place this retinue of assistants into the egg at the right time and in the right place. If the mother carries the gene grandchildless she gets it wrong. Everything else works fine, but the bacterial assistants aren't in the right place in the daughter's own egg when they're needed, so they don't do their job.

Between the 1920s and the 1980s geneticists built up a huge library of genetic aberrations in fruit flies. (They did the same for many other organisms, too.) Some of the effects are relatively minor—lack of pigment in albinos; more, less, or different hairs on the body; eye color—but some are very significant, such as vestigial wings. Many changes cause development to grind to a halt altogether, because a crucial instruction is missing or wrong. These are called lethals. Many of these effects were understood long before molecular biology came on the scene. Others have been found since, but haven't changed anybody's mind much. However, some genes were discovered that have very important systematic effects upon development.

Some variants of *Drosophila* were found in which some antennae were replaced by legs (a gene called antennapedia), or whose second wings—which are usually highly modified—had developed as normal wings (bithorax). The genes responsible for these effects weren't the common kind that determine characters—antennapedia can produce a perfectly normal leg, but in the wrong place. They are known as homeotic genes, and they are evidence that the developmental process involves variations on a theme, because they change one variation into another.

There were biologists who had already become convinced, through comparative studies of development, that the various appendages along the length of the fruit fly were *all* variations on a theme. To them the fruit fly was like a Swiss Army knife—a body plus a set of attachments, which could be arranged in many different ways during the manufacturing process without difficulty. Other biologists had concluded that early ancestors of the fruit fly had four wings, and that one pair had since been modified by evolution, to make tiny gyroscopes that help the insect stay the right way up. Both types of biologist were greatly heartened by the discovery of homeotic aberrations, just as paleontologists, who think that fossil creatures with more toes than usual were ancestors of the horse, are heartened by the occasional birth of a modern horse with too many toes.

Homeotic genes change our picture of how the DNA blueprint works. Imagine an aircraft factory which suddenly starts to turn out airplanes with perfectly formed landing gear where a wing should be. It is possible that a draftsman got drunk and systematically drew an entire plan for landing gear in the wrong place. It's more likely that there was a whole stack of numbered blueprints for various pieces of the plane, plus a master blueprint of the whole machine consisting of labeled instructions of the type "fit item number 555 here." A secretary mistakenly typed "575" instead of "555," and landing gear (blueprint 575) ended up where a wing (blueprint 555) ought to have been.

In short, it looks as though homeotic genes affect the "administration" of development, not the nuts and bolts. They program not individual pieces, but the way those pieces are organized. They program crucial switch points in development.

When scientists discover something as interesting as homeotic genes, they want to dig deeper. Obviously they will want to know whether there is something special about such genes' DNA. They discovered that all homeotic genes contain a particular DNA sequence, called a homeobox. It's fairly short, and it may be a warning label, a signal to the chemical factory: "I am a genetic switch." Moreover, the same special DNA sequences, more or less, turn up in all sorts of other creatures: fish, mice, people.

Moreover, the way in which homeotic genes function is to some extent understood on the chemical level. In *Drosophila* at least, and to some extent in other organisms, homeotic genes are the ones that turn other genes on and off. Their products fit the turn-on sequences of some particular set of genes, so a change in a homeotic gene will either fit nothing, in which case development grinds to a halt, or it will fit some other set of genes. In short, DNA sequence number 555 specifies an antenna, sequence 575 a leg, and the homeotic gene allele antennapedia specifies the wrong number. We mentioned the existence of junk DNA—for example, sequences of bases that are relics of evolutionary ancestry, no longer used in normal development, "turned off" permanently. A change in a homeotic gene could activate such a genetic fossil, and a fruit fly can suddenly develop features that haven't been seen since its great-great- . . . -great-grandmother was alive.

Maternal effects, homeotic genes, and lots of other anomalies and developmental oddities have made molecular developmental biology into one of the most active research areas. The hope is that further understanding will let us make purposeful changes in the development of animals—and perhaps even of ourselves—by editing, or even rewriting, parts of the genetic blueprint. There are ethical issues here, as well as technical ones;

but the prospect of eradicating or otherwise dealing with genetic diseases, or of producing pest-resistant plants, is attractive. What used to be magic is becoming a biochemical technology: genetic engineering.

■ RECURSION IN DNA

The phrase "genetic engineering" trips easily off the tongue. People talk about it as if it's just a matter of making a few changes to the DNA blueprint, and then you get a new organism. The problem is that the blueprint image is a huge oversimplification. It suggests that the developing organism just reads its way through its "Book of Life" sentence by sentence, building bits and pieces, rather like a child assembling a toy car from a plastic kit or a factory building a real one. But the DNA blueprint is not like this. In particular, it is recursive—self-referential. In everyday life we learn to distrust self-reference—for example, the dictionary definition of "the" as "*the* definite article." But computer scientists find self-reference enormously useful—for the same kind of reason that the definition "A wall is a row of bricks with a wall built on top of it" does actually tell you how to build a wall. (It's just a short way to say "A wall is a row of bricks with a row of bricks built on top of it with a row of bricks built on top of it with a row of bricks build on top of it with . . .") What it doesn't do is tell you how to *stop,* but that can be taken care of in other ways. Because DNA chemistry is like computing, it also makes use of recursion.

Car blueprints specify how to make complicated little components like gearwheels, how to combine them into a gearbox, what to hook the gearbox to, and so on. Biological blueprints do both more and less. They do less in that their main task is to direct the manufacture of tools—proteins that go on to make other changes as a result of their own peculiar chemistry. It is as if a car blueprint specified which factory workers were to be activated, but only the workers—various specialists—knew what they had to do. So the blueprint says "Activate Fred," and no more; but Fred knows that his job is to put the gearbox in.

However, the DNA blueprints also does more than the car blueprint. Car blueprints do not specify exactly how every component should be made: Nuts and bolts, for instance, are taken off the shelf from the storeroom. But DNA blueprints have to specify the precise chemical structure of all the "worker" molecules they activate. The car blueprint doesn't have to specify designs for the tools used by the factory workers, such as power screwdrivers; but the DNA blueprint not only has to specify designs for the tools, but also for the tools that make the tools. And it must provide instructions for how to read the instructions.

We want to spend the rest of this section showing you just how incredibly complicated the DNA blueprint really is. Please don't be put off. The whole point is that the system is so complicated that it's almost impossible to understand. *We* certainly don't, and we don't expect you to either. What we want you to take away from this discussion is meta-understanding: comprehension that the workings of DNA become fiendishly complicated if they are catalogued in detail. We have to show you some of those details to convince you, but we don't want you to remember them.

The DNA blueprint operates on four different levels. Much of the cell's tool kit is made from RNA, and the RNA structure is read directly from parts of the DNA sequence. The ribosomes—the tiny automated tools that read messenger RNA and assemble proteins—are made from two kinds of RNA: large and small ribosomal RNA. The "disposable" linkage molecules that collect the various amino acids and bring them to the ribosome for assembly are transfer RNA; those, too, are read directly from DNA subsequences. The enzymes that do the job of reading the RNA tools from the DNA subsequences are proteins, known as RNA polymerases. Like all proteins, they have to be made—and they need the rest of the system, *including themselves,* for that to happen. So do the important proteins of the protein-assembly system, the ribosomal proteins. The entire structure is self-referential.

But there is yet another level of recursion: when to make each of the bits, and how much to make. These decisions are controlled by proteins specified by yet other parts of the DNA chain by way of messenger RNA. The controlling proteins fit onto special DNA sequences on either side of the transfer RNA, ribosomal RNA, and protein-coding sequences; and they enable or inhibit RNA polymerases sticking to that protein-coding sequence. The production of the control proteins is of course controlled by other (occasionally the same) control proteins.

There are special sequences on the RNA, called ribozymes, which act like enzymes to cut and paste the RNA sequences that the polymerases copy from the DNA, and pick out the messenger RNAs inside them; then these messenger RNAs are pasted together to make the compound gene sequences that code for proteins. There are special enzymes and other proteins that escort these molecules into the cell nucleus, where the DNA is, or into the cytoplasm, where the ribosomes are. There are even special proteins, called histones, that pack the DNA away when it's not in use.

The image of DNA as a blueprint sweeps all of these complexities under the carpet. But it's not such a bad one provided you bear in mind how a really sophisticated factory uses blueprints. A factory blueprint is not just a list of instructions such as "Put bolt R141 into hole B775." It contains, implicitly or explicitly, lots of extra instructions—where to find bolt R141 in the storeroom, when to order more stock, and so on. Modern factory technology is also recursive: The machine tools used to make cars are themselves made by machine tools.

In this sense a living cell is very like a factory, and basically the operation of the cell is all chemistry—but chemistry-as-molecular-computation more than chemistry-as-colored-liquids-in-test-tubes. The complicated chemicals do such unlikely things, keep on doing them, and interact with so many other chemical systems doing even more unlikely things, that we have to use special biological words, not chemical ones, to describe them.

"Genetic engineering" reaches into this incredibly complex network of chemical computations, makes a few changes, and observes that something interesting or useful results from them. All without really understanding how the chemical factory works. Would you be happy doing that to a real factory—even if you couldn't detect any side effects? Or would you be worried that although you've replaced a few random bits of machinery and are now making teddy bears instead of aluminum cans, one day the production line will screech to a halt as fluff from the teddies builds up and clogs the cooling system?

TROUBLE WITH TRUBBLES

We've seen that it is possible to communicate chemical needs to an alien intelligence, by treating chemistry as complicated "structured" physics. Could we also communicate biological needs, by treating them as complicated "structured" chemistry? Does genetic engineering on Zarathustra follow the same rules as on earth? Or is their fuel our cosmetic this time?

STANLEY: Captain, I have serious news. The hydroponic gardens have been damaged. Without them we will run out of food and oxygen.
CAPTAIN ARTHUR: How did that happen? It was a perfect landing.
STANLEY: Yes, but the local gravity is different. A robot gardener misjudged its position, hit a watering line, and exploded, burning the entire stock to a frazzle.
NEEPLPHUT: Do you have the genetic greenprints on record?
CAPTAIN ARTHUR: Pardon?
NEEPLPHUT: On my world all living organisms are determined by linear replicating sequences of simple chemical molecules. I am no biologist, but if your biochemistry is similar, then perhaps our molecular engineers will/will not (delete whichever is inapplicable) be able to grow new food plants for you.
STANLEY: I'd give my right arm for some soyasynthasteak!
CAPTAIN ARTHUR: You may have to, Stanley. Tell Neeplphut about DNA. [*He does so.*]
NEEPLPHUT: Interesting. Zarathustran genetic material, ZNA, employs a code too, but it uses pairs selected from *eight* bases, in accordance with the Principle of Octimality. Are you *sure* your DNA uses triplets of four bases? That is most strange.
STANLEY: Octimality?
NEEPLPHUT: Nature prefers the number eight. I explained that when you told me that full electron shells have eight electrons in them.
CAPTAIN ARTHUR: So *that's* why you got so excited! We didn't understand a word you were saying.
NEEPLPHUT: The Principle of Octimality is our most treasured natural law. I am encouraged that you find 64 distinct "triplets," as you seem to insist they must be, because 64 is the square of 8. Indeed, ZNA provides the same number of possibilities; there ought to be some structural correspondence. But only four bases? I am astonished.
STANLEY: It's been thoroughly verified, Neeplphut.
NEEPLPHUT: Then I will/will not (delete whichever is inapplicable) take your word for it. I suppose that 4 can be viewed as the ⅔ power of 8. But ⅔ is a very curious number, not at all octimal. . . . [*Reluctantly:*] It is true that 4 is a fractional power of 8; something that a rather unorthodox scientist of ours calls a fractimal. Terrans have fractimal genetic material. . . . I must tell Brondelmat; he's bound to find it significant. [*The discussion proceeds.*

ZNA replication processes are very similar to those for DNA, and so on. Somewhere in the fine details of messenger ZNA, however, inconsistencies appear. Finally Neeplphut brings along his colleague Fimplut, a biochemist, with a bundle of textbooks, and he and Stanley compare the molecular structure of the terrestrial DNA with the Zarathustran genetic material ZNA.]

FIMPLUT: It is *enormously* different. I am not sure that our techniques will work at all.

CAPTAIN ARTHUR: Does it matter? The DNA blueprint is a complete description of the plant. Can't you just work from that in the same way that you do from ZNA? Or maybe translate from one to the other?

FIMPLUT: Well, um, that depends/does not depend (delete whichever is inapplicable). What I mean is, we do not do it that way.

CAPTAIN ARTHUR: *What* way?

FIMPLUT: We do not have a complete understanding of how to go from ZNA base sequences to plants. We have to insert engineered ZNA into existing organisms—we mostly use trubbles—and let them do the job for us.

CAPTAIN ARTHUR: Then try it with DNA instead.

FIMPLUT: I fear that DNA will either have no effect on trubbles, or it will kill them. I am eager/reluctant (delete whichever is inapplicable) to try the experiment, but neither result will help you make a salad.

CAPTAIN ARTHUR: But the information you need is all present in the DNA!

FIMPLUT: The information in Phlegbart and Snulligan's *Perispherical Potentate* is present on any commercial hypercube, but without a hypercube player, the chant of the three little school-fnergs must remain purely implicit. Not only that; there is the problem of maternal ZNA. Trubbles supply start-up information to the developing organism from their *own* ZNA—fortunately, it is very flexibly programmed. Do you have a flexible supply of DNA maternals with you?

STANLEY: No, all the hydroponic plants were destroyed.

FIMPLUT: Then you have a problem. We cannot start up a genetic environment without the requisite start-up information. It is like trying to program a smell-recorder without a manual. Or with one, for that matter.

■ CONTINGENCY PLANS

Fimplut has pointed to a serious difficulty. The metaphor "genetic code" suggests that the entire structure of an organism is implicit in the DNA "message." But is it *enough* to know the code message, even given that you know what each coded symbol represents? Let's think about another area of science where we encounter codes: computer programs. In a computing metaphor, DNA contains the instructions, neatly encoded in "digital" form (with the four digits A, T, C, G). The cell acts like a computer, carrying out the instructions specified in the program. The attraction of this metaphor is that it recognizes more explicitly than does the blueprint one that the developmental process is flexible. Much of the power of computer programs lies in their ability to change the actions taken according to circumstances, an ability known in the trade as conditional branching. The DNA program also has this ability.

Nowadays, when you buy a new computer, you can't just switch it on and expect it to work. It comes with a separate start-up disk, which installs various useful features. As well as the program code in the new computer, you need the start-up disk, which is produced on a previous computer. The same goes for the DNA code. An egg (computer) needs not only its own code (DNA sequence in its genes) but a start-up disk (mother's provision of RNA information). The early stages of egg development are controlled not by the egg's genes, but by the mother's. We've already seen how the mother snail provides her babies with start-up disks that tell them which way to coil. Because Mendel's character "color of pea" refers to the pea plant's "eggs"—its seeds, peas—it looks as if that particular gene is a maternal-effect one. Of course, the maternal RNA start-up disk is programmed from her DNA, so geneticists don't worry much, because they can argue that DNA code still controls the whole process. The fact that it's displaced by one generation doesn't bother them.

■ CONDITIONAL BRANCHING

The developing egg is affected not only by maternal genes, but by its physical environment. Chemical systems are generally sensitive to temperature. Reactions usually proceed faster, and often proceed differently, when the chemicals involved are warmer. Bread rising is an example: The gas that causes it to rise is produced more quickly in a warm kitchen than in a cold one. Chemical reactions are also sensitive to the medium in which they occur, the presence of impurities, and many other factors. The chemistry of development is not immune to such effects, but it has learned to cope with them. There are at least two approaches.

Some developmental systems deal with the problem by ensuring that all conditions are carefully controlled, just as a factory making polyethylene will control temperature, humidity, and air purity. Mammals are exponents of this approach: Their eggs develop inside a mother, whose own thermostatic systems keep her body—and hence the eggs that she is nurturing—at a suitable constant temperature. The chemistry of mammalian development is very sensitive to temperature change. If mammalian eggs are grown *in vitro,* as is done for part of the development of "test-tube babies," then all conditions in their immediate environment must be held within very narrow limits, otherwise the embryos fail to develop. A variation in temperature of 3°F is enough to wreak havoc.

Amphibians, such as frogs, use a different method: contingency planning. Their eggs, like those of most nonmammals, develop outside their bodies, where control is impossible. English and North American frogs lay their eggs in ponds in early spring, and the temperature can vary from 80°F in the daytime to just above 32°F—near freezing—at dawn. They cope by providing a selection of enzymes for development: "fast" ones for use when it's cold, "slow" ones for when it's hot. Many organisms that are required to function over a wide range of temperatures use the same trick.

Quite a lot of frog DNA is used to program these alternatives. Which strategy makes more effective use of the DNA program? Does the mammalian mother use more DNA to program the development of her own body, with its thermostats and controllers of chemical concentrations within the womb, than the frog does in programming alternative chemical pathways? No: Frogs have longer DNA sequences than mammals, even though mammals are arguably more complex creatures. But it's not clear whether shorter DNA sequences are more effective, or merely shorter. Moreover, the question is mixed up with others. Some amphibians protect their eggs by making them bad-tasting or even highly poisonous, and even provide warning colors to show this. The female mammal has long-range eyes and a sophisticated brain to detect and avoid predators. The DNA sequence must also take care of such items. We don't yet know enough about the way the use of DNA is optimized to decide which approach is best. Nature seems happy with both.

The metaphor of DNA as a program has another appealing feature. Programs are software, as distinct from hardware—the actual electronic components—and as such are easily modified. Modifying blueprints is possible, but it's not a feature of blueprints that springs to mind immediately. The evolutionary process, about which we say more in the next chapter, can readily modify the DNA sequence. Indeed, to some extent the problem for developing organisms is to prevent this from happening too often. Any copying process is prone to errors. If you photocopy a document and there is a smudge on the glass, the copy is imperfect. Medieval monks made many errors when copying manuscripts. We all know the game of Telephone, in which a message is sent through a series of people and "Send a dozen tanks; the town is being attacked" ends up as "Send Cousin Frank; the clown is feeding the cats."

Copying errors in the DNA program are called mutations, and they can have far-reaching effects. They can change the sequences in the genes that determine proteins, they can change the control sequences that decide when to switch genes on or off, they can change the proteins that should do the switching on, they can change the receptors that regulate the amounts of hormone secreted by glands. . . . All of these chemical interactions can be tweaked so that the developmental path of the descendant is different from that of its ancestor. Because some genes control the timing of the switching on of other genes, the whole developmental process can be accelerated or retarded; individual enzymes (like the set in a frog embryo that compensates for variations in temperature) can be tuned up or down. This can result in some systems developing more quickly, or others being delayed.

SYSTEMATIC CHANGES IN DEVELOPMENT

In most mammals the development of the heart and the circulatory system has been accelerated, so that they can be used to transport nutrients from the lining of the womb to appropriate parts of the embryo. In human embryos, the heart is the first functioning organ to develop. In contrast, some organs are so retarded that some of us don't have time for them to develop at all—for example, wisdom teeth. Eventually, with luck, we will retard wisdom teeth so far that we do away with them altogether.

Most animals stop developing around the time they become sexually mature—they can't stop developing before that stage, since sexual maturation is itself development—so anything not developed by then loses its chance. This opens up two very different options, both of which enable the adult form of the descendant to resemble the juvenile form of its ancestor. They are called progenesis and neoteny, and we take them in turn.

Where it is to the creature's advantage to reproduce rapidly—for instance, when supplies of food are ample and the name of the game is to grab as much as possible—then those mutations that make puberty occur earlier, resulting in precocious sexual behavior, will be fostered. This is progenesis, and it can be seen in many insects that exploit temporary gluts of food. Mushroom flies may breed when still maggots; some cadaver beetles breed as grubs. Progenesis is the preferred explanation for the small size of individuals in isolated populations of mammals, such as may occur on islands or in other cases where there is strong competition. The adult Maltese elephant, now extinct, was only three feet tall.

The opposite effect, neoteny, is also very common: when life as a larva is really good, the larva delays transforming to the adult form. There is a Mexican salamander which, like the frog, begins as a tadpole and changes, or metamorphoses, into an adult form. However, the salamanders known as axolotls do not metamorphose at all, but breed while still tadpoles—enormous, well-developed tadpoles, but tadpoles nonetheless. Humans are extremely neotenous primates; we retain all kinds of juvenile characters such as curiosity, hairlessness, head shape, and cuddliness into adulthood. Women are perhaps further along this road than men. Other species hover on the verge of doing away with their adult form altogether; they can nearly breed as larvae, but still need the glandular changes that occur when they metamorphose into adult form to kick their sexual apparatus into operation. Thus the mayfly lives for three years as a larva in a freshwater stream, and has reduced its adult life to one day of reproductive activity.

All this suggests another question. Differences and change aren't the only things that require explanation. If small changes in the developmental program can make an enormous difference so easily, why is everybody the same? There must be lots of differences in the environment, if not in the genetics, so why don't we see big differences within a species? Nongenetic influences can be important for the behavior of genes—for example, the drug thalidomide, administered in the late fifties, produced major defects in developing human embryos. Again, the question is not so much why that happened as why lots of other drugs, or mutations, don't have similar effects. If development is all chemistry, then it should respond to outside chemical influences.

VICTORIAN VERSATILITY

We let a Victorian lady answer this question for us. She is Augusta Ada, Countess of Lovelace, daughter of the poet Lord Byron and a founding figure in computer science. She developed many of the programming ideas related to Charles Babbage's famous but unbuilt "analytical engine," a mechanical computing machine. She appears here because we need the help and advice of someone versed in computing who knows Victorian natural history backward and forward but is unaware of current theories of biological development. Like you, she has read this far and thus knows many things of which the original Ada could have no inkling. She is talking to Wallace Lupert, a fictitious figure who is intended to represent a consensus of modern biologists.

AUGUSTA ADA: It would appear to me that substances like thalidomide, which have a direct effect on a developing organism, should be very rare. It is only to be expected, because the pattern of development of most creatures is very stable.

WALLACE LUPERT: But bees and termites develop differently on different diets. Some fish can adopt different forms if you starve them, but still breed. And what about shepherd's purse?

AUGUSTA ADA: I see what you mean. In a crack in the pavement it can produce just two leaves, one flower, and one purse with a seed, in about ten days, but in a flower-bed it can produce a cubic yard of plant with thousands of purses, each with several fat seeds, every week for three months.

WALLACE LUPERT: Quite. Are you saying that *food* is a rare substance?

AUGUSTA ADA: It does not seem especially surprising to me that reducing an organism's food supply might lead it to follow a complete developmental path, but in a more restricted manner. At each stage, as the genetic switches are thrown, it does as much as it can with the available supply of energy and material. Bees and termites are rare special cases, and no doubt their very flexibility is programmed genetically. The analytical engine, too, can change its computations in response to outside signals—provided I write the program that way. But in any case such behavior is also rare. The diet of most animals does not affect their development greatly. Think of stick insects, or those breeds of butterflies that mimic leaves. If they change their development, they lose their disguise and are eaten by predators.

WALLACE LUPERT: They have a very "tight" DNA program.

AUGUSTA ADA: That is correct. An error made in instructions to the analytical engine may result in the cessation of its calculation, not in a completed calculation with an incorrect result.

WALLACE LUPERT: But some creatures are much more versatile. What about rats, cockroaches, or lawn plantains, which can live in a huge range of habitats without any developmental changes at all—or at least, nothing very great?

AUGUSTA ADA: In such cases, evolution has selected for versatility. Humans are unspecialized as a result of neoteny, their development halted before specialization sets in.

WALLACE LUPERT: Ah! The badger's tale!

AUGUSTA ADA: I beg your pardon?

WALLACE LUPERT: It's part of *The Sword in the Stone*, a book that was written after the age in which you lived. The badger is writing a treatise that explains why Man has become the master of the animals. All embryos look pretty much the same, you know. The badger's theory is that when God manufactured living creatures, he first made them as embryos, and called them all before him, asking them to consider what they wanted to be. He offered them a gift: to alter any parts of themselves into something that would be useful in later life. The embryos thought it over, and one by one they made their requests. The badgers asked three boons: a thick skin, a mouth that could be used as a weapon, and garden forks for arms. Everybody specialized in some manner—until God got to the last embryo, Man. "Please, God," said embryo Man, bootlicking as usual, "I think that You made me in the shape which I now have for reasons best known to Yourself, and that it would be rude to change." God then congratulates Man on being the only creature to solve His riddle: that adaptability is superior to specialization.

AUGUSTA ADA: Remind me to read that one day. The irony, of course, is that Man specializes too.

WALLACE LUPERT: What do you mean?

AUGUSTA ADA: Man specializes in being versatile. Human genes specify the character of *versatility*. That is a character that also requires a tight program.

WALLACE LUPERT: You think human genetics is very rigid, then?

AUGUSTA ADA: All genetics is very rigid. That is why species differ so strongly. All individuals in a species have virtually identical genes. Except for a few mutations, of course—either valuable characters on the way in or bad ones on the way out.

WALLACE LUPERT: You mean everybody gets the same genes from both their mother and father?

AUGUSTA ADA: Of course! Everybody gets the same genes, period. Those are the only genes there are. Look, if genes could vary very much, we would all be different shapes and sizes. And everything would get mixed up as a result of reproduction. No, genes have to be mostly the same. You can get the odd mutation from one parent, of course. Not from both—that would be highly improbable.

WALLACE LUPERT: Okay. Nowadays, if you've inherited the same gene from both parents, we say you are homozygous for that gene. When you get different alleles from your parents we say you are heterozygous for that gene. You're saying that any individual is homozygous for almost all genes.

AUGUSTA ADA: I am saying far more. I am saying that most genes are the same in *all* individuals. But I suppose that any differences in the population would make some individuals heterozygous for some genes, after a generation or two, because of random breeding; so my statement is equivalent to yours in practice. You can tell by animal-breeding experiments. You seldom see extensive mutations.

WALLACE LUPERT: Hmmm. What would you say if I told you that people have studied the numbers of kinds of protein in about 100,000 genes, without doing breeding experiments, and discovered just from those chemical differences that any given individual is heterozygous for about 10 percent of its genes? In every species, humans included?

AUGUSTA ADA: I would be exceedingly surprised.

WALLACE LUPERT: But it's true. In fact, about a third of all genes exhibit differences somewhere in the population.

AUGUSTA ADA: Then why do we not see major differences in people?

WALLACE LUPERT: The genes are different, but the organisms that they produce all look much the same.

AUGUSTA ADA: That is merely restating my question. In any case, I understood you to say earlier that *Drosophila* is a popular experimental animal because changes to its genes produce many changes in its appearance.

WALLACE LUPERT: Yes, I did say that, and it's true.

AUGUSTA ADA: Then you have propounded a paradox.

■ CANALIZATION

The paradox was anticipated in the 1950s by Conrad Waddington, now seen as one of the greatest embryologists. He devised experiments to investigate it and suggested a concept to resolve it. He called his resolution canalization, and if we're going to understand how different genetic programs can produce the same animal (or the same program can produce the same animal at different temperatures and under different conditions) we must come to grips with it. Earlier embryologists called the effect regulation and used it to explain twins. It has no natural explanation in terms of simple—or even quite complex—chemistry.

In *Drosophila* experiments, you observe the effect of mutations in some particular gene, while the rest remain unchanged. The clue is that wild *Drosophila* do not show the effects of single gene mutations to anything like the same extent as the inbred, domesticated laboratory stock. Accordingly, their genetic makeup is said to be wild type; it is typical outside the laboratory. The development of wild *Drosophila* is very resistant to changes either in environmental conditions or developmental program. Waddington showed that the effects of what he called "different genetic backgrounds" is to give the developmental program different sensitivities to environmental effects. For example, if you take a lot of developing *Drosophila* pupae and warm them, a few of them produce flies that lack one particular vein in their wings, called the cross-vein because it cuts across the direction along which the main veins lie. Waddington took flies that lacked the cross-vein after warming, and bred them selectively to increase the proportion that lacked it. When about 70 percent of the offspring failed to develop a cross-vein when warmed, a few flies also lacked it *without* any warming.

Some biologists thought that Waddington was claiming to have found inheritance of acquired characters, a version of evolutionary theory that was (and is) out of favor. In fact, Waddington was demonstrating the selection of flies that possessed a low threshold to an environmental stimulus. In his view, evolution would rarely, if ever, select for the overt effects of a single gene mutation. The members of a species differ in all kinds of ways, but the developmental program is sufficiently resistant to these differences that it can reach the same result despite them. (We can also interpret his ideas in terms of the DNA program, but the detailed workings of DNA were not then known.) Species achieve this stability by accumulating a repertoire of "buffering" tricks, such as those used in frog eggs to counteract temperature changes. The cross-veinless flies had lost some of their buffering capability during Waddington's program of selective breeding, so some of them were affected even by normal temperatures.

Waddington's mental picture of development was what he called the epigenetic landscape—a gently sloping surface, like an elastic sheet, with interconnecting ridges and valleys that run down the slope. Development is represented by ball bearings rolling down the valleys of such sheets and falling off the bottom edge. Which part of the edge they drop off determines the characters of the developed organism. The genes determine the depths and positions of the valleys, and the heights and positions of the ridges. Think of them as pulling the surface from below, molding it into different shapes. Meanwhile the environment presses from above and also molds the landscape. The ball bearing can roll around inside the valleys in all sorts of ways, but unless it crosses a ridge, it can't change valleys. Moreover, different valleys might converge farther down the slope. The final "destination" is unaffected by many variations of the possible paths.

In such a picture, the selection of cross-veinless flies has reduced a ridge somewhere to the point at which little more pressure from above (warming the pupae) makes the ball bearing cross what is left of the ridge and take a new path. Once Waddington's stock of flies had been selected to have a low ridge, then in some of them the genetic pull was enough to depress the ridge without involving any unusual environmental pressure.

Wild creatures have well-established valleys with few low ridges, so development is resistant to the pull of the genes, or "canalized." Nor is it greatly affected by differences in the environment; a big change is needed to remove a big ridge. Domestication, especially inbreeding for one specific character without attention to the rest of the program, produces low ridges and shallow valleys, far more vulnerable to changes either in genetics or in environment. This is what happened to laboratory *Drosophila*.

Whether or not Waddington's mental picture resembles actual biology, canalization is what produces the magical aspects of development. The grooves in the developmental landscape are so resistant to differences that two half-eggs can each make a perfect baby—one of a pair of identical twins. But laboratory *Drosophila* have been selected to have landscapes that are very far from resistant; this makes them great for experiments, but useless for living in the wild. Their development is responsive to changes in the environment and that fact enables scientists to pursue the chemistry of development without being frustrated by canalization, which prevents the very changes that they want to study.

■ INHERITED WEALTH

What brought canalization into being? Here's one way in which the development of complex animals can be made more resistant to genetic or environmental differences—so much so that it may look to us as if that development is seeking some predetermined goal. Parents can pass on not just genes, but a lot of the equipment for dealing with them. The maternal DNA program doesn't just hand over a DNA copy; it also provides the egg with a lot of energy and material. Humans exploit the results of this maternal provision by hens and grasses when they eat eggs or cereals. But, in addition, the maternal cell that will become the egg possesses a complete set of working chemical tools for reading the program and acting on it. The egg doesn't have to construct its own DNA reader as a consequence of its own DNA program; it gets a free one from Mother. Mother's program also provides a full set of maternal gene instructions, and it can program her ovaries to make the eggs with a very specific architecture, getting them off to a flying start. This is a fairly universal phenomenon; most animal eggs don't rely on their own DNA program until they are some way into development. The first part of such development is automatically resistant to change because it is under the mother's control, and she herself is resistant to change because *her* mother was, and so on. The embryo's self-controlled development need be resistant to change only after the point at which the embryo itself first takes charge.

Species are grouped into classes of various sizes; one of the largest is called the phylum. The stage at which an embryo takes charge of its own future is called the phylotypic stage because all animals in the same phylum resemble each other most at that point. Earlier stages have different amounts of yolk and different shapes of egg; later stages are heading off, like the creatures in the badger's tale, toward different adult forms. But all the types of animals in a given phylum seem to have "agreed" to retain a common structure at that vital period when the new genetic program is starting to run, when the baton is being passed from mother to child in the developmental relay race. The species in a given phylum converge onto the phylotypic stage from very different eggs; they diverge away from it toward their adult forms; but they all pass through that one crucial stage. Only after the phylotypic stage can defects in the embryos' own genetic programs show up. That is the point at which most of the hundred million larvae produced by one mother oyster die, as their own genetic defects become apparent.

The central point is that the DNA program in the embryo does not initiate development. To some extent it only maintains earlier development, and adds bells and whistles to it at later stages. It looks after the timing of puberty, tunes the neoteny, provides the correct eye color and the right number of whiskers. It must also organize the offspring's egg cells to have the right bits and pieces to continue the process into future generations. It is in many ways less than a blueprint, less than a computer program, and that is where Captain Arthur came adrift. Zarathustran science, like ours, lacks the technique to handle the initial stages independently of parental organisms. Neither civilization has the technology to make an egg.

The DNA sequence is a long message, containing information about which proteins should be made and when, and which enzymes to use at what temperatures. The egg does not start life with a clean slate. It begins inside a cellular computer system that is not only already programmed, but already running. Suppose we could make an egg, get all the molecules in the right places. Could we start it up? When you buy a new computer it doesn't work unless you go through a lengthy installation procedure. As we said, nowadays that's done automatically, with a start-up disk made in another computer. Real eggs get their start-up disks from Mother, along with everything else that gets them going. We'd be in trouble unless we could provide some surrogate start-up disk.

Even if we could start it up, would it work properly? Could we give it the right environment for development? We hear a lot about genetic engineering, which might more realistically be described in computing terms as hacking the DNA code. Must we now also consider hacking the surrounding operating systems, as well as the DNA program itself? Is it wise to hack the program when the operating system is an almost total mystery?

Or does it make no difference? After all, that computer system was set up from the mother's DNA. Is the manufacture of a viable egg just a matter of enough R&D? If Congress voted enough money, could we push the Artificial Egg Project through to a successful conclusion? Can we view development as just a mixture of chemistry, cybernetics, and information theory?

Only if we can include maternal-effect genes, canalization, the hierarchical structure of genes, and a thousand other things that are currently unknown. That's rather a tall order; already it's a lot more complicated than just assembling a toy organism from a super-Lego protein kit by following the DNA blueprint. And if making something that complex is hard, how about taking an entire system of organisms, letting them interact with each other, and then persuading the whole system to change toward ever more complicated states of organization and diversity? For that must be the next item in our reductionist program.

We call it evolution. Darwin didn't, as it happens—but the very last word in *The Origin of Species* is "evolved."

4 THE POSSIBILITIES OF EVOLUTION

Two dinosaurs—small, gentle plant-eaters—were browsing on the leaves of a small bush when one of them suddenly froze in terror.

"Eeeegh! Look what's just arrived!"

The other one turned round to see what the fuss was all about. "Aaaaagh! A Tyrannosaurus rex! And it's seen us! We're dead meat!" When it turned back, the first dinosaur was heading for the horizon as fast as its short stumpy legs could carry it. "Don't be silly!" the second dinosaur yelled at the retreating figure. "You can't outrun a tyrannosaur!"

The first dinosaur looked back over its shoulder and grinned wickedly. "No!" it shouted. "But I can outrun you."

We now see that much of the complexity of living organisms is the working out of complicated, but structured, chemistry. The connection between chemistry and biological activity is relatively straightforward for a virus, which is little more than a segment of chemical software wrapped up in a protective package; but the virus cheats by borrowing far more complex chemistry from bacteria or cells. Bacteria in turn may be viewed as chemical workshops, and cells as chemical factories. Organisms that do not just exist and reproduce, but develop, pose more difficult problems. By referring to them as industrial complexes we have tried to make it clear that biological development, too, can be interpreted as complicated chemistry, just as a car can be interpreted as a complicated assembly of wheels, cogs, and electrical gadgetry; cars, though, are simplicity itself when compared to even the simplest living creatures.

On the other hand, it would be wrong to say that living organisms are "just" complicated chemistry. They are a very special form of complicated chemistry, with very unusual properties; those properties cause us to ask questions that would not occur to anyone who thought in conventional chemical terms. If we were to think in purely chemical terms about living creatures, every question that we asked would have an answer that consisted of an enormously long sequence of chemical changes. We would soon tire of such explanations, and look for a simpler answer on some other level.

One of the most important questions about living creatures is, Where did they come from? It's wonderful that they work because of intricate chemical mechanisms; its fascinating to tease out the structure of those mechanisms—but how did such amazing mechanisms arise in the first place?

This question leads directly into the theory of evolution. Virtually all human thought about biology, over the past hundred and fifty years, has been tied to the ideas of evolution. Evolution offers the hope of understanding things that otherwise appear totally baffling. It is no wonder that many people, faced with the astonishing manner in which living creatures operate, fall back on a theological explanation, that life was "just" created. Evolution, properly understood, provides a rational alternative: It shows how the operation over huge periods of time of simple principles (but within an increasingly complex environment) might lead to the remarkable organization that we see all around us. Evolution, as a theory, has itself evolved; it has become rather subtle and not a little complex itself. To some extent this reflects the complexity of terrestrial life, which any theory of the development of life must come to terms with.

Few of our daily experiences equip us to think sensibly about evolutionary systems: We tend to act on simplified models of the world; we seldom think about the effects of small changes over huge periods of time; and we almost never try to tackle anything remotely as complex as the totality of life on earth. Many aspects of evolution run counter to our intuition.

REPRODUCTION ISN'T

We've used computer images of biological development: the mother computer building a computer egg that changes its structure under the influence of a built-in program. We've also seen that at least the early stages of

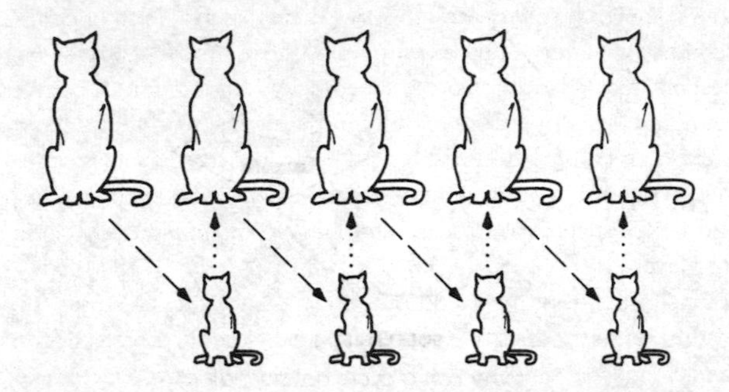

FIGURE 13

How not to make a copycat. Here ↘ represents "gives birth to" and ↑ represents "grows up into."

these changes involve a start-up disk provided by the mother, and not just the built-in software of the egg. Until the egg gets going properly, it can't access its own software. This is a very different picture from, say, cats "making" kittens that grow into more cats. That is, reproduction isn't just the simple process that we conventionally think of, turning cats into copy kittens into copycats into copycopykittens . . . as shown in Figure 13.

Real feline reproduction is more like Figure 14; it is eggs that are reproduced, and these grow cats (with eggs inside them). Indeed the word "reproduction," certainly to modern ears, has the wrong overtones. We talk of a photocopier reproducing a document. But a son or daughter cat is not an

FIGURE 14

How cats are reproduced—but not copied.

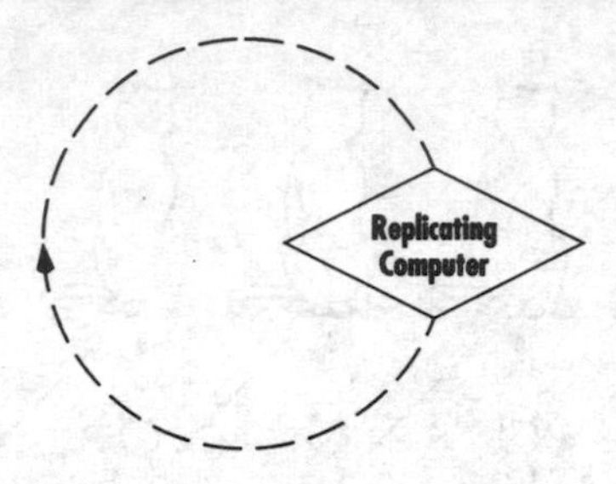

FIGURE 15

How can a cycle get started?

exact copy of its mother. We must therefore distinguish replication—the production of exact copies—from biological reproduction, which permits modification of the structure as well.

In our computer metaphor, suppose for a moment that the computers just replicated, so that the next generation of computers ended up, when fully developed, by being an exact copy of the previous generation (Fig. 15). In such a system the computers *can't get better*. Moreover, it's hard to see how such a cycle can get started. If you need an identical computer to make a computer, where did the first computer come from?

Clearly we're missing some important aspects of the process.

NATURE'S SCAFFOLDING

Here's a similar puzzle. Men with steel pickaxes are digging iron ore from a mine. You need iron ore to make steel pickaxes, you need steel pickaxes to dig iron ore. How can such a process start? Put this way, the answer is obvious enough. There are other sources of iron than deep mines; alternatively, you can dig mines without iron tools if you have to. The moral is that you can get into a cyclic process from outside. The hard thing is to get out of it again; you can do that only by changing the process. Again, once an arch is built, it holds together because the weight of each stone is supported by its neighbors. But you can't put the stones in place one by one with nothing to support them. So you provide some kind of scaffolding. In the same way, even if you are observing a computer that builds perfect copies of itself, that doesn't mean that the process can't get started. The first

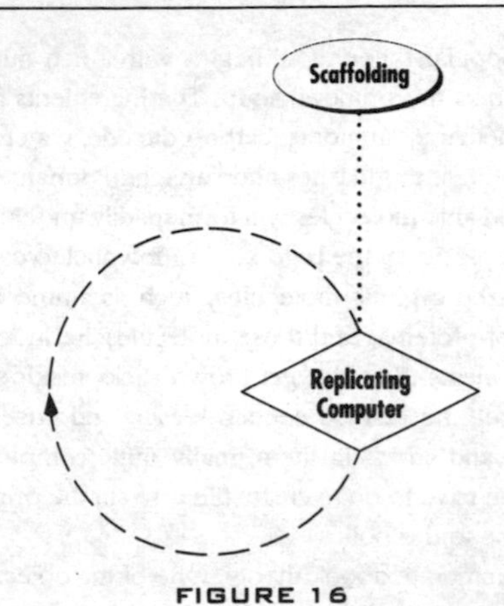

FIGURE 16
Start with scaffolding and then take it away.

such computer might be assembled by some totally different process (Fig. 16).

The maintenance of life is a cycle involving the chemistry of DNA and its realization in biological development. And we've seen that DNA is highly self-referential. How can such closed cycles get started? The answer is that you don't start them fully formed. The origin of life is not a loop; it is a tale of chemical processes that result in the formation of a loop. The main *chemical* step is the appearance of RNA and DNA. We know that carbon tends to form big molecules; that kind of complication is "downhill" to carbon, in the sense that going downhill is the easy direction. But the organized "data-processing" complexity of RNA and DNA seems to be quite a different matter. It's not at all clear that data processing is downhill to carbon chemistry alone. It must, however, be downhill to something; otherwise it would never have happened.

One thing that is downhill to DNA is replication. Once you've got some DNA, then it doesn't take long to have an awful lot of it; so there's no problem in explaining the origins of lots of DNA. It's also not too hard to see chemical reasons for short DNA chains being able to join up into longer ones. The question is, Where does the first little bit of it come from? The answer must be that it comes from the accidental—that is, unpredictable but inevitable, like traffic accidents—combination of simpler molecules. According to one popular theory it all begins with a rich mixture of organic chemicals, known as the primeval soup. The ingredients for the soup are indeed simple. Methane, ammonia, carbon dioxide, water. Energy, probably in the form of lightning flashes and sunlight. Formaldehyde, cyanides, and other small organic molecules will form readily under such conditions. And because complexity of the Lego kind is downhill to carbon, so too are many medium-sized organic molecules, such as amino acids, the basic building blocks of proteins. And those molecules are indeed produced in laboratory experiments. You also get brown sticky residue on the glass of the reaction vessel; the residue contains resins and tars, molecules with around ten thousand atoms in them, really quite complex organic compounds. So all you have to do to create life is to stir the primeval soup, and wait for it to come to the boil.

This is the common textbook theory. One of the objections commonly raised by opponents of evolutionary theory is that molecules like DNA, or even just hemoglobin, are too complex to arise through chance interactions of smaller molecules. The analogy that they put forward is the old cliché about monkeys typing *Hamlet*. In principle, if a monkey hits keys at random, eventually it *will* type out *Hamlet,* and anything else you care to name. But it takes extraordinarily long to do so. Given that the complexity of life has to arise through random mutations, the universe hasn't been around long enough for molecular monkeys to have much chance of typing hemoglobin. We'll leave this objection hovering for the moment (but see chapter 10 for one possible answer).

There are many scientists who—for similar but subtly different reasons—think that it can't be quite that easy. The molecular structure of DNA and RNA, they maintain, is a little too complicated to have arisen chemically from primeval soup. There needs to be some sort of scaffolding, some simpler replicating chemical process upon which the DNA replicating loop can build. The scaffolding used to build the loop is long since gone, but we can speculate on what might have happened.

One of the more fascinating theories is that of Graham Cairns-Smith, who focused on the "high-tech" nature of DNA replication and felt that it needed explaining in terms of something lower down the technological scale of chemistry. Are there any natural replicating chemical systems that could act as scaffolding? There are: crystals. If a tiny seed crystal is dropped into a solution of the same chemical, it will grow, replicating its atomic structure. The growing crystal will also, from time to time, become so large that it breaks into pieces under its own weight; now you've got *more* crystals. The macroscopic object "crystal" replicates, and not just its atomic lattice. Not only that, there is a kind of heredity at work. Flat crystals produce more flat crystals; lumpy ones produce more lumpy ones.

However, crystal replication alone is a little too simple. What is missing, as yet, is sufficient variability. Cairns-Smith realized that this missing ingredient can be provided by clay. Interesting chemical reactions tend to take place on surfaces, because surfaces are places where the individual reactants can be brought into contact and held until they combine. Surfaces are catalysts. The surface of clay is highly complex, with intricate needlelike crystal structures all over it. Those structures contain irregularities, dislocations, at which the regular lattice structure of the crystal is broken. Because crystals like to grow by extending their atomic lattices, the dislocations are repeated the next level up. The process is a bit like building a brick wall. If you make a mistake in the first row of bricks, then subsequent rows have to repeat the same mistake in order for the bricks to fit. The pattern of dislocations on the surface of clay is highly variable, but whatever it is in any given instance, it replicates. There is even the possibility of mutation: New dislocations may form spontaneously here and there.

Suppose now that some type of clay has the right kind of chemistry to catalyze the formation of organic molecules on its surface—molecules that might actually stabilize the replication of that kind of clay, in a sort of chemical symbiosis. As the clay replicates, you get more clay with the same property, hence more of the same molecules. We now have a replicating cycle that produces more and more copies of a molecule, *even though that molecule cannot replicate unaided*. The precise cycle is different from that used by life, however; life puts in a generation-gap phase shift, but clay does not (Fig. 17). However, the kinds of molecule that are helped to form by using the surface of clay as a catalyst include proteins, sugars, and nucleic acids. A replicating cycle of nucleic acids—RNA, perhaps even DNA—could get started by using the clay as scaffolding.

Maybe the Bible was right. Maybe man *is* made from clay.

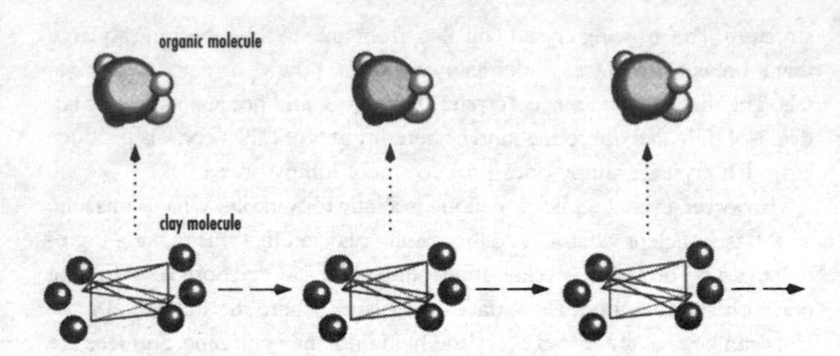

FIGURE 17

Clay scaffolding for replicating an organic molecule. Note the difference between this and Figure 13. Now → means "copy" and ↑ means "build on."

■ THE ASYMMETRY OF SELECTION

A cyclic process of perfect replication can't improve its products, because they're all the same. But the difficulty is worse than that, because in a cyclic process of imperfect replication the quality of the copy generally degrades. Put a document into a photocopier and make a copy. Then copy the copy, copy the copy of the copy, and so on. Modern copiers are pretty good, but after a few hundred steps, the quality of the document gets quite bad, and it doesn't take long before you end up with an illegible mess. It's true that occasionally a photocopier may actually improve the legibility of the document that is being reproduced. Dot-matrix printing, which forms characters with an array of disconnected dots, often looks better after being photocopied, because the copying process tends to fill in the gaps between the dots. But such cases are rare. They can happen by chance, if an odd speck of dust happens to be in just the right place, or sometimes more systematically, as with dot-matrix output, but most of the time a copy is worse than the original.

The downhill direction for photocopying—the easy way to go—is toward worse and worse copies. Life, in contrast, seems to go uphill. The story of life on earth is one of repeated improvements, with life tending to become more complex and more highly organized. This "progress" is not uniform, and it is sometimes set back by, say, a major catastrophe—but in general life perversely seeks out the uphill direction.

Many people are so puzzled by the nature of living creatures that they explain it through creation stories, in which some preexisting being of enormous complexity is responsible for the lesser complexity of life, and deliberate design changes are responsible for improvements. The underlying intuition is "conservation of complexity"—the idea that simple causes can't have complicated effects. We told you in chapter 1 that this intuition is wrong. It's not really necessary to invoke such a tremendous concept as a preexisting being to find a source of complexity. It's actually rather easy to make an imperfect copying process produce improvements. However, you don't do this by making single copies one after another. Two new ingredients are required. First, you have to make a *lot* of copies. Second, you need some kind of selection process that throws away anything that gets worse, but keeps anything that gets better. With photocopies you need a person to make this idea work, and it's not terribly practical. With life you don't, for life can perform the selection though its own activities. If creatures compete with each other for resources such as food or light, and if only the winners get to reproduce, then any random changes that increase the chance of winning tend to be reinforced in the population, and any changes that increase the chance of losing tend to die out.

A simple experiment in "artificial life" confirms this. We'll invent an imaginary population of six worms of various lengths, along with some reproductive rules. It takes two worms to breed, and the result treats getting longer or getting shorter symmetrically. The offspring are equally likely to be longer than their parents, or shorter; the average lengths of the offspring are the same as those of the parents. However, natural selection acts asymmetrically, and favors longer offspring. Here are the precise rules. At each stage, take the worms and divide them into three pairs at random; these are breeding pairs, but they breed in an unusual fashion. First, they fight to the death—and the longer one always wins. (If they are the same length, the winner is random.) Then the winner produces two offspring, whose lengths are determined by tossing a coin. If the coin lands heads, then the offspring are a pair that "mutate" to be respectively one and two units longer than the winner of the fight. If the coin lands tails, then the offspring are a pair that "mutate" to be respectively one and two units shorter than the winner of the fight. The winner then dies, leaving only the off-

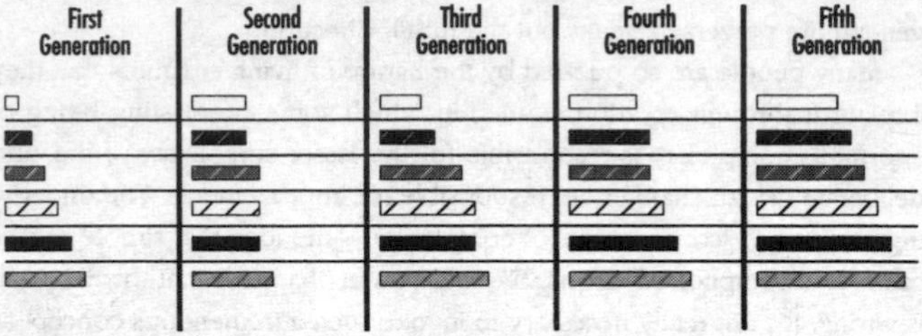

First Generation | Second Generation | Third Generation | Fourth Generation | Fifth Generation

FIGURE 18

Successive generations of artificial worms. Mutations are equally likely to be longer or shorter, but selection for greater length eliminates most short mutations.

spring. The next generation therefore also has six worms, and the process can repeat.

For example, suppose the initial population has lengths 1 2 3 4 5 6. Three random pairs are (1,3) (2,4), (5,6). Of these, worms of lengths 3, 4, and 6 respectively win their fights. If worms 3 and 4 throw heads, but worm 6 tails, then worm 3 produces offspring (4,5), worm 4 produces offspring (5,6), and worm 6 produces offspring (4,5). The next generation therefore has lengths 4 4 5 5 5 6. Figure 18 shows the first five generations in such an experiment. Observe the steady increase in length, even though the coin-tossing ensures that reductions in length are just as likely as increases. It happens because there is a built-in bias: The longer worm in each pair—the victor in the battle for survival—is the basis for the change decided by the random coin-toss.

This combination of random changes followed by competitive selection lies at the heart of evolutionary theory.

THE EVOLUTION OF EVOLUTION

Evolutionary ideas had been around for a long time when young Charles Darwin, at Shrewsbury School in England, learned his biology, his philosophy, and his natural philosophy—the nearest thing in those days to what we would now call science. Erasmus Darwin (Charles's grandfather) and others, such as Jean-Baptiste Lamarck, believed that animal and plant species are not fixed for all eternity, but are subject to transformation. Two of the biggest problems for the species transformers were how life began in the first place, and how an animal—indeed, an ape—could be transformed into man. But even then, some thinkers of evolutionary persuasion had tried to grasp these nettles.

What Darwin did was explain the transformation of species, not just assume that it must occur. He provided a self-evident mechanism that makes species change look inevitable. A standard story is that he read Thomas Malthus's *Essay on the Principle of Population,* whose central point is that an exponentially expanding population (one that multiplies by a fixed amount at each generation) will outgrow its linearly growing food supply (one that adds a fixed amount at each generation). Malthus argued that, as a consequence, "natural increase" of population will eventually outgrow food supply. It is normally assumed that the lesson Darwin took from Malthus is that among naturally breeding wild animals there is always competition for food. There are many small differences among these competitors (even within a species), so that those better able to find food, store it, or use it effectively will on average breed more of their own kind.

But there was a deeper message for Darwin in Malthus. This is that nature's rules, even simple rules like the one dictating that exponential growth will outstrip linear growth, bind human plans and desires into inevitable patterns. Although the earlier work of Isaac Newton carried a similar message, the overt objects of Newton's attention were planets and tides, not politics, so his message wasn't widely appreciated. Malthus opened Darwin's eyes to the fact that human beings, too, are bound by the laws of nature. He took that lesson to show that natural selection—leading to, among other things, species change—is an inevitable consequence of the competitive breeding of organisms subject to limited resources.

The mechanism is basically the one that applies to our artificial worms, but the rules are more complicated. In order to breed, creatures must survive to breeding age. They do so by competing with other creatures, and only the winners get to breed. The results of their breeding possess similar characters to their parents, but subject to random mutations, which may lead to "better" creatures or "worse" ones, in the sense that "better" creatures tend to survive the competion better, whereas "worse" ones tend to do less well. The mutations are random, but, as Darwin argued, competition tends to eliminate the worse ones. This is his principle of natural selection.

Darwin did not simply declare that evolution has occurred; he set up an argument that makes sense of the hierarchical classification of animals, because natural selection has *driven them to diverge*. Similar animals have common ancestors in the recent (by geological time scales) past; different animals have more remote common ancestors. And what drove the two kinds apart, making them recognizably different, is that they specialized toward different ways of competing for the same resource. Then their specializations began to free them from the competition. Dogs and cats, for instance, have taken up different ways of being carnivores, and hunt different prey differently. They no longer compete in any vital fashion.

Darwin worried about his explanation, because it required heredity to behave in ways that he thought were incompatible with each other. He was happy about characters that weren't part of some continuous spectrum, say the presence or absence of a particular set of feathers in a bird, but he was puzzled by characters that vary continuously throughout a population, but don't vary continuously among the progeny of particular individuals. He knew that features like large or small noses can persist in families for many generations, but he didn't see why that should happen when intermediate sizes existed elsewhere in the population. This is often explained in textbooks by saying that Darwin believed in blending inheritance, as mentioned in conjunction with pink hair in the previous chapter. That's wrong. He was far too familiar with domestic pigeons, dogs, and horses for that; and he was surrounded by people saying, just as they do today, "Doesn't he have his father's nose?" or "Big ears run in the family, you know." He knew that such regular differences exist within and between breeds and in families, and are incompatible with blending. But he couldn't explain them.

Darwin was also aware that some differences, such as the muscular arms of the blacksmith and his sons, result from differences in upbringing. Apprentice blacksmiths develop muscular arms because they spend a lot of time lifting heavy tools and materials (and horses' legs); they don't have to inherit them. Darwin worried about whether such acquired characters would appear in the next generation. Lamarck believed that they could be passed on to progeny, but Darwin disagreed.

In successive editions of *The Origin of Species* Darwin had a variety of tries at these problems, but his worries weren't really sorted out until Mendel's ideas were merged with evolutionary theory in the 1930s. On their own, Mendel's characters explain only obvious traits such as albinism, red hair, or wrinkled peas. The choice between a small number of factors (recall that the modern term is "alleles") doesn't help with *graded* differences such as height, weight, or intelligence, because there are too many possibilities to be controlled by just a few factors. In the early days of the development of statistics Darwin's cousin Francis Galton also worried about how graded characters could occur. Only in the 1930s was continuous variation brought under the Mendelian umbrella, when Sewall Wright introduced the idea of "peaks of adaptedness"—most effective values among a continuous range—to provide the missing link between continuous variation and discrete characters. His image was that of a mountain range, in which there are continuous graded changes of height, but nevertheless a few peaks stick out as the highest places around, and evolution causes graded characters to select values corresponding to those *separated* peaks. We'll explain these ideas more fully in chapter 6, but they deserve a mention here.

■ LET THE SCALES FALL

It is easy to get confused about evolution. There is a very seductive, but unhelpful, set of comparisons which we all tend to make. We set up three scales (maybe more) which run from the simple to the complicated, and then we get them all mixed up together in our mind. Even professional evolutionary biologists do this. However, they are actually very different scales, with different uses and philosophies. Indeed the third, which has the longest and most honorable history, has no useful philosophical basis at all.

The first scale is an individual one. You began life as an egg, which made an embryo, which made a fetus, which made a baby, which . . . You will even continue your journey into complexity as you read this book, and your mind is (we hope) taken to new places and learns to think in new ways. This scale is called your development, or ontogeny. We dealt with this at some length in the previous chapter. The DNA explanation of development puts it in a very different light. When a car is assembled in a factory, it gets more and more complicated as extra bits are added on. At first sight the same goes for the developing embryo: It starts out pretty simple-looking, then grows various things like skin, a spinal cord, eyes, arms, legs, and a brain. But those complications are not being added on from outside by factory robots. They are implicit in the DNA blueprint; the complexity has been there all along, and it emerges in its full glory as the embryo develops and the implicit becomes explicit. Five feet of DNA inside one human egg, carrying molecular-sized information. Think about it.

The second scale is the evolutionary one. Two billion years ago the most complicated living things were bacteria. About one billion years ago DNA became segregated into the nucleus of cells, instead of just being strung around inside as in bacteria; soon after, creatures that resembled today's amoeba became multicellular and invented a new trick, development. By five hundred million years ago most of the major kinds of life that currently exist had appeared in some form. Life on land opened up entirely new patterns for these forms. By 120 million years ago insects, dinosaurs, and mammals had evolved—and not just a few different kinds, but a vast range of species; now there was enormous biological diversity. Warm-blooded mammals and birds have been the dominant life-forms on land for only forty million years; big-brained mammals have been around for only twenty million. Our own specialization, *really* big brains, has been around for only a couple of million years. This represents a highly nonlinear increase in complexity, as measured by braininess (or some other criterion, such as independence from environment, or how much privilege progenitors pass on to descendants, of which more later).

The third scale is the Ladder of Life. This consists of the organisms that exist now, arranged according to their complexity and without regard for evolution. Many zoology books are arranged like this, with *Amoeba* and other protozoa at the front in chapter 1 and the primates (us) at the end in chapter 13. Many natural-history museums still have a series of displays with "primitive" animals at the left and "more advanced" ones toward the right.

Mammals, the group to which we belong and with which we best empathize, sit at the top end of the second and third scales, but that doesn't mean they are necessarily "the best." We happen to have specialized in brains, which give us the ability to compare ourselves with other creatures but also lead us to use braininess as a criterion. A canine Ladder of Life would no doubt put dogs at the far right and use the criterion of sense of smell, in which we privileged apes are sadly deficient. Modern kinds of fishes, called teleosts, evolved later than mammals and have been much more successful, by many criteria other than braininess. For example, they have many more species and live in more diverse habitats. We could invent other scales. If we used the fishes' innovation, the air bladder, as our benchmark, we would again find much better air bladders now, in modern fishes; and we could also set up progressive series of fish species in which there is a steady improvement in air bladders, starting from their invention about eighty million years ago. Or we might use the complexity of blueprints, measured (somewhat naïvely as we shall see) by the amount of DNA in the nucleus. Some newts and lungfishes come at the top of that scale today, and we certainly don't. Does that bother you? Size isn't everything, but again there has been a general increase over the last thousand million years. Living creatures are like cars in this respect: As evolution proceeds, there are more kinds of them and they get more complicated, by whatever reasonable measure of complexity you choose.

It is rather obvious that the individual developmental scale is different from the other two; after all, it's about what happens to *one* organism. It still gets confused with the others, though, because of a series of puzzling resemblances between stages in the developing embryo and stages in evolutionary history. Early on, human embryos develop gill pouches. Why? One suggestion was that our development reflects our evolutionary history—that "ontogeny recapitulates phylogeny," as Ernst Haeckel put it. It is now thought that if this is true at all, it is true only in a very indirect sense. But for many years it was the mainstay of embryology and evolution, because it offered the alluring prospect of deducing a creature's long-dead evolutionary ancestors from its living embryos. (Wrongly, but that's another story.)

The other two scales have a seductive similarity, but they cannot possibly be the same. Dinosaurs are well represented in the evolutionary tree, but none are alive today, so they can't appear on the Ladder of Life. Moreover, there is no way in which any animal now living can be the ancestor—in an evolutionary sense and on an evolutionary time scale—of any other animal now living, not without a very fancy time machine. The confusion of the second and third scales is evident in statements such as "Humans are descended from apes." Not so. Modern humans and modern apes are descended from some common ancestor in the evolutionary tree.

Perhaps to our eyes that ancestor looks more apelike than human, which is why we get confused, but today's apes could equally reasonably claim to have descended from humans. The misunderstanding allows some cheap shots at evolutionary theory—"Just *which* ape do you claim to be descended from, Mr. Darwin?"—and it's so much simpler to stay confused and have an apparently easy target than to use your brain and try to understand a new idea.

Another source of the confusion is that there is a very strong feeling in many people, even many zoologists, that there is something primitive or ancestral about *Amoeba* or *Hydra* or worms in comparison with mammals. Haeckel's image of the embryo climbing its own family tree has its attractions. However, it is misleading. We may be tempted to compare the gill pouches in a human embryo to those of a modern fish—to think of them as something that developed during the "fishy stage" of our evolution—but we would do better to compare gill pouches in a human embryo to the kind of pharynx that *all* vertebrates have at that stage of development, which makes gills in fish and a lot of different talking and swallowing equipment in us.

Embryos don't climb their family trees, *Amoeba* isn't our ancestor, however distant, and the modern Ladder of Life does not show modern but still primitive representatives of important stages of our evolution. Lungfish are players in their own story, not in ours. Yes, there are three persuasive and different scales on which organisms can be arranged—in illuminating ways—from simple to complex, but you should never confuse them with each other.

■ WHY HAVE SEX?

In creatures that reproduce asexually, the genetic material of the offspring is, apart from mutations, identical to that of the parent. Asexual organisms make photocopies of themselves. Sexual organisms have a new trick available to them: They can mix genes from both parents. This isn't done in a haphazard way, dumping all the parental genes into a bag and choosing at random. Instead the mechanism of reproduction combines parental genes in a way that ensures that exactly one version of each gene is passed on. Remember that the cell's DNA is packaged into chromosomes, and that chromosomes from the parents "cross" and exchange genetic material in corresponding positions. So a child may very well inherit "her father's nose." It's not as simple as such phrases suggest, though. No single DNA sequence will contain the code that determines the nose as such. It's not at all clear that DNA "knows" that noses exist. The child might have a nose whose shape is largely determined by genetic material from her father, but with some of her mother's mixed in; her skin, which covers her nose as it does the rest of her, might be mostly the result of genes from her mother; and so on. You can see how mixed-up a nose could get. Because DNA sequences do not correspond directly to observable characters, it's important not to think of the offspring as an assembly of selected parental characters. However, many characters—for example, blood groups—do correspond fairly well to specific regions of DNA, and for such characters the image of a DNA strand that directly dictates them is not such a bad picture.

Why did sexual reproduction, with all its complications for DNA, ever get started? You can't answer such a question by elaborating on the molecular engineering involved. An evolutionary view is needed. What's in it for sex? In 1964 H. J. Muller pointed out that asexual "photocopier" reproduction has the same defect that we've already noted for repeated photocopies: Successive operations tend to degrade. He likened the process to the irreversible behavior of a ratchet, which can move on a notch but never back. We remarked that occasionally a photocopy may actually improve on the original, but Muller argued that because genetic material comes in discrete packages, there is a mechanism that irreversibly degrades the genetic material of asexually reproducing creatures.

The basic point is that most mutations are bad, in the sense that they lessen the organism's chances of surviving. Random changes to any computational system mostly lead to trouble, and DNA is no exception. Now, mutations happen all the time. In a population of creatures of some particular species, there will be some with no bad mutations, some with one, some with two, and so on. The proportion having no bad mutations will be relatively small, and over a long period of time chance events that affect reproductive success will cause that segment of the population to die out. Once it has died out, creatures with no bad mutations will not reappear—or, more accurately, will reappear so infrequently, by a mutation that "undoes" a bad one, that to keep the discussion simple this event can be ignored. *Click* goes Muller's ratchet, and once it moves on a notch, it can't go back. In the resulting population, the fewest bad mutations now occur in those creatures with exactly one bad mutation; and by the same argument, random fluctuations will eventually remove them from the gene pool. *Click*. You see how it goes. Muller's ratchet implies that in creatures that reproduce asexually, the number of bad mutations will necessarily keep increasing, and eventually so many bad mutations will build up that the creatures die out completely, because too many bad mutations are fatal.

Sex is a way of getting around the mechanism of Muller's ratchet. With creatures that reproduce sexually, some offspring may still inherit a bad mutation—say, from the father—but others will not, because they get the relevant DNA from the mother instead. Those offspring that have fewest bad mutations will reproduce most successfully. The new broom of sex sweeps the population clean of bad mutations. Like all brooms, it's imperfect: Some offspring with bad mutations will survive to reproduce, and if both parents share a bad mutation, so will all the offspring. But the chances of that are also very small, and again this event can be ignored.

This is a genetic backup mechanism, not just for odd strands of DNA but for entire genes or gene complexes. Every sexually reproducing organism carries around two versions of every gene, but expresses only one. The other is a backup, and it remains in the gene pool even though it is not expressed. So one major evolutionary advantage of sexual reproduction is that it throws a wrench in Muller's ratchet.

The complications of animal sexuality would fill a great many books—indeed, they already have, from the *Kama Sutra* to *Wuthering Heights*. We mention just one striking example here, for later use: the anglerfish. In humans, sex is determined by the sex chromosomes, which come in two types called X and Y. Males have the pair XY, whereas females have XX. The two chromosomes differ considerably: Some 115 genes have been identified that occur only on the X chromosome, and there are probably many more. But the sex of an anglerfish is not determined by its genetics at all, even though male anglerfish are less than a thousandth the volume of females and spend their lives *attached* to females. If an anglerfish larva encounters a mature female, then chemicals produced by that female switch it into "male" mode. It attaches itself to the female, stops growing when still very small, becomes a parasite, and concentrates on sperm production. All other larvae turn into females (and most get eaten before they become much larger). It's a cunning system, which ensures that every male has a mature female companion without wasting any extra effort on surplus males.

TWO KINDS OF CHARACTER?

Even into the 1970s, people tended to assume that animals and plants had two very different kinds of character. First, there are the characters emphasized by botanists and zoologists interested in the systematic classification of organisms—characters that don't change, and that provide information about the organism's remote ancestry. They are the deep, important homologies, mappings of an organ or system of organs onto another superficially different organ or system. For this reason we shall call them deep characters. In this view the various parts of a flower—petals, sepals, and the anthers that contain the pollen—are "really" leaves; insect antennae and mouthparts are "really" legs; the whale's limbs resemble our arms or bat's wings far more closely than they do fishes' fins; all vertebrates have gill pouches and a brain. And so on. Here we are not referring to functional similarity, such as that between a fly's head and ours, but about similarities in the developmental program.

The second kind of character is more trivial, the kind that geneticists work on: color of seed coat or of mouse hair; wrinkliness or smoothness of a pea; small differences in size; patterns of bristles on *Drosophila;* eye color or blood group in schoolchildren; whether you can roll up your tongue. These we shall call simple characters.

A few characters messed up this nice polarity. Homeotic genes, in particular, seemed much too important for geneticists to be able casually to change them back and forth in successive generations—antennae becoming legs and then antennae again. *Drosophila* with the bithorax gene are four-winged, like their remote ancestors, whereas the modern version has two wings and two tiny gyroscopic balancers. To convert a modern fly into an ancestral one by changing just one gene was nearly as shocking, to those in the know, as breeding two ostriches and getting a dinosaur would be today.

We now understand why some gene changes—those most easily found in living, developing progeny—have trivial effects, while a few, which mostly act early in development to turn whole groups of genes on or off, have much more dramatic effects. We are even beginning to see links between the genetic blueprints of different (but closely related) organisms and the way they develop. We can't "read the blueprint" and work out what kind of developmental processes it controls, but we can see that genetic differences—differences between the blueprints of related organisms—have resonances with the differences in their development. Provided we don't succumb to the temptation of having embryos climbing their family trees again, this should give us confidence in our understanding of evolution. It is genetic sequences that evolve, and different sequences result in different adults.

Look at that thought from another direction. Adult organisms can't evolve. All they can do is die. Eggs can't evolve either; all they can do is develop. If neither adult nor egg can evolve, what can? Only the blueprint. The blueprint is the creature's DNA sequence. So we are led to the viewpoint of the neo-Darwinists: The raw material upon which evolution operates—the stuff that mutates randomly—is DNA. Their view, expounded beautifully in *The Selfish Gene*, is that evolution is all about differences in DNA.

Mutations, the genetic changes that evolution operates upon, are changes to the chemical structure of DNA. They can't be any old changes, though, because damage to a chemical—a few atoms knocked off here, a few added there—usually gives a different chemical. A mutation has to keep the whole system working. The simplest way to do this is to change the sequence of DNA bases: to insert or delete a base or a short subsequence, or to replace one subsequence by another. The coded message changes, but the symbols themselves remain intact. A mutation is a chemical typo.

Pursuing the same line of thought, a gene must be some set of DNA code words that produces a coherent developmental effect. It might be a particular subsequence, or a collection of subsequences; the precise layout presumably depends upon how the organism goes about interpreting the code. A mutation in a given gene—a typo in a given paragraph of the Book of Life—changes the resulting organism.

Natural selection, however, operates on creatures, not directly on DNA. This makes the whole process rather subtle. An oversimplified caricature—unfortunately, one that is often put forward without being identified as an oversimplified caricature—goes like this. The form and behavior of an organism are determined by its genes, which produce particular characters such as blue eyes, sharp horns, or powerful legs. It is characters that determine whether the organism has a competitive advantage. Random mutations change the DNA code, hence change one or more genes, hence change one or more characters. Selection eliminates organisms whose characters reduce their chances of survival, and thereby eliminates the corresponding genes. A "good" mutation that increases the organism's survival chances will itself survive; a "bad" one won't.

For example, most of you will have seen albino animals—rabbits, mice, fish. They are pale, with pink eyes. Albinism is caused by a lack of pigment. That pigment is made by a protein, the enzyme tyrosinase, which is itself made by a particular DNA sequence. If a mutation occurs in that sequence, then the protein doesn't get made, so no pigment gets made, so you get unpigmented animals, albinos. Here we can trace the links between DNA sequence and character with a great deal of confidence. There are many other characters for which a similar story can be told, and neo-Darwinists tend to assume that every evolutionary change occurs by a more or less subtle variation on this kind of mechanism. We will criticize this caricature as our tale unfolds; we will explain just how subtle the link between DNA and organisms actually is, just how flexible all the links in the causal chain really are. But for the moment the caricature will suffice.

In particular, it immediately resolves the problem of having two different kinds of character (plus some awkward intermediates). With DNA as the ultimate stuff of heredity, organisms don't have to worry about different kinds of character, malleable or rigid. Those differences are determined by the developmental program specified by the sequence of bases in the DNA. Some DNA sequences just make a protein, some have a more organizational role and turn on groups of other genes, some turn *those* on, and so on. There's a hierarchy of genes, like the managerial structure of a big company. Genes controlling malleable characters are down at the bottom of the hierarchy, shop-floor workers pushing buttons on the protein machine. Genes controlling unchanging structures such as the spinal cord or gut are somewhat around the level of senior management.

■ TWO KINDS OF MUTATION?

Mutations in DNA are generally small—a change to one base, or the insertion or deletion of a base. Occasionally entire strands of DNA may be inserted or deleted, but let's think about single bases and look at a typical mutation. The sequence

. . . CAT·CAT·AAG·TGA . . .

specifies the amino acids

. . .*valine.valine.phenylalanine.threonine* . . .

If we change the initial C to A we get

. . . AAT·CAT·AAG·TGA . . .

which means

. . .*leucine.valine.phenylalanine.threonine* . . .

One amino acid has changed, so the protein that is produced has changed too. The result of this single change to an amino acid may or may not be crucial, but it often is. Compare with a mutation that changes the third base T to C:

. . . CAC·CAT·AAG·TGA . . .

Because CAC and CAT both specify valine, this mutated sequence specifies the *same* list of amino acids:

. . . *valine.valine.phenylalanine.threonine* . . .

as before. Most biology books say that if a change to the code produces the same amino acid, then it makes no difference, but biology itself is more subtle than the textbooks. Transfer RNAs for different triplets for the same amino acid may be present in very different amounts in different kinds of cell. There could be a lot of transfer RNA for the triplet CAT but not very much for CAC, or the other way round. Then the apparently neutral mutation from CAT to CAC could result in a lot of this particular protein being produced in muscle cells, but very little in skin cells, say. Moreover, a *second* mutation could change CAC to TAC (methionine), whereas no single mutation can change CAT to any triplet for methionine. Indeed the only such triplet is TAC, and you need two changes to get it from CAT. So the biology books are oversimplifying the situation.

What effect can a single DNA mutation have? The above mutation just changed a tiny bit of one protein. Does a tiny change to DNA always produce a tiny change in the resulting organism? Not necessarily. It depends on whether the change affects a simple character or a deep one.

The same distinction, for the same reason, occurs in the bit strings that make up a computer program. A particular binary sequence, such as 10011101, has no intrinsic meaning. In a word-processor program, for example, it may encode the letter "d" in the welcoming phrase, "Welcome to

SuperWord," or it may determine some operation to be carried out by the computer deep within its processing program. If that sequence is changed, in the first case it will simply change the welcoming message to "Welcome to SuperWorm" or something similar. In the second case it could do almost anything—wipe the hard disk, switch the operation to a totally different program, delete all "R"s, print in italic all paragraphs with an even number of words, or play the opening bars of the Hallelujah Chorus. Most probably it would crash the machine. There's no difference in the bit strings in these two cases, but a huge difference in what they mean, in what actions they trigger.

Although mutations in DNA sequences are random, in the sense that any particular base is as likely to mutate as any other, this does not imply that all mutations have equally significant consequences. We've already seen that some regions in the DNA contain junk, while others code for the production of proteins or control the switching on or off of other genes, and so on. A mutation in junk DNA will have no effect (if the DNA is junk and if the mutation doesn't turn it back into nonjunk) and no overt effect—because that's what we mean by junk. A mutation in a DNA sequence that codes for production of a protein will build a different protein. The effects of this will depend upon the nature of the error, but they will generally impair the functions for which that protein is used. A mutation in a control sequence of DNA will switch on the wrong genes. The result will be a coherent sequence of events, but not the sequence intended—just as *Drosophila* may develop a perfectly formed leg where an antenna should be. This implies that certain large-scale coherent effects may be "downhill" to evolution by random mutation. It may be easy to bring them about. Indeed, there seems to be no limit to the complexity of the change that a single DNA mutation might trigger. A mutation is a genetic switch, and one switch can control a flashlight, or the entire electrical system of the United States. "Structural engineering" features of DNA chemistry, as well as "program interpretation" features, have a strong and highly sensitive effect on the implications of a single mutation. The idea that all mutations are on the same footing is quite misleading. "Nearby" sequences may have profoundly different developmental consequences.

The research programs that have sprung from this insight—that the meaning of the DNA sequence in the developmental program, and not just its form, matters—are many, various, and productive. It genuinely seems that only the details of how, in each particular case, the DNA determines the developmental steps, remain to be worked out. In the language of computer science, the only task left is that of "disassembling" the DNA code, turning cryptic symbolic instructions such as CGATAA back into their chemical and organizational implications.

THE RHYTHM OF LIFE

In addition to "molecular computation," there is another major feature of the chemistry of life that is very different from conventional test-tube chemistry. Living chemistry is much more dynamic. The reactions of conventional chemistry go to completion—you mix silver nitrate and sodium chloride solutions and watch the silver precipitate out as silver chloride, because the "law of mass action" drives the reaction toward a single product. In the thirties it was realized that the chemistry of life is quite different: Many of the cell's reactions occur in cycles. The classic one is the Krebs cycle, in which the cell's foodstuffs are converted to carbon dioxide and water in a series of steps that cleverly regenerate all the intermediate substances used, such as high-energy phosphates. Companies that sold biochemicals began to produce vast charts with "all" the cell's biochemical pathways on them, to be pinned on laboratory walls and look impressive. Right now there is one on the wall of the Nonlinear Systems Laboratory at Warwick University, bearing the label "The Mathematician's Nightmare."

People had realized that the law of mass action must also apply to these cycles, leading to one specific product, and that the concentration of the resulting product must somehow feed back to the initial reaction rates, to keep the cycle in balance. However, one property of these reaction chains and cycles was not appreciated until the seventies and eighties, although Brian Goodwin had published a book about it in 1964. Namely: If you take a signal, delay it for a fixed length of time, and then feed it back where it started from, you effectively have a clock. But this one is a molecular clock, a tenuous, flowing network of chemical change.

A quite different breed of biologist, interested in animals and plants instead of chemicals, had developed a vocabulary to describe the rhythmic aspects of their behavior—heartbeats, sleep/wake, breeding cycles. But Brian Goodwin's research group of biochemical adepts at Sussex University had the words "We got rhythm" written over the lab entrance. Delays in feedback produce oscillations, and Goodwin had realized that even the cell's most basic biochemistry—the transcription and translation of the genes that code for enzymes—could not proceed uniformly and smoothly. So the enzyme, and perhaps the resulting products too, must be produced in fits and starts. The same went for all the other reactions in those impressive but tidy cycles on the wall charts: The whole cell had to be in a constant state of push-pull-bang-whistle-bleep, with every component varying wildly.

This wasn't the calm, sedate chemistry envisaged by the chart makers.

We've gotten used to this idea by now, but it took a long time before it became part of the thinking of the typical biochemist. The main drive came from endocrinology, the study of chemicals (mainly hormones) produced by glands. A very important factor in the clinical measurement of hormone levels is the time of day. Adrenaline is low at night, and cell divisions can proceed faster. During the day prolactin varies one way, at night another, and both depend on other circumstances. Clinicians found that the hormones that make the pituitary gland stimulate the patient's ovaries or testes had to be given at thirty-minute intervals, or they didn't work. Both the sensitivity to stimuli, and the stimuli themselves, were rhythmic. In the fifties we had learned that the pituitary is the conductor of the endocrine orchestra; in the eighties we discovered that *all* the glands were playing music, and quite complicated music at that. This "temporal organization," as Goodwin called it, pervades all of life's chemistry, right down to the most trivial reactions in the cell. Not only that, but it links up to the external cycles of day and night, and the seasons, with the twenty-four-hour cycle being the dominant one. The four-minute cycle of gene transcriptions and Krebs-style oxidization; the four-day estrous cycle of the mouse; the annual breeding cycle of the sheep; the five-to-six-year plagues of lemmings; the synchronized seventeen-year life cycle of certain cicadas—all life's got rhythm.

■ SELF-ORGANIZATION CAN BE DOWNHILL

It is late evening in the *Thighbone*'s cabin. The conversation began with the relative merits of the Whifflpoffl Lungbats and the Dimpfot Octimists, two competing teams in the Zarathustran phugball league, and—amply assisted by a case of Zarathustran ocktails brought on board by Neeplphut—has moved on to the big questions of Life, the Octiverse, and Everything.

STANLEY: I've never understood why living things seem to be able to complicate themselves without outside help. Where does the extra complexity come from?

NEEPLPHUT: I wonder whether mutating the seventeenth "S" to a "C" would lead to a winning position?

CAPTAIN ARTHUR: Neeplphut, stop playing with the *Thighbone's* computer!

NEEPLPHUT: I am sorry/get stuffed (delete whichever is inapplicable). I am experimenting with artificial life. I have written a program—in octal arithmetic, of course—that sets artificial organisms loose in the computer's memory.

STANLEY: *What?* You're setting bugs loose in my computer?

NEEPLPHUT: Do not alarm yourself, Stanley. They are merely strings of symbols. Look! [*The screen displays a list:*]

S	start
E	end
Fs	find preceding start
Fe	find next end
C	copy everything between "start" and "end," including the symbols S and E themselves, and put the copy into the space that begins immediately after "end."

CAPTAIN ARTHUR: They don't look much like organisms to me.

NEEPLPHUT: Well, for example, a primitive organism might be a sequence of instructions such as

S Fs Fe C E
↑

where the arrow shows the initial position of the program pointer. The evolutionary process is begun by setting the program pointer to the first symbol, and then carrying out the instructions in turn as the pointer moves along, one instruction at a time.

STANLEY: And what happens?

NEEPLPHUT: Something very interesting, Stanley. This particular organism finds its own start S, then its own end E, then copies everything in

between—namely itself—in the spaces after its final E; and the pointer moves to the next space, giving

S Fs Fe C E S Fs Fe C E
↑

The program has *replicated*!

CAPTAIN ARTHUR: That's interesting. Of course, the programming language does include a "copy" command, so it's not a *total* surprise. Does the process continue?

NEEPLPHUT: Indeed, as you can easily verify. The replications continue indefinitely, and soon the computer becomes filled with copies of the primeval organism:

S Fs Fe C E S Fs Fe C E S Fs Fe C E S Fs Fe C E . . .

STANLEY: That's perfect replication—but in the real world, errors sometimes occur.

NEEPLPHUT: Of course. I have programmed those in too. Suppose, for example, that one of the E symbols mutates into an S. Now you have

S Fs Fe C S S Fs Fe C E S Fs Fe C E S Fs Fe C E . . .
|
mutation

STANLEY: What does *that* do?

NEEPLPHUT: Computer memories do not occupy unlimited space, Stanley. When the computer's instruction pointer gets to the end of the memory space, it automatically starts again at the front, and eventually it runs into the first "S" of the mutated sequence. When it does so, a new life-form is created, namely the sequence between the first "S" and the next "E," which is now further up the sequence:

S Fs Fe C S S Fs Fe C E

Amazing! There are precisely eight instructions between the "S" and the "E." Octimality triumphs yet again! And this organism is not only longer than the first one; it is more complex—it contains *two* "C" commands. The first leads to

S Fs Fe C S S Fs Fe C E **S Fs Fe C S S Fs Fe C E**
↑

and the mutant organism has replicated successfully!

STANLEY: Yes, but the second "C" instruction creates a copy of the primeval organism, the second half of the mutant; and this [*italics*] overwrites the front half of the new copy

S Fs Fe C S S Fs Fe C E *S Fs Fe C E* **S Fs Fe C E**

↑

and kills it!

CAPTAIN ARTHUR: And then the primeval organism replicates and kills the second half of the mutant, before resuming its own replication as before!

STANLEY: It's a rather silly mutant, Neeplphut.

NEEPLPHUT: Well, I'm just trying the simplest things I can think of. There are many other possibilities. Suppose, for instance, there is a second mutation, so that the sole surviving mutant organism now becomes

S Fs Fe C S S Fs Fe **E** E

|

mutation

Then I have produced a slightly shorter mutant

S Fs Fe C S S Fs Fe E

that copies itself *without* making a copy of the primeval organism, and so survives to take over the whole of memory space—

STANLEY: That one has only seven instructions between the "S" and the "E," Neeplphut. It looks rather septimal to me.

NEEPLPHUT: [*Looks irritated but pointedly ignores him*]:—and of course a slightly different program, one that writes the new copy to a random place in memory, say, would lead to a first mutant that did not kill its own offspring, and to a complicated competition between program organisms with memory space as the resource.

CAPTAIN ARTHUR: I can see it's fun to play with, Neeplphut, but what's the *point*?

NEEPLPHUT: The point, my good captain, is that even in this extremely simple example, you find self-replicating mutants of increased complexity, parasitizing the replication program of the primeval organism, and a rudimentary case of immunity when the primeval organism fights back. This kind of behavior is typical of computational entities.

STANLEY: And, thanks to DNA, life is the computational entity par excellence. So we shouldn't be surprised if the self-organizing and self-complicating features of life seem counterintuitive.

NEEPLPHUT: That is correct, Stanley. It is your intuition, not life, that is faulty.

MECCANO ANIMALS

There are further complications that arise when you try to understand changes in organisms by way of changes in *DNA*. Waddington's experiments on genetic assimilation showed that a lot of the characters of an organism are not expressed in the form or function of the organism. They are potential. Of course they are; we knew that. Male characters show up when one hormone is present; female characters show up when a different hormone is—no matter what sex is specified by the genetics of the cells in the body. The developmental pattern of both sexes is present in each embryo, and a single switch, a hormone (ultimately determined by a single short DNA sequence), decides which of the two potential patterns of anatomy and physiology will actually be realized. Similarly, environmental cues, such as warming a *Drosophila* pupa, can switch development into another path, cross-veinless, in flies that are genetically susceptible to that treatment.

On the other hand, there is also a great deal of stability; the system has many givens. If a mutation makes an animal grow a lot bigger, perhaps by providing more growth hormone, it doesn't burst out of its skin when it gets too big. Skin isn't just a bag of predetermined size; it seems able to grow to match whatever body it contains. Equally, blood vessels and nerves seem to spread right through. Animals don't develop a new organ, then have to wait for matching mutations in the instructions that govern nerve growth so that nerves grow into it. They're not like Meccano creatures, to which evolution adds a new electric motor but no battery to power it, or a new wheel but no axle for it to pivot on.

Actually, that last statement isn't entirely true. There are a few animals that do seem to be built using the Meccano approach. They always have exactly the same number of cells in a given organ, and their entire development runs along very rigid lines. Cut a nematode egg in half and you don't get twins: you get two different, often nearly perfect, half-nematodes, one built with half the collection of Meccano pieces, the other built with the other half. It's like cutting a fire-engine egg in half and getting a truck and a ladder. The nematode *Caenorhabditis* is the most famous of these, and we probably know more about the relation of its genetics to its developmental program than we do for any other organism. (We know more about the genetics of the bacterium *E. coli,* but that doesn't develop.) Precisely because of this developmental rigidity, it seemed to Sidney Brenner that *Caenorhabditis* was the ideal animal for investigations aimed at mapping the DNA program (in the computing sense) point-for-point to the developmental program. By "mapping" we mean that each part of the DNA sequence is to be associated with a specific step in the organism's development: *This* is the sequence that makes cell number 97 grow in the right place; *this* is the sequence that determines the overall structure of the nervous system; and so on.

Surprisingly, more than half of the "pattern-forming" genes in this nematode seem to be of the maternal-effect type. Most of the "housekeeping" genes that produce the enzymes that perform standard tasks in all cells make use of maternal photocopies. The baby worm's own genetics take over once the cells are up and running. And this is just the start of a surprisingly complicated story. This archetypal Meccano organism may just build itself from the pieces in the box, but the assembly instructions are a great deal more subtle than "Fit screw A into hole B."

Thus there are at least three levels of genetic control development. Changes in low-level genes affect only single characters—whether pigment is made, how many bristles one has, how much growth hormone is produced. Changes in "managerial" genes produce a coordinated suite of developmental changes, either because they act very early (and may be maternally donated) or because they act on other genes (homeotics) or on whole suites of hormone-sensitive tissues (sex-determining genes). Finally, some genes are parts of integrated "developmental kits." Skin must have a lot of built-in properties to respond to what other genes make the body do as it develops, and so must muscles, bones, and joints, which must all develop in concert if the animal is not to fall apart when it tries to walk. Perhaps there are some parts of the genetic blueprint that oversee the relationships of muscles, bones, and joints, or the connections between sense organs and the brain.

■ SOURCES OF INNOVATION

We've now traced the core of the evolutionary process to DNA. Individual mutations are just changes in the DNA code. But natural selection doesn't act directly upon DNA codes. It's not a quality-control inspector who looks at code sequences, calculates their fitness, slaps on a sticker, and rejects those sequences that fail to come up to scratch. Natural selection acts at one remove: Mutations in DNA produce differences in the form or function of the organism (via development), and it is upon those differences that natural selection acts. The results feed back to the DNA, because to become replicated it needs the organism. DNA that produces losers in the battle to reproduce will die out; DNA that produces winners will replicate and thus continue to produce winners (until evolution elsewhere changes the whole game). Living creatures fight proxy battles, but DNA controls the boardroom decisions. Richard Dawkins's book *The Blind Watchmaker* describes these battles extremely well.

The way this feedback cycle operates is subtle. Sometimes we can convince ourselves that we understand particular instances; we can see the survival advantage in some adaptation. For example, carnivores usually have their eyes at the front of their heads so that they can keep their prey in sight as they chase it; herbivores have eyes on the sides of their heads to provide a wide field of vision, keeping a lookout for predators. Such understanding must be subjected to critical scrutiny, however appealing it may seem. Thus conventional wisdom has it that the giraffe's long neck lets it reach leaves higher up the tree than its competitors can. This may well have some truth in it, but it's only part of the story. The giraffe also has long front legs, which let it run faster, another obvious survival factor. Suppose you were an animal with long legs but a short neck; how would you drink? You could kneel, of course, but if your legs were long and awkward when folded that might not be a bright idea. A long neck would come in handy.

There are many strange evolutionary novelties, whose origins seem very puzzling. They deserve attention, because it is the tricky cases that provide the strongest tests of the theory of evolution, and it is upon these cases that thoughtful critics of the theory sensibly tend to focus. Stephen Jay Gould has written many beautiful essays that pick up these "Just So Stories" and show how a plausible series of ancestors can be inferred—or invented—to account for natural selection of really odd adaptations. There are three major routes to innovation.

The first route we have already touched upon in chapter 3; it makes use of the twin processes of progenesis (early breeding) and neoteny (omitted adulthood). In both of them the sequence of development is (at least to begin with) unchanged, but the timing is altered, so that particular features may be suppressed or emphasized. Our previous example was the evolution of the axolotl, which, you recall, breeds as a tadpole.

The second route involves an apparent discontinuity in the chain that leads from ancestral forms to their modern descendants. "They must have died out at this stage because the organ can't have been any use when it was that small/rudimentary/different." Think about the elephant's trunk. What advantage does it convey? Well, you can get at high leaves with it (like the giraffe's neck) and drink without bending down (like the giraffe's neck) and heave logs around easily (quite unlike the giraffe's neck). Now, somewhere along the line of development from an ordinary nose and upper lip to a full-sized trunk, what you get is a short dangly thing on the front of your face. It's no good for grabbing leaves from tall trees, no help with drinking, and totally useless for heaving logs around. Therefore, the argument goes, there's no advantage in having it, so the evolution of a trunk can't get started.

There are several answers to this objection. An extreme one is the catastrophist position: Occasional evolutionary jumps can occur, so you don't have to go via a short dangly thing. But that doesn't explain how the jump occurs. It might be some accident of DNA evolution, a sort of elephantine antennapedia—or trunknosia—but how did the blueprint for a trunk get into the DNA, ready to be used, without being field-tested first? A much simpler explanation is that there are all sorts of uses for a short trunk. The tapir gets excellent service from one today. It's good for grasping things—not tree trunks, but bunches of leaves or grass. It's not bad for poking into niches that ordinary animal mouths can't reach, to seize plants growing between rocks or ringed around with thorns. In short, an organ can evolve through rudimentary versions, provided those versions convey some advantage. It need not be the same advantage that a fully developed organ would confer. A very rudimentary wing, for example, might be useful for cooling purposes (flap it feebly to create a draft or to cool the blood inside it); a partially developed one could permit short gliding flights just long enough to confuse a predator.

The third route to innovation invokes collateral necessities: "You can't get there from here, but if you went *that* way first . . ." It leaves odd groups of characters hooked together in ridiculous ways. All land vertebrates evolved, it is said, from the creature known as *Eusthenopteron,* which de-

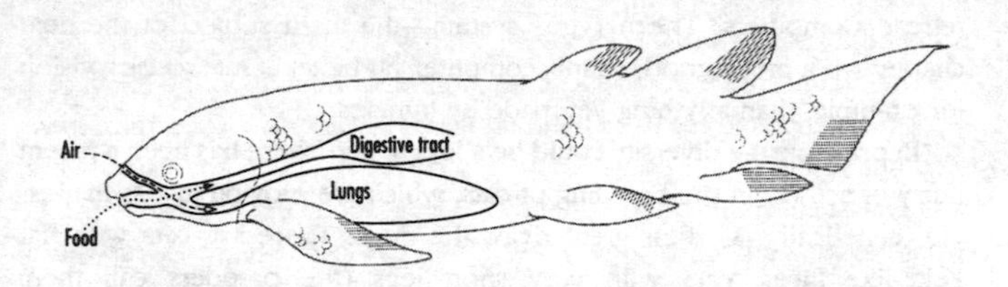

FIGURE 19

Eusthenopteron's unfortunate crossover

cided to set up home on the land. Long ago, a fish developed lungs. Its descendants had lungs above, below, or around the gut. *Eusthenopteron* probably had its lungs below the gut (Fig. 19), so that its airway had to cross its foodway. As a result, all of *Eusthenopteron*'s descendants, including us, have our airways crossed with our food passages. The advantages of having these two things at all seem to have outweighed the problem of an occasional individual choking to death. If only *Eusthenopteron*'s lungs had been above its gut . . .

Our excretory and reproductive systems are also mixed together—an accident that, duly processed though the human psyche, may well be the ultimate origin of the phrase "dirty books."

■ DIVERSITY RULES

When we look at life on earth today, the most obvious and impressive thing about it is its enormous diversity. The selfish gene employs a tremendous range of vehicles for its successful replication. If we accept Waddington's message that the potential genetic versatility in every species is enormously greater than what is expressed, then the diversity is even more impressive. If you can think of it, there's a life-form somewhere on earth that has it. Radar? The bat's sonic echolocation comes close enough. The wheel? There are no wheeled creatures of mammalian size, but the flagella of some single-celled organisms have a rotating axle and bearings.

Metal detector? Sawfish use sensors for electric fields to detect prey buried in sand. Jet propulsion? Squids use it to confuse predators and beat a hasty retreat. Computers? The nervous system—the main subject of the next chapter—is a pretty good organic computer, far better at many tasks, vision for example, than anything yet made by humans.

In principle the diversity could be a lot greater. There has been a recent flurry of activity in the breeding of cats, which are proving just as malleable, genetically, as their great rivals the dogs. There are cats with flat Peke-like faces, cats with very short legs (the breeders call them Munchkins—ugh), unusually large cats, and furless cats (very wrinkly). These strange (and mostly ill-advised) breeds have not been produced by subjecting cats to radiation beams or mutagenic chemicals; they've just been selectively bred from the existing genetic repertoire of ordinary catdom. They don't usually occur because they're not adapted well to survival (which is why the breeds are ill-advised). The cat with a flat face, for example, has terrible sinus trouble and its eyes have to be wiped all the time.

But even ignoring the potential diversity and concentrating on what is actually expressed, there's an awful lot of life around. That's because there are an awful lot of possible DNA strings, and mutations provide the diversity of developmental programs that enable new structures to develop. The developmental programs that produced more viable organisms prospered.

A naïve view of natural selection would seem to rule out diversity. If only the fittest survive, why don't we end up with huge quantities of one superfit animal—the perfect survivor—and nothing else? This indeed tends to happen with breeding systems selected by humans on the basis of criteria such as productivity. There are remarkably few strains of wheat in the world. However, fitness isn't a numerical quantity like price or temperature. Given two organisms, neither need be fitter than the other. It depends on what environment the fitness is related to. A dolphin is fitter than a human for survival in the sea, and less fit in a forest.

■ BURGESS BUSHES

As well as increasing diversity, all the common images of evolution, such as ladders and trees, carry overtones of progress: ever onward, ever upward. Lamarck placed humankind at the pinnacle of the Ladder of Life, and he was not alone in seeing the human race as not just the product of evolution but its goal. The story of evolution that we have been telling is a very deterministic one. Mutations may be random, but only the winners survive, and winners are those with superior characters. If you ran evolution again, with different but equally random mutations, on the whole you might expect the same winners to emerge.

Or would you? According to Stephen Jay Gould, in *Wonderful Life*, the answer is a resounding no. We owe our presence on this planet to a long series of accidents, very few of which would likely be repeated if history were rerun. Gould's argument that contingency—randomness—plays a major role in the results of evolution, as well as in its internal workings, is based upon a layer of rock known as the Burgess shale. Fossils of soft-bodied creatures are very rare because they tend not to be preserved. The Burgess shale, however, has beautifully preserved fossils from the time when multicellular animals were first starting to evolve. On the tacit assumption that the diversity of life today, plus its known ancestry, is all there is, previous generations of paleontologists had shoehorned the Burgess shale animals into previously known compartments—hung them on existing limbs of the evolutionary tree. Then Conway Morris and other young paleontologists took a less prejudiced look and discovered that most of the animals of the Burgess shale are totally unlike anything we have ever seen before. So different are they that they led Gould to a new view of evolution. There is not one evolutionary tree, but many. Instead of a treelike structure for the relations between species, think of something that more closely resembles a moor with lots of grasses, a few bushes, some saplings, and only one or two tall trees. Interpreting this metaphor for organisms, we see that nearly all new species fail immediately, leading to a localized loss of diversity; a few survive and diversify and have a fair amount of success for a while, but then fall foul of adverse conditions; and very few indeed hit upon a significant new evolutionary "invention," a body plan that lets them keep diversifying into towering timber. There are also a few experiments that didn't diversify—isolated branches of ancient bushes that have grown up into the canopy like a liana. They include horseshoe crabs, the coelacanth (a species of fish thought to have been extinct for millions of years until one was caught in 1938 off South Africa), *Peripatus* (a creature halfway between arthropods and worms), and the tuatara (a New Zealand lizard).

Evolution should be viewed not as one tree, then, but as a moor, with mostly stunted growths. The tree that we see now includes those creatures whose descendants still survive in some form, but those are only a tiny proportion of the total that have lived at some time in history. Trial and error involves a lot of trials and 99.999 percent error, so Gould sees the evolution of humanity as being accidental, purely contingent. There is a Burgess shale creature that may well have been the ancestor of the vertebrates (and hence us); certainly it is the most likely creature in the Burgess shale to have evolved in that direction. It appears no better adapted than any of the other peculiar creatures with which it coexisted.

Yet most of those creatures became extinct. The Burgess shale fossils were created by a mud slide, a random event that killed off all the inhabitants of that particular pool. There were other mud slides, other pools. The mix of creatures would not have been identical in every pool. Survival could have been, to a great extent, a matter of chance. Suppose that the species of creature that evolved into today's vertebrates had become extinct, as most of its fellows did . . . then no vertebrates, no Burgess, no Gould, and no you. Equally, if the K/T meteorite, which (many people now think) crashed to earth and thereby brought about the end of the Cretaceous period, killing off the dinosaurs, had missed, then the dinosaurs would have continued for another 150 million years. The shrewlike ancestors of us mammals would have stayed tiny and continued clambering around in trees, and again there would be no Burgess, no Gould, and no you.

Gould's theory has been disputed on many grounds. Some people question the supposed diversity of the Burgess shale fauna. At least one creature, interpreted as having a totally new body plan for animals, seems to have been reconstructed upside down; its right-side-up version appears rather more prosaic. Others question the contingency; perhaps it would have been possible to predict which creatures in the pool had the greatest chances of survival. But the central point—that when you have a big diversity of animals in relatively small populations, chance effects become much more important—deserves to be taken very seriously. There is no reason why evolution should effectively be deterministic; *small* populations can easily be wiped out by sheer bad luck, and then survival is a matter of fortune rather than fitness.

However, Gould may have overstated his case. We've discussed the existence of two types of character: simple characters directly produced by the protein-building capabilities of DNA; deeper characters that depend on DNA sequences that control the action of other sequences. Diversity may be a deep character of this kind. It is possible that the Burgess shale was formed around the time that evolution discovered the trick of flexible programming. If so, its creatures may be the result of natural selection operating on deep characters. Because they are deep characters, we should be very wary of drawing conclusions about which organism would be most successful simply on the basis of external form. Perhaps the superior survival abilities of the creature that evolved into us actually would have been predictable in advance. This is an intriguing question, and we will come back to it in chapter 10.

Darwin himself had rather different worries about the origin of humankind. While it's very plausible to us that little dinosaurs became birds, or even that replicating clay molecules led to nucleic-acid replicators, we tend to feel that our own origins can't be as simple. Our physical form is one thing; we have convincing fossils of our recent, prehuman, and becoming-human ancestors. The problem is our mental capabilities. We can think, reason, do sums, write books about the collapse of chaos, and argue about the religious significance of evolution. None of these abilities seem to be explicable in the same terms as our physical attributes. How does intelligence fit into the reductionist picture?

5 THE ORIGINS OF HUMAN UNDERSTANDING

A very ugly and conceited young man, wanting to get married, went to a matchmaker and asked for help in finding a wife. "I want someone utterly beautiful and totally exceptional."

"I have just the girl," said the matchmaker. "She's rich, intelligent, and absolutely stunning."

"Hold it," said the young man, suddenly suspicious. "Why is she still single?"

"I admit," said the matchmaker, "that she has one tiny problem."

"I thought so."

"No, it's nothing very terrible. It's just that— Well, one day every year, she goes a little bit crazy. Doesn't cause any trouble, just a bit weird. After it's over, she's fine for an entire year."

"I can live with that," said the young man. "Where is she?"

"Not right now," said the matchmaker. "If you want her to marry you, you'll have to wait a bit."

"Until when?" the ugly and conceited young man asked eagerly.

"Until the day she goes crazy."

Many people are happy with the idea of animals evolving and having evolved, but draw the line at human beings. There are so many aspects of human beings that seem totally different from anything we find in animals. We have language, creativity, artistic tendencies, mathematics, the written word, culture. Above all, we are intelligent. How can such dramatic differences from the rest of the animal kingdom ever have evolved? Of course, our view of the special nature of human beings is not terribly objective, because, unavoidably, we live inside one of them. We have no idea what it is like to be a dog or cat, assuming that such a statement makes any more sense than asking what it would be like for a tree to be a rock, so we have no very good basis for comparison.

The paleontological record is clear enough that no serious scientists dispute the basic point: Humanity did evolve from animals. *Homo sapiens* is a branch of the ape family, a branch so confident of its superiority that it names itself "Man the wise," flatly denying the evidence of most of its own daily experience. Cynicism aside, it's also undeniable that we are a special, highly unusual kind of animal. For a start, we *can* be cynical, and notice it, and make fun of ourselves. And then write about it. And then write about— But perhaps we should stop before we get caught in an infinite regression.

Most of the abilities that we parvenu apes cherish as signs of our superiority to the rest of the animal kingdom are present, at least in rudimentary form, in some species or another, as Jared Diamond recently hammered home in *The Rise and Fall of the Third Chimpanzee*. However, some of the links he discusses—for example, that between human agriculture and the ants' use of aphids and fungus farms—are somewhat strained. Moroever, there is a certain inconsistency of viewpoint in maintaining both that humans are extremely unusual and special and that all of our apparently unique features have animal precursors.

Nearly all of the really dramatic differences between human beings and other animals are the result of our brains. For an animal of our size we have an unusually large brain; while size isn't everything, size plus structure that can make good use of size is. How did our brains and their remarkable abilities evolve? Brains are built from networks of nerve cells, which transmit electrical signals and form "electronic circuitry" that processes data from our sense organs and sends instructions to our muscles. We have to explain this whole package. As you might expect, the most important link in the evolutionary chain is the development of the nerve cell. Brains must have started out as rudimentary groups of a few nerve cells, so before we can tackle brains, we have to ask why a tiny group of nerve cells might offer any evolutionary advantage.

■ COMPLEXITY INCREASES

Before tackling nerve cells, though, we must establish a general principle: that evolution mostly leads to increased complexity in individual organisms. Complexity is downhill to evolution. The examples of artificial worms and Neeplphut's computer organisms are simple cases of this at first sight puzzling tendency in evolutionary systems. Now we wish to argue that it is a very general tendency, and for general reasons.

There are two rather difficult ideas to be grasped. The first is that it is "easier" to add stages onto an already effective sequence than it is to modify earlier steps in the sequence. So most innovations that offer a competitive edge are refinements that complicate (and often enlarge) the *adult* stage of organisms. The second idea concerns the kind of innovation: It is more likely that competitive advantage will be gained by adding something rather than removing it. Neither idea is universally valid, but both are true much more often than they are false. Of course the exceptions are interesting in their own right, but let's concentrate on the main line of thought, taking the two ideas in turn and developing them a little.

First, that it is easier to add new stages to an effective sequence than it is to change it. The reason is quite simple. When you add a new stage you can build on what already exists. The old line of development will continue to work just as before (except that your extra stage may wreck something that would otherwise have worked well). Don't forget, we're talking of modifying a process here, not just changing the product. The process works fine until you get to the new stage, because it worked fine before. If you tinker with an early stage, however, you may well mess up everything that happens afterward. Think of early frog development: the formation of a hollow spherical blastula, which engulfs a bit of itself to make a gut. Sometimes the blastula exogastrulates—forms a gut on the outside instead of the inside. It gets its development wrong—though when it does so, like any machine that behaves in an abnormal manner, it is still obeying the laws of nature; they're just leading to a different result. This mistake doesn't just cause problems for the gut of the developing animal. In normal development the nervous system forms when the epidermis switches to nerve-forming mode. When the gut is on the outside, this switch never gets tripped, and no nervous system develops. Stomach on the outside implies no nerves anywhere. It's like those assembly instructions that tell you to screw the wires into the sockets but have omitted to tell you to fit the wire through the hole in the outer casing first—only this time your modification means that there isn't a hole in the outer casing, or maybe that there's no outer casing at all, or no sockets, or the wrong size of wire.

Computer scientists write programs in terms of small, self-contained sequences of operations, known as procedures (or subroutines, objects, or demons). If you know that each procedure works on its own, you can concentrate on how to hook them together. Biological development is a bit like that, but because the procedures are carried out in a fixed sequence, you can't modify one in the middle without affecting everything that follows it. Sticking something new on the end is far more likely to work. However, just occasionally you might get lucky if you tinker with something in the middle; we are discussing a general tendency, not a rigid rule.

The second idea is that improvements generally involve extra gadgetry rather than less. This is not a rigid rule either, but again it makes a lot of sense. Suppose you decide to remove something. It was presumably there for a reason, so you'll lose whatever advantage it originally conveyed. The advantage you gain by simplification either has to be so good that you don't mind, or (more likely) the thing you've removed has already become obsolete. Both of these methods for improvement are uncommon—though striking when they work. Most of the time, complicating development is, paradoxically, easier than simplifying it. If a predator has good ears but poor eyesight, you don't remove its ears, you improve its eyes. In a complex structure, tweaking and fussing, adding quality control or bells and whistles, has a much better chance of giving a competitive edge than does simplification. There is generally more stuff under the hood of a modern car than of a veteran. A silicon chip, despite its small size, is more complicated than a vacuum tube: a modern desktop PC can do more things than an old mainframe computer could because it's more complicated.

One of the most interesting cases where a major developmental simplification did occur was the invention, by mammals, of the technique of keeping embryos within the mother's body until they are fairly well developed. Genetic investment in temperature control for the mother then rendered obsolete masses of DNA contingency plans for different temperatures in the embryonic environment. That's the main reason why human DNA is shorter than frog DNA, and it's an example of privilege, of built-in advantages due to improvements in the process of development. Silicon chips also involve a kind of technological privilege; the effort is invested in the chip-producing plant, not in the chip itself. Chips are a dime a dozen (fairly literally), but vacuum deposition equipment or photographic mask aligners cost a fortune, and chip designs are guarded like state secrets. Notice that the simplification of mammal DNA happened as a quality-control improvement first, being followed by the loss of obsolete material. There are evolutionary pressures behind that loss, not because the outmoded DNA occupies unnecessary space—evolution doesn't seem to care much about that kind of efficiency—but because mutations in sequences of DNA that are never used (since they code for a contingency that never happens) do not have an adverse effect on survival. So a creature with the mutation survives just as well as one that doesn't, and the unmutated DNA sequence is less able to ensure its own replication. Something more drastic must have happened in mammals, but it's not at all clear what.

IT'S BETTER TO BE DIFFERENT

Creatures do not evolve in isolation, but as part of an ecology—a coevolving system of organisms, interacting strongly with each other. Ecologists like to talk of environmental niches, which are possible "lifestyles" for the organisms in a particular environment. These niches are not fixed, not God-given; "niche" is just a convenient metaphor for describing a creature's relation to its environment, including other creatures. Until dogs have evolved, the niche for dog fleas is not unoccupied; it is nonexistent. The story of an ecology is usually one of increasing complexity, with organisms evolving into rather more specialized niches, and thus becoming more specialized organisms. The Burgess shale is an ancient example; two modern ones are African lake cichlids and Darwin's finches, which we now describe.

Darwin realized that the ecosystems of small oceanic islands would be very different from those of large landmasses, because the effects of geography would be very limited, because the ecology would involve many fewer species, and because an isolated ecology would develop differently from one perpetually subjected to outside disturbance. "In the Galápagos islands nearly every land-bird, but only two out of eleven marine birds, are peculiar," Darwin noted. By "peculiar" he meant specialized. The same is true today. There are thirteen species of finch on the Galápagos islands, all evolved from a common ancestor, probably during the last five million years or less. They are all black or brown and from three to five inches long. They are astonishingly specialized. There are three tree finches, two cactus finches, four ground finches, a warbler finch, a mangrove finch, a woodpecker finch, and a vegetarian finch. In the absence of the usual competition, the original finch species evolved to occupy a whole set of niches (a phenomenon known as adaptive radiation). Each finch has its own characteristic shape of beak, adapted to its chosen niche. The vegetarian finch, for instance, has a short, thick beak for cracking seeds.

The effect of natural selection here is not that of "Nature, red in tooth and claw," as Tennyson put it; animals are not competing desperately for the same food supply. On the contrary, it is that of animals learning to live with their fellows by finding an alternative food supply. This is not to say that competition does not occur; after all, it is competition that results in natural selection. However, in practice you hardly ever see the competitive element of natural selection. It happens so fast, on a geological time scale, that all you see is the eventual winners.

However, on a human time scale we can sometimes see the changes as they occur. Darwin's finches are still evolving. From the middle of 1976 there was a drought on the island of Daphne Major. The plants produced few leaves, so there was a dearth of caterpillars. A few pairs of cactus finches bred, but their offspring failed to survive. Medium ground finches didn't even breed. By the time the rains came back, early in 1978, only 15 percent of the population of medium ground finches remained. Those that did had been selected for size: The large birds fared best. Their beaks, in particular, were conspicuously large. This seems to have been the factor that influenced their survival. When seeds become scarce, birds that can exploit larger seeds, not normally eaten at all, have an advantage.

An even more extreme case of effective isolation of an ecology occurs in the cichlid fishes of the African Rift Valley lakes. Cichlids are perchlike fish between two inches and two feet long; there are about twelve hundred species worldwide (excluding those in the Rift Valley lakes). The three main lakes in the Rift Valley are Victoria, Nyasa, and Tanzania, and they have been there for between half a million and ten million years. During that time the landscape has changed, usually slowly, with changes in the connections of rivers and pools, which is how the fishes got into the lakes to begin with. The cichlids in each lake seem to have started from a single founder species, a different one for each lake. There are now some 750 species in Victoria, two thousand in Nyasa, and three thousand in Tanzania—more in either of the latter two lakes than in the rest of the world put together.

Despite starting from different species, each lake has produced *the same* ecological spread, the same set of environmental niches. They all have standard fish, which eat smaller animals, and fish that eat algae; but they also have specialists, such as fish that eat only eyes, or only fins, or only other fishes' babies. In each lake there is a small blue-black fish with three broad gold longitudinal lines, which occupies the same niche in each case; but the skeletal structure shows that each evolved from a different ancestor.

When, recently, catfish were introduced into these lakes by local fishermen looking for a more salable product, they competed very effectively indeed within the entire complex cichlid ecology. The catfish didn't distinguish between one species of cichlid and another; they were just better at grubbing around on the bottom of the lake than any of the cichlids were. Until the introduction of the catfish, the fittest bottom-grubber was, by force of circumstance, the fittest bottom-grubbing cichlid. With the arrival of the catfish, all that changed, and the bottom-grubbing cichlids died out. Fitness, we repeat, is fitness relative to the existing competition, and new competitors change the nature of the game.

These two examples show how, even in an isolated ecology with limited raw materials, evolutionary pressures lead to diversity and the occupation of ever more specialized niches. But this process of continuing complication can't go on forever. Living creatures are forced by evolutionary pressure to operate right at the limits of what they are capable of, to perform a delicate balancing act on the edge of disaster. There may come a time when the "style" of an organism—its system of organization—starts to get top-heavy. Having chosen to specialize, all it can then do to improve is to become more specialized; it's trapped in an evolutionary dead end. Many prehistoric animals, such as saber-toothed tigers and titanotheres, grew so large and lumbering that they were poised right at the limits of what their environment could sustain. Their size was such an advantage that it made it worth their while, in evolutionary terms, to grow so huge, but there was a price to pay. Operating right at the limits made them very vulnerable to environmental changes—in particular, any reduction in food supply.

In such circumstances it is evolutionarily worthwhile for some of the competitors to cut out the later, complicated part of their life history; this excision results in neoteny (omitting the adult stage) or progenesis (breeding as a larva), depending on whether stability or exploitation is the background rhythm. If it is advantageous to stay the same, to continue to occupy a well-developed and canalized niche, then you get neoteny; if there is a new niche to be developed out of the old one and exploited to the full, the result is progenesis. By leaving out the later, complicated part of development, a tested-but-simplified earlier state can be used as the new adult type in a currently underexploited part of the ecology. Among the marine plankton there is enough food for everybody, and many sea creatures send their larvae there for school lunches. However, when the evolutionary squeeze is on, the plankton may lose their adult stages altogether and become predatory larvae, who not only bully the other schoolchildren but eat them.

From this new basis, a new set of competing complications can be established. Advance, retreat-and-consolidate, advance again. These "simplifications" add to diversity, because the original creatures usually hang around too. There is a texture within such a diversity, as well. All of the organisms have parasites as well as predators, and these parasites form evolving subsystems of their own, which track their hosts, diversifying with them and continually trying out new tricks of their own.

■ GENETIC KITS

The standard textbook story is of a random mutation in the DNA providing a range of developmental variants (many of which are lethal: The developmental system crashes). However, a single-gene mutation rarely has a predictable effect upon the development of an organism. Waddington showed this clearly: The extent to which development is affected either by an environmental kick or a gene difference depends upon what other genes are present. In particular this is true for the "wild-type" version of the appropriate gene, which we saw in chapter 3 is typical outside the laboratory: When the mutation first happens there is *always* a normal version available on the equivalent chromosome from the other parent. Almost all mutations produce a loss of function of the proteins that the gene produces; such a mutation won't show up if the normal gene is present. Mutant genes of this general type are thus recessive; if Dad gives you a mutant eye-color gene, you've still got the normal one from Mom, and that one wins, so your eyes are the normal color.

More subtly, in canalized development, where the entire system has stabilized itself against environmental and genetic changes, there are all kinds of alternative routes to each stage or function, often because the original gene sequence has been duplicated or indeed multiplied. It is very common to find that a particular protein, such as hemoglobin, is represented by many variant DNA sequences. In us, one of these makes special embryonic hemoglobin, which is so avid for oxygen that it can take it from the mother's blood; fetal hemoglobin is used later, and finally there is adult type. There are many other hemoglobinlike sequences in the human genome, though, which aren't used anywhere. There are good reasons for keeping such copies; they're not just superfluous junk. All of us keep a lot of clothes in a wardrobe or drawer; we don't wear everything at the same time. This isn't wasteful; we find it very useful to have "one to wash, one to wear, and one to keep in a drawer, in case." Similarly, anyone who uses computers learns to keep backup disks with copies of all their important software and documents. These are not used, but stored. When (it always *is* "when," not "if") the computer breaks down and chews up its hard disk, on which the working copies reside, you can put everything together again from backups. The same goes for upgrades of software to allegedly superior versions: Sensible users keep copies of the older versions, just in case the shiny new improvement turns out to have unexpected flaws. (Again, it usually does, so the flaws aren't really unexpected at all.) Life acquired the DNA-backup trick at some fairly early stage of evolution, mainly for the "upgrade" reason. Most mutations occur in unused parts of the DNA, it seems, resulting in evolutionary decay unless that DNA adopts selfish-gene tricks. When, occasionally, those bits are used—as when two recessive genes come together in one creature—then evolution can experiment, without losing things that work already if the experiment doesn't lead to anything useful.

There is an even more subtle reason why most DNA changes (and most environmental differences) don't affect the developmental program in any obvious way. This is because the systems concerned are versatile. Like the subprogram for skin, which permits any size of embryo to grow within it without bursting, most of the developmental programs of most animals and plants have lots of if-then contingency plans. In chapter 3 we explained how the frog's embryo develops at the same rate, no matter what the temperature, by using different enzymes at different temperatures. Let's clarify that a bit. It's unlikely that there are labels that really follow instructions of the type "Use when temperature exceeds 80°F" on bits of messenger RNA—gene photocopies—which the egg received when it had a maternal nucleus. The egg would need a thermometer for such an instruction to make sense. Presumably what happens is that the appropriate bits of messenger RNA possess quantitatively different labels, and the embryo makes use of progressively faster ones as some reaction loses ground relative to others. For example, some necessary enzyme product may dwindle as the temperature drops. In that sense the embryo does have a thermometer, but the process of accessing the correct enzyme runs automatically, because the embryo's DNA programming is indirectly temperature-sensitive. Such built-in versatility, in which the ends are predictable but the means can be varied, is characteristic of life. When a bee leaves the hive in search of nectar, you can predict that it will shortly settle on a flower—but you can't predict *which* flower, or what route it will take to get there.

Textbooks emphasize the homeostasis of adult animals—their built-in ability to maintain the correct temperature and the right amount of stored food. But the embryonic program must do more than maintain; it must progress. Waddington gave the name "homeorhesis" to the mechanisms that achieve this maintenance of developmental pathways. The term covers various kinds of developmental versatility; not only the hot frog's egg, but the way that two halves of a frog's egg make two perfect tadpoles, or the way that shepherd's purse can make either a huge shrub or a tiny dwarf weed. Many people assume that all living creatures have this kind of versatility, that it's some side effect of, or necessary condition for, being alive. Not so. As we've already remarked, the nematode worm *Caenorhabditis* is a Meccano-built organism, barely versatile at all. All such nematodes have the same number of cells in each body organ (except gonads and skin), and they can't repair wounds. Versatility needs explanation and mechanisms; it's not just a property of living material.

■ COMMUNICATIONS SPECIALIST

Versatility depends on having access to appropriate information. You can't repair a wound, for example, unless in some sense you "know" that there *is* a wound, and where it is. It might be possible to respond to some generalized chemical signal emitted by wounds, but this is unlikely to be a foolproof method. In the same way, a versatile developmental program can't respond differently to differences in the environment unless it "knows" about that environment. Again, this can be done in a relatively crude fashion. In the early development of an embryo, a lot of information is provided by internal feedback mechanisms, such as rough measures of the ratio of cytoplasm to nucleus. This answers questions such as "Shall we stop cleaving yet?" and "Are the cells small enough?" but it can't handle anything subtle.

A much better solution is to develop a specialist class of cell that can detect information about the organism or its environment and convey it to a place where it can be acted upon. Many kinds of multicellular organism have developed just such a communications specialist: the nerve cell. A nerve cell is very long and very thin—it takes just two nerve cells to connect your toes to your brain—and it conducts electrical signals by what is essentially a chemical process. Nerve cells occur earlier in evolution than brains, and in simpler animals. *Caenorhabditis* does have nerve cells, but apparently it doesn't use them to convey developmental information; it just assembles the Meccano, piece by piece, and the nerve cells are among the pieces. Nerve cells are used to keep an eye on the development of a creature as simple—relatively speaking—as *Hydra,* and in most other animals. Adult hydras, which have a network of nerve cells over the surface of their bodies, can coordinate their overall movements in a way that appears to display "intention": They bend toward prey.

There is an entire mathematical theory of neural nets, idealized structures that mimic some of the known features of nerve cells. You can think of them as elaborate switching circuits, where the behavior of any particular switch depends upon the signals that it receives from all the others to which it is connected. If this sounds like a cellular version of electronics, that's not a bad image. Electronic circuits can be designed to carry out almost any desired function, including complicated computational tasks, and so can neural nets.

Nature, however, doesn't design things. It lets them evolve. The network of nerve cells, the connections, and the chemical "computations" carried out by each nerve cell that govern the signals it sends out and its own response to the signals that it receives—all presumably encoded somehow in the organism's DNA—can be modified in innumerable ways by mutations. Some of those modifications lead to greater reproductive success, others do not. Those that don't, die out; those that do, survive and are replicated. Presumably the "intentional" movements of the hydra came about because those ancestral hydra that came up with this particular nerve net and its associated chemical rules fared better than their competitors. Rules that caused the hydra to bend away from prey, for example, would have been a disastrous failure.

The abilities that stem from a network of nerve cells give a major twist to the evolutionary tale. They change all the rules of the game. Instead of outrunning a predator, you can outcompute it. You can sense it coming and make a prudent retreat. Instead of grubbing around blindly for prey by feel, you can detect some characteristic feature, such as a chemical that it emits. You can use chemical signals to ensure that you find a suitable mate. Predators can eavesdrop on your signaling mechanism and give you a nasty surprise. And so on. All of this means that there is a huge evolutionary advantage in experimenting with improved nerve cells, neural nets, sense organs, and computational devices that extract useful information from what the sense organs report. The same battery of equipment is useful in controlling biological development and generating versatility. Nerve cells can be active monitors of a creature's developmental state, and can react quickly and at a distance to important changes.

NEURAL CIRCUITRY

A nerve cell, or neuron, consists of a roundish cell body, from which extends a long fiber, the axon. At the end of the axon, and sometimes elsewhere along its length, are structures that form junctions, called synapses, with other nerve cells. Branching structures of different kind, called dendrites, also extend from the cell body; these are receivers for connections from other neurons. Neurons communicate by electric impulses, generated within the cell body and transmitted along the axon, roughly like messages being transmitted over a telephone line. A great deal is known about the electrochemistry of the transmission process, but very little is known about the way in which neurons "interpret" incoming pulses and decide what pulses to send out in response. We understand some basic features of how the neural telephone system works, but virtually nothing about the meaning of the messages that flow through it.

The theory of neural nets got started in 1941, when Warren McCullough

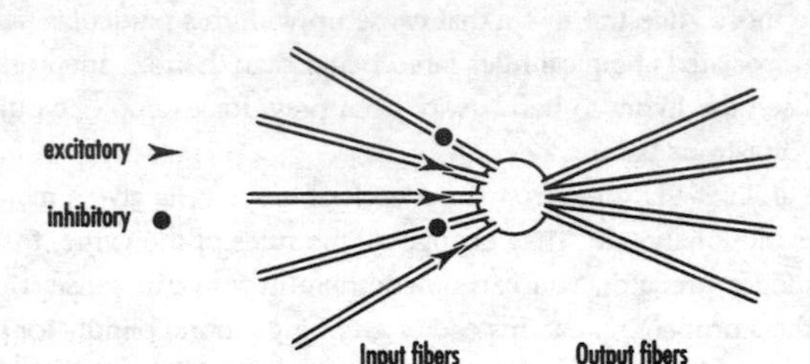

FIGURE 20

Inputs and outputs in a 1940s model neuron

and Walter Pitts developed a simple mathematical model of neurons and their interconnections. The modern setup is different in detail, but it is easier to follow the general idea in the original version, so for a few paragraphs we'll be old-fashioned. A Pitts-McCullough neuron is a blob (the cell body), equipped with a number of output fibers (Fig. 20). These connect it to other neurons, being thought of as input fibers to the neuron in which they terminate. There are two kinds of input fiber: excitatory and inhibitory. What's the difference? When a neuron fires, it sends a signal along each of its output fibers. By convention, this signal takes the same time, one unit, to reach the neuron at the other end. Each neuron has a threshold, a value which determines whether it fires in response to incoming signals. Inhibitory inputs have a total veto: If any signal arrives along an inhibitory input, the neuron will not fire. If no inhibitory signals occur, the neuron counts how many excitatory inputs are carrying a signal, and if that number exceeds or equals the threshold value, then it fires.

These idealized neurons differ from real ones in many respects: The synchronous timing is a major difference; the simple "threshold" internal mechanism is without doubt far too simple; and so on. Modern versions overcome these limitations. But even these golden-oldie models are very versatile. For instance, you could fit them together to make a computer. Instead, we'll tackle a much simpler task, something that might be of use to an organism developing a rudimentary sensory system. Figure 21 shows a very simple Pitts-McCullough neural network, with a mere seven neurons, shown by large circles. (The human brain has *ten billion* neurons.) Excitatory connections are shown by lines terminating in arrows, and an inhibitory one is indicated by a line terminating in a small circle. The number written inside each large circle is the threshold for that neuron. This particular network connects to the outside world via an input connection, and produces an output that can be used to make some decision based on the processing that goes on within the network. It is thus a very simplified model of a sensory process in a living creature, detecting some feature of the outside world and responding to it on the basis of some kind of neural processing.

What does this network do? You might like to think about that before reading on.

Suppose that at some instant of time the left-hand neuron receives an input signal. It then fires, because it has a threshold of 1 and there is no inhibitory input. One time unit later, three neurons receive this signal: the one to its immediate right with threshold 1, and the two above and below with thresholds 2 and 3 respectively. The neuron to the immediate right,

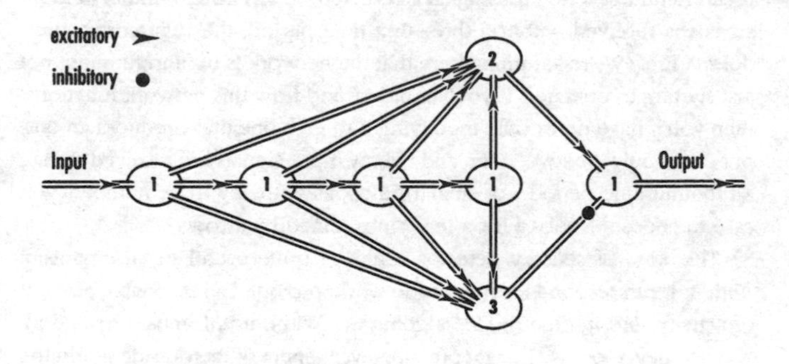

FIGURE 21

A Pitts-McCullough network. What does it do?

with threshold 1, behaves in just the same way, as do the next two neurons in the chain of four threshold-1 neurons that runs from left to right. The fifth neuron that produces the final output also has threshold 1, but is not directly connected to this chain. The chain of four neurons acts as a "delay line," a short-term memory. Any input received by the first neuron in the chain is sent on to the next neuron one time unit later and to the third neuron one time unit after that, reaching the fourth neuron after three time units.

The neuron with threshold 2 will fire at some instant if and only if at the previous instant—that is, one time unit earlier—at least two neurons in the delay line fire. The neuron with threshold 3 is similar, but it fires if and only if at least three of the neurons in the delay line have fired at the previous instant. Now think about the right-hand neuron, with threshold 1. The connection from neuron 2 is excitatory, but that from neuron 3 is inhibitory. In order for this final neuron to fire, producing an output signal, neuron 2 must have fired one time unit earlier, but neuron 3 must not have fired. In other words, one time unit earlier, exactly two neurons in the delay line must have fired. (It takes at least two to make neuron 2 fire, but more than two will also make neuron 3 fire.)

In short, if the leftmost neuron receives exactly two input signals within any period of three time units, then the rightmost neuron fires (two time units after the second input signal is received). If any other number of input signals is received within a three-unit time period, the rightmost neuron doesn't fire. (We're assuming here that the network is up and running, not just starting to operate.) If you've understood how this network functions, then you'll have no trouble modifying it to give one that produces an output if and only if between ten and twenty input signals are received during a fifty-unit time period—or similar functions with any three numbers you care to choose. Not bad for a few blobs joined by arrows.

This kind of sensory detector could be quite useful to an organism. With it, it can respond to the outside world provided a reasonable amount of activity, *but not too much,* is going on. Without it, it must respond always or never, or else just ignore whatever aspect of the outside world the input is detecting. Imagine a marine creature that feeds by putting out a tentacle. An occasional isolated stimulus to its sense organ could be anything: a crab brushing past and already long gone; a pebble dislodged from above. Too much activity might mean that a predator is nosing around, ready to pounce as soon as the unwary animal puts out its feeding tentacle. Intermediate levels might signal the presence of a small, slow-moving creature: food. The neural net just described wouldn't make the right decision all the time, but an overactive animal would rapidly be eaten by predators, and an underreactive one would starve. Survival would select for the most effective neural network.

There are at least two distinct ways for neural networks to evolve. The most gradual is the fine-tuning of thresholds in an already established network, a rudimentary kind of learning process. Our tentacle feeder could evolve to one that made the most effective choices possible with a particular delay-line network, by fine-tuning the thresholds of the neurons responsible for the least and most activity within a given period so that the chances of survival increase. In 1949 Donald Hebb said that nerve cells that fire together grow together. More drastic is a change in the geography of the network itself—new connections, changes from inhibition to excitation, even new neurons. Over millions of years, using a combination of these processes, random mutation and natural selection will lead to increasingly complex and sophisticated sensors and analyzers.

■ ENTER BRAINS

The development of neural sensors and analyzers seems to have been a historically important survival trick, because most complicated animals have nerve cells. In the free-living larva, and in the adult, they form a sensorium, an array of sense organs; and they connect to a motorium, a motion generator, composed of muscles, glands, and other effectors. Most animals are more complicated than hydra and have some kind of central switching system, which determines what sensory inputs evoke what responses. This central switching system is almost always at the end of the animal that first encounters any new problem—the front end. It thus defines a "head" by attaching the major sense organs to that end and placing the main computational devices—the brain—nearby.

The human brain has a number of distinct regions that evolved at different times and that perform different functions. Brains are built from a small number of big circuits with fairly specific functions; big circuits are built from smaller ones; small circuits are built from individual nerve cells; and nerve cells evolved from other kinds of cell. So the problem of brain formation reduces to the general question of biological development. The precise way in which the brain works is another matter.

It's important to realize that you don't have to possess a brain to be versatile. Chemical gradients, for example, can lead you to food, and you don't need a nerve cell to sense those. Amoebas, which have no nerve cells, can take different routes to a piece of food or away from some noxious chemical. But if you want an adaptable way to react to the environment, you need a brain. To a developing animal, variety or versatility in behavior is just as useful as variety in form. By segregating behavioral complications into computing circuits in the brain, or data stored there, you can serve more functions within the same structure. If a horse needed a separate set of legs to trot rather than walk, it would be in deep evolutionary trouble. Instead, as the horse evolved, so did a neural control circuit for its gait, which chooses the best pattern of leg movements for a given speed and terrain. Hardware is specialized; software is adaptable.

As a result of this adaptability, the brain is a very successful evolutionary trick. If you can eat several kinds of prey, or recognize several predators, you're better off than your rivals who can do less, so you'll probably outbreed them. You gain access to simple survival tricks: "Run away from anything bigger than you are." No, you don't need a sophisticated method to assess size: "If you have to bend your eyestalks upward to see it, then scram. Fast." There are tricks involving memory: "That taste made us sick last time we ate it—let's try something else this time." There are tricks of pattern recognition: "Another black-and-yellow stripy thing—wave your arms furiously and drive it away."

Once brains are around, it pays to develop better brains. An evolutionary arms race then develops between predator and prey. Imagine a herd of rather slothful creatures with poor eyesight and little sense of smell, preyed upon by a creature not much better equipped, but having sharp teeth. It pays the prey to evolve greater speed to run away, and better eyesight or smell to spot the predator coming. (That is, those prey animals that develop in those directions have a greater chance of surviving to hand on the genes responsible for them. Creatures don't deliberately evolve in some direction, but to avoid cumbersome circumlocutions we will sometimes use that kind of metaphor.) If the prey evolves any of these things, the predator must do so as well, or else change its preferred prey. As Dawkins has observed, there is an essential asymmetry in these evolutionary arms races. The prey is running for its life; the predator is only running for its lunch. As a result, predators tend not to *over*evolve, becoming so successful at hunting that they wipe out their prey. Indeed, that kind of "success" as a predator is not at all a survival characteristic, as the world's fishing fleets are belatedly beginning to understand. And so predators tend to weed out the old and the feeble, or the young who have strayed from the herd; the fit adult prey mostly survive.

Brains are essentially involved in this process. You can't run efficiently without a brain to control and coordinate the movements of the muscles. You can't run fast without senses to make sure you're not about to fall over a cliff or crash into a tree. So the development of senses and sophisticated movement must go hand in hand with that of neural circuitry to make them work. In technology, you can see the same pattern setting in—for example, consider the development of airplanes. Early airplanes used the pilot's brain, and little else, to set the positions of control surfaces and to make sure the plane wasn't going to hit anything. Today all passenger aircraft are equipped with radar, a machine "sense" that can outperform eyesight when it comes to "seeing" through cloud or fog, and they use increasingly complicated electronics to operate the control surfaces and adjust the engines. Without its computer the space shuttle would fly little better than a brick: It is aerodynamically unstable in normal flight and an unaided human pilot can't control it.

The end results of such an arms race can be very subtle indeed. A male moth can recognize its mate by a single chemical signal, a pheromone; a gazelle will need rather more sophisticated analyzing circuitry to recognize a particular pattern of smells as "lion." A female firefly can be attracted by a flashing light of the right color and frequency; a gazelle needs more sophisticated visual circuits to spot a lion partially hidden behind a bush. Predators have evolved some kind of mental model of mechanics—just how much effort to put into a jump, and at what angle to launch itself, to bring down a running wildebeest. So, too, have monkeys that leap from tree to tree. We have a similar model somewhere in our own brains, which is why we can catch a ball thrown into the air.

As the neural network analyzed above shows, data processing takes time. Complicated data processing takes a lot of time. If a complicated process is to be useful, then the processing must be speeded up. A useful evolutionary trick for this is to sheathe the nerve axon in the substance myelin, which is in effect an insulator to stop the electrical signal from escaping. This offers a kind of instant evolutionary upgrade to any neural network that uses it; everything works faster, but otherwise functions much as before.

The alternative is to stick with more primitive detection systems, as we do to protect our eyes. If something is heading toward our eyes, a very rapid reflex takes over, closes them, and causes us to turn our head away. We don't wait and try to figure out what the object is and whether it's harmful. If you have your nose pressed to a windowpane and a dog on the other side leaps at it, you pull away, even though you know intellectually that the dog can't actually bite your nose through a sheet of glass. Our brains contain all sorts of traces of their evolution. As any competent engineer knows, quick-and-dirty solutions can often be the most effective.

INFORMATION THEORY

Every so often we've dropped the word "information" into the discussion, but we've never really explained what we mean by it. One way to see brains is as information-processing devices, so we'd better remedy the omission now. Electronic engineers have developed a quantitative measure of information, which goes back to Claude Shannon in the 1940s. Any message can be coded in the binary system, as a string of 0's and 1's, and each such digit is called a bit. The amount of information in a message is the number of bits required to represent it—that is, its length. The answer to any yes-or-no question increases our knowledge by one bit of information. The important feature of information, so defined, is that in a message it can't be created, only destroyed. If you know how much information goes into a message, you also know the most you can get out of it. (Some information may get lost, for example if the receiver breaks down or there are hisses and crackles on the line.)

This quantitative theory of information has important applications to error-detecting and error-correcting codes. Another application is to the question of data compression, which will become important later on. If a message contains redundant elements then it can be shortened; the information content can be reduced. For example, suppose that the message is a telephone directory for a town with a hundred thousand subscribers. For simplicity assume that each subscriber has a ten-digit phone number, that the average name contains twenty characters, and that addresses are omitted. The directory thus contains three million characters. (Each can be converted to binary form by some fixed code like A = 00000, B = 00001, and so on, but think of one character as a fixed unit of information. A character is then roughly equal to five bits, but that's not important here.)

Because some human names are more common than others, the data in the directory can be compressed by encoding the information more cleverly. As a simple illustration, in an American directory there might be around a thousand subscribers with the surname "Cohen," and probably none at all with the surname "C." If each "Cohen" is abbreviated to "C," four thousand characters are saved.

If there happen to be a few people with the surname C, then a more esoteric scheme can be used—say, "CC" in place of "Cohen," a saving of three thousand characters. But it would be better to encode like this:

Cohen → C
C → CC

saving 3,997 characters. (Unless there are people with the surname "CC.") Moral: Different codes compress the data by different amounts. Information theory places precise limits on how much compression is possible for any message containing a given amount of redundancy; the limits are expressed in terms of a quantity that is confusingly called entropy. (Confusingly because it's not the same as the corresponding concept in thermodynamics.)

Particular messages, however, can be compressed a great deal further (in part because restricting the range of possible messages increases the effective redundancy). To take this to extremes, we could encode the entire telephone directory by the message "A," where the code is

Albert Aardvark 071-433-9152
. . .
Zola Zarathustran 020-366-6664 } → "A"

Decoding such a compressed message, however, requires having access to the entire original directory in plaintext. So the extreme compression is obtained by a cheat.

The essence of the cheat is that in order to read a message, you must have access both to the ciphertext (the message in code) and a description of the code. So the true quantity of data required is the length of the ciphertext plus that of the codebook. Changing every "Cohen" to "C" and every

"C" to "CC" saves 3,997 characters in ciphertext, but requires an extra twelve characters to specify the code:

Cohen → C; C → CC

assuming "→" is an available character. Thus the true saving drops to 3,985 characters, but it still is a saving. The extreme case, where the entire directory is encoded as "A," actually increases the number of characters. In place of

DIRECTORY

we have

A

A → DIRECTORY

so the number of characters grows by three: the "A" that forms the entire ciphertext, and "A →" in the code recipe.

All discussions of information in real systems, such as brains or DNA, must take into account not only the length of the messages, but also the "hidden" information in the decoding machinery. Unfortunately, people often forget this point, as we will see later.

■ EYES ARE NOT CAMERAS . . .

Nearly all animals complicated enough to have several organs have invested in eyes—sensory organs that respond to light. Visible light is that part of the electromagnetic spectrum that can penetrate watery fluids, which is what your eyes are filled with. Eyes come in a huge range of patterns. The simplest are single cells with a light-sensitive area at one end and a fiber like a nerve axon at the other. At the other extreme of complexity are the cameralike eyes of vertebrates, like us, and mollusks, like the octopus. Insects and crustaceans have enormous compound eyes with arrays of tiny lenses, more like some television cameras than like a Minolta. Different models of eye must communicate, to their brains, very different aspects of the light that hits them. So the brains must have very different circuitry to decode the visual information and make sense of it.

Your brain gives you the impression that "you," your conscious self, is looking out of your head through a window—or, more accurately, two windows, but the images are combined, and you only find out that it's two windows if you shut one eye. When you use binoculars or microscopes, or wear glasses, what you see is just what you'd expect if you fitted those holes with lenses. You've been taught that your eyes are like cameras, with a lens that focuses incoming light and an iris that adjusts the amount of light that enters the eye, but it doesn't seem like that from inside your own head. Your brain is working very hard to create the illusion that "you" are looking out of holes in your head where your eyes should be. But you're not. Your retina, at the back of your eye, stops all the light. It is the nerve cells from your retina that carry the picture into your brain, and they carry it as discrete electrical impulses, not as continuous variations of light and shade.

The standard, simplified image of the visual process is that the optic nerve is like a bundle of optical fibers, so that each receptor cell in the retina is wired up to a corresponding nerve cell in the brain. The area of the brain on the receiving end of this bundle of fibers is called the optic tectum. It's true that there is a very good correspondence between small illuminated areas of the retina and patches of nervous activity in the tectum, so the visual image really is mapped onto the tectum, much as a picture on a computer screen is mapped to individual pixels, phosphor dots. This description suggests that the brain then directs its attention at the dot image in the tectum, as a proxy for the image actually received by the retina. Perhaps other parts of the brain have neural links along which they can interrogate bits of the tectum: "Any light shining on your bit of the retina, Neuron OT-5776?"

If you don't think too hard about what's involved, that model of vision may seem satisfactory. But the system must be more complicated. For example, we recognize a flower as a flower, no matter what part of the retina is receiving its image, whether or not the flower is tilted, whether it is close up or far away, and even when it is waving in the breeze. We can recognize it in bright light, moderate light, and deep shade. These changes in size, position, and illumination make a huge difference to which cells in the tectum are activated, but we don't notice. So the brain processing must be a lot more subtle than just asking which bits of retina are receiving light from where. Indeed, it is so complicated that we currently have no very good idea of just how it works, and that's one reason why robot vision remains in a very rudimentary state.

We do know a certain amount, and that lets us guess at the general nature of the processes involved. One surprise is that the "front end" of the visual system is much more complicated than we might expect. In the early

1960s David Hubel and Torsten Wiesel showed that the nerves from the eye to the brain are not just passive carriers of information about the presence or absence of light. Some are, but about half are not. There are many layers of cells in the retina, and the receptor cells are connected to nerve cells in the next layer, and so on. Many of the cells in the lower layers also contribute fibers to the optic nerve. These fibers carry quite complex information: "My area has just gone dark slowly"; "a vertical line has just passed across my area"; "Something fast enough to be a fly has just passed nine o'clock high." We don't mean that actual phrases are conveyed along these nerves, but that's what the impulses that are conveyed along them mean to the brain. In frogs, for instance, the tongue flicks out toward nine o'clock high far too quickly for there to be any decision-making mechanisms between sensorium and motorium.

Thus your retina is not at all like the film in a camera. Detailed pictures from both retinas do get passed to the two sides of the optic tectum, half of them crossing over as they do so, so that the right half of both retinas goes to the left half of the tectum. So each half of the tectum receives information from both eyes, and that's part of the mechanism that lets us "see" a three-dimensional picture even though each eye receives in two dimensions. But most of the signals that pass along the optic nerves contain high-level information: Impulses in nerve fibers signal the presence of vertical and horizontal lines, patches of light or darkness, grids of special spacings and angles.

There are very local effects, too. You can distinguish two objects that are closer together than neighboring receptors in your retina. Your eyes flick from one to another, rather irregularly but not totally randomly. These movements flicker the edges of images across the receptors, causing local circuits to "increase the contrast" and—probably—tell the brain that they've done so. Color contrasts may also be enhanced by simple, patterned, local circuits.

You may think you see out of your eyes, but it takes enormous processing power to produce that illusion. The world that you see "out there" is really an ever-changing pattern of electrical impulses inside your own head, the one place you *can't* see. But all of that processing is "transparent" to your conscious visual senses—that is, you don't have any conscious awareness of the details of the processing. You can get hints, though, from various visual illusions that fool the image-processing programs into producing bizarre outputs. While writing this section we noticed that Figure 21, on page 147, does precisely that. The circles at the extreme left and right look bigger than the others, because the converging arrows make the other circles seem smaller. A more startling example is the Cheshire cat illusion (Fig. 22). Sit a cat in front of you on a table. Hold a mirror at 45 degrees to your line of vision so that your right eye sees the cat but the left eye sees a blank wall reflected in the mirror. Now wave your left hand beside your head, so that it moves across the image presented to the left eye.

The cat . . . disappears. If you concentrate your attention on the cat's mouth before you wave your hand, then only the grin on the Cheshire cat remains. What you see during the illusion depends on what regions you were paying attention to beforehand.

It is only by examining very carefully just what happens in the optic nerve when simple images fall on the retina, and figuring out (by comparing them with computer circuits) how the complicated neural connections in the retina function, that we can begin to understand how we see an image. Some neurologists, using new techniques to trace all of the hundreds of connections of each nerve cell, are beginning to understand the processes in the various parts of the brain that are involved in vision. You can

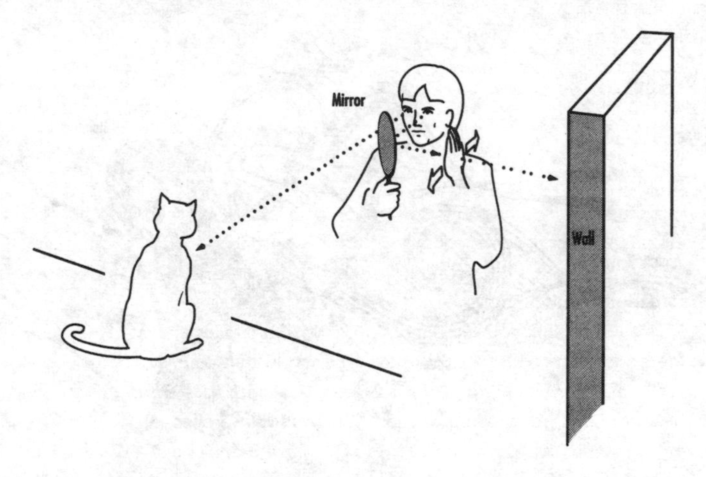

FIGURE 22

How to create the Cheshire cat illusion. The grin is invisible from behind.

read the words on this page without apparent difficulty. Science is only just beginning to understand how you do it, but we do know that it is the computational abilities of networks of nerve cells that makes it possible, and that those computations are inordinately complex.

■ ... BUT EARS MIGHT BE MICROPHONES

From eyes to ears. Our ears are quite different from our eyes, not only in form and function but in the way the brain processes the sensory information they provide. The behavior of the sensory organ itself is easier to understand, and there seems to be less in the way of high-level preprocessed signaling to the brain. Hearing, and our senses of balance and acceleration (which also reside in our ears), depend upon the physics of the ear (Fig. 23).

The perception of sound may be simpler than that of light, but it's still pretty complex. Sound waves are guided into your outer ear and impinge upon the eardrum, which they make vibrate. The eardrum has evolved so that it does not resonate with any particular frequency of sound—unlike, say, glass, which can be shattered by a powerful soprano—and it passes its

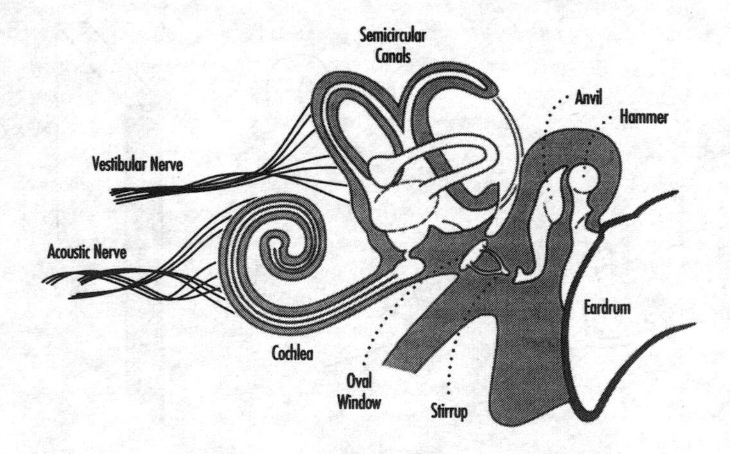

FIGURE 23

Anatomy of the ear

vibrations on, with very little distortion, to a tiny bone at its center, fancifully known as the hammer. This in turn passes the vibrations to two other small bones, the anvil and stirrup, across the middle ear and on to the inner ear, through an opening called the oval window. This is the bottom of a helical spiral, the cochlea, which has a channel that passes up one face, across the tip, and down the other side, letting what remains of the vibration out again into the middle ear. A membrane called the basilar membrane is strung between the up and down channels, and in it are sensory hair cells.

Various regions of the spiral react in different ways to different frequencies of the incoming sound, and stimulate corresponding parts of the basilar membrane. Its hair cells then send signals to the brain. The ear is basically an analog device, responding to signals that vary continuously, whereas the eye is much more a digital one, responding to on/off switching signals. The cochlea is a tuned receptor, turning pitch into position: The basilar membrane, with its associated vibration sensors, vibrates in different places when the eardrum vibrates at different frequencies. Low frequencies cause the tip of the cochlea to transfer energy through the "tuned" basilar membrane, but high frequencies pass through the sensitive barrier at the base of the spiral. We know some of what happens because of an industrial disease. Boilermakers, whose ears were subjected to very loud noises at very restricted frequencies, used to develop selective deafness. It was found that the sensory cells in very specific, tiny regions of the cochlea had degenerated.

The actual movements of the basilar membrane are extremely small; for conversational levels of sound they move much less than the diameter of an atom. The special hair cells that receive the vibration and turn it into a nerve impulse are individually tuned to specific frequencies, making it possible to discriminate pitch with enormous accuracy. Different pitches are sensed at different distances along the cochlea.

Our sensation of sound waves has an interesting feature: It is logarithmic. That is, twice as much incoming sound energy produces the same increase in perceived loudness. A signal with an energy of 0.002 watts sounds twice as loud as one of 0.001 watts, and a signal of 20 watts sounds twice as loud as one of 10 watts—even though the increase in energy is far greater. This is a general property of our sensory systems, equally valid for light energy and apparent brightness; it lets us compress a huge range of sensory input into a much smaller range of responses, so that we can hear a moth's wings fluttering without being deafened for life by a thunderclap. Our sensitivity to pitch is also logarithmic, but this is probably a result of the geometry of the cochlea and its tuned membrane. Some thirty thousand nerve fibers in the auditory nerve carry all the information about the mix of amplitude and frequency to the brain. So effective is the system that we can derive stereophonic information—depth and directional sense—from the disparity in sound received by our two ears. Stereo hi-fi creates the illusion of depth of sound by sending different signals to each ear; that's why it needs two (or more) loudspeakers.

The ancient Greek cult of the Pythagoreans—the first known wholists, whole-number enthusiasts—discovered some curious numerical oddities in our perception of musical sounds. Sounds of different pitch appear harmonious when their frequencies are in simple integer rations, such as 2:1 or 3:2. This was one of the pieces of evidence that convinced the Pythagoreans that the universe revolves around whole numbers. The most basic interval of all is the 2:1 ratio, the octave. Piano keys repeat the white notes CDEFGAB and return to C on the eighth step—hence the term "octave." Our twelve-note musical scale, with its extra black notes, the sharps and flats, is a cunning compromise based around these ratios. One mathematical reason why our ears prefer these ratios is that sounds whose frequencies are not in simple integer ratios are "out of step" and tend to interfere with each other, producing phenomena known as beats, which we perceive as a horrible low-frequency buzz. However, this can only be part of the story: Our hearing has many nonlinear features for which such an analysis breaks down.

Our ears contain other sense organs, too. All of them depend on one kind of sensitive hair cell, like those that turn sound vibrations into nerve impulses. They have a bristle on top that senses when the hair bends, braced against a bunch of ordinary bristles that constrain the sensitive bristle's movement, so that it can only sense bending in one particular direction. In our fishy ancestors such cells were used in the organ called the lateral line to detect both the flow of water past the fish and slow watery vibrations. Modern fishes and land vertebrates have refined some of their functions to include hearing and balance, but they use much the same pattern of sensors, called vestibular organs (Fig. 24).

The most obvious bits of vestibular organs are three semicircular canals,

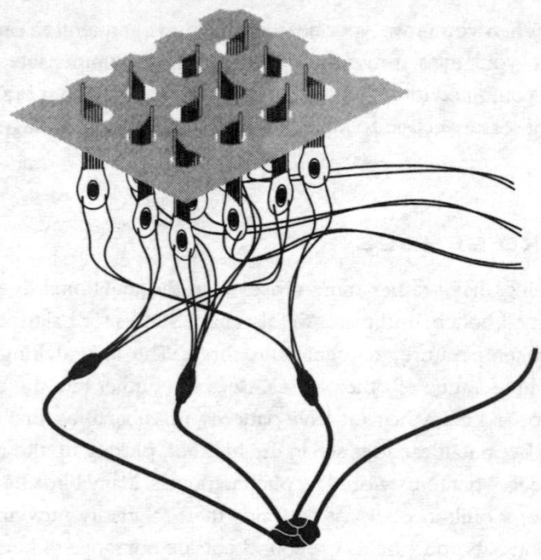

FIGURE 24
Vestibular organs

at right angles to each other. They are filled with fluid, which can circulate through a common reservoir. When we turn our heads, the fluid doesn't move much, so we drag the little sense organ—it looks like the bristles of a tiny paintbrush—through the fluid. When we stop turning, the fluid, which by now has begun to move, flows past the bristles, drawing them the other way. This is a rather coarse system—like dead-reckoning navigation, in which each day's new distance and direction are estimated and used to update the previous position on the map—and like all such systems it loses track because errors accumulate. You may notice this when you drive round a traffic circle: It often seems like a lot more than a 360° turn.

We also have tiny, dense, calcium-containing rafts anchored to groups of hair cells in the reservoir. As our heads tilt, they drag the hair cells past the fluid, and different ones signal to the brain: "We're tilting down toward the left front." During the development of this system, all the hair cells with the same directional sensitivity are connected to the same nerve cells. These send simple messages to the brain: "My group, the east-southeast gang, is under some stress here."

The direction- and attitude-sensing capabilities of your ears are crucial to vision. When you move your head, but keep your attention on a particular object, your eyes move the opposite way to compensate. All your senses are combined into a consistent overall picture by your brain, which takes the necessary action to keep the incoming sensory messages making sense.

WEIRD SENSES

Human beings have rather more senses than the traditional five of sight, hearing, smell, touch, and taste. We also have senses for balance, acceleration, and temperature, to name but three. The animal kingdom has evolved a huge range of other senses. Bees see further into the ultraviolet than we do, and many flowers have patterns, invisible to us, that guide approaching bees. Rattlesnakes see in the infrared, picking up the body heat of a mouse in what to us would be pitch darkness. Many birds have a magnetic sense, a built-in compass that aids their migratory movements. For echolocation bats and whales use sound outside our range of hearing. Talk to a blind person about his or her world picture, and you will appreciate how different it is from your sighted one. No doubt many new senses remain to be discovered; there is no reason whatsoever to imagine that we have discovered all the possibilities. Animals are probably exploiting features of the physical universe that we haven't even thought of.

Here's an example, a sensory system that until very recently went totally unrecognized. (Sorry, but we can't, by the nature of things, tell you about one that hasn't yet been discovered—though we hear that Dr. Delius dreams in undiscovered colors.) Sawfishes (and more generally sharks and their relatives the rays) have a system of long tubes known as ampullae of Lorenzini, which open on their snouts. The tubes are full of jelly and have an obvious sensory cell, with a nerve fiber at the inner end leading to an important part of the animal's brain (Fig. 25). The tubes are nearly an inch long in newly hatched sawfish, which are themselves only about six inches long, and run all along the spiky nose of the adult sawfish. So quite a lot of biological effort seems to be invested in them, but what can they be detecting?

To answer this, we must find out what kinds of change makes them respond by sending signals from the sensory nerve cell. It turns out that they respond to temperature changes; the rate at which they produce electrical

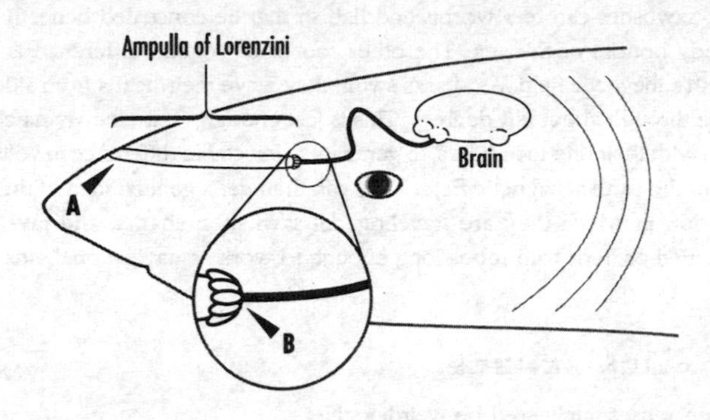

FIGURE 25

An ampulla of Lorenzini. The voltage A at the snout end is the same as the voltage at the inside end, where it is compared with the voltage B deep inside the fish by sensing the voltage difference across a single nerve cell.

pulses alters when the temperature does. So are they thermometers? Not necessarily: They also respond to chemical changes in the surrounding water, after delays of minutes or hours. We can't infer that a particular organ acts as a sensor for a particular effect just because it responds to that effect. Your nose will respond rapidly and painfully if you dip it into an acid bath, but that's not what noses are *for.*

The responses of ampullae of Lorenzini to both temperature and chemistry are equally misleading: Their primary function is to detect electrical voltages. The sheath of the tube is a good insulator, and the jelly inside is a good conductor, so the voltage at the inside end is the same as the voltage at the snout end. Here a single nerve cell can compare this voltage with the voltage deep inside the fish, where the nerve cell is situated. The tube thus detects the voltage difference between points many inches apart. It is sensitive to a difference of a few millivolts, so it can pick up very shallow voltage gradients in the surrounding water. These come from two sources. The first arises when other organisms produce voltage differences by pumping salt around their bodies, especially during muscle and nerve activity. Sharks can feel the electrical activity of other creatures' nerves. By waving their noses over the sand like a treasure-hunter with a metal detector, sawfishes can feel worms and flatfish that lie concealed beneath the sandy bottom of the sea. The other source of voltage difference is the earth's magnetic field. As sharks swim, they wave their heads from side to side through about ten degrees. This is just enough, in the newly hatched fish with their one-inch tubes, to generate a detectable difference in voltage from the earth's magnetic field. They can then get a general idea of the direction in which they are traveling. All sawfishes, sharks, and rays are hatched or born with tubes long enough to work as navigational aids.

■ ALIEN SENSES

Alien senses might well be weirder still:

NEEPLPHUT [*who has been grubbing happily around in the ship's attic, and appears in the cabin carrying a long black box*]: What is this device, Captain? It has the name of one of your ancient gods written on it. Is it of religious significance?

CAPTAIN ARTHUR: Let me see that. No, Neeplphut, Yamaha isn't a god.

NEEPLPHUT: I am sorry. I have an exquisite mastery of your written language, except for the vowels and the consonants.

CAPTAIN ARTHUR: It's an electronic keyboard, Neeplphut. We use it to play music.

NEEPLPHUT: I notice a pattern in the letters marked above the keys— A B C D E F G repeated over and over again. A cycle of period seven. This yamaha is a horribly septimal device, Captain. But what would one expect of Terrans anyway? Why the repetition of the letters? Are they duplicate keys?

CAPTAIN ARTHUR: No—each successive repetition of the same letter signifies the same note, but an octave higher.

NEEPLPHUT: What are "octave"? They sound promisingly octimal. We also group sounds into sequences of eight.

STANLEY: See, Captain? *Some* features of perception are fundamental and universal.

CAPTAIN ARTHUR: An octave is the interval between two notes, one of which has twice the frequency of the first.

NEEPLPHUT: Why do you call it an octave when the frequency multiplies by two? That seems fractimal to me, not octimal.

CAPTAIN ARTHUR: It's called an octave because, if you number the letter A as 1, B as 2, and so on, then the next repetition of A is number 8.

NEEPLPHUT: That explanation smacks of crypto-septimism to me, Captain! And in any case I fail to see why doubling the frequency should have any special importance.

CAPTAIN ARTHUR: The structure of our ears means that sounds one octave apart combine harmoniously. Let me show you by playing some notes. Listen: here is middle A and here is the A one octave higher. Do you hear how similar the two notes sound?

NEEPLPHUT: No.

CAPTAIN ARTHUR: Well, let me play a tune, and you'll probably understand what I'm getting at. [*Plays a few bars of the space anthem "The Star-Sickled Hammer."*]

NEEPLPHUT: Stop! Please, *stop!* My auditory organs are in pain!

CAPTAIN ARTHUR [*a trifle miffed*]: Oh, very well. Nobody ever appreciates my playing.

STANLEY: It was no worse than usual, Captain. I'm surprised it distressed Neeplphut as much as *that*.

NEEPLPHUT: I suspect we have encountered a perceptual incongruence, Captain. How do your auditory senses work?

CAPTAIN ARTHUR: [*Explains about basilar membranes and all that.*]

NEEPLPHUT: Captain, your description is that of an analog device. But Zarathustrans hear sound digitally. Our sense organs contain a series of neural circuits that react to incoming vibrations by switching on or off a variable number of neural channels. Sounds appear pleasant to us if the resultant stream of digits displays mathematical patterns. Let me play Zartmo's *Topologgiatura,* and you will understand. [*He inserts tape into player. An appalling cacophony emerges, with sounds that leap up and down across enormous intervals in a discordant and arrhythmic fashion.*]

CAPTAIN ARTHUR: Stop! Please, *stop!* My ears are ringing!

STANLEY: That reminds me: When we first met you, we heard the same kind of weird noises. We don't anymore, because everything we hear is filtered through the translator. But you Zarathustrans speak like that too, don't you?

NEEPLPHUT: Of course. Like you, we make sounds that are meaningful to our auditory organs.

STANLEY [*interrogates translator*]: You knew this all along, didn't you?

TRANSLATOR: When I tried to explain to you about the grammatico-logical axioms of the nitrogen breathers of Omega Aurigae you instructed me to go blow my fuse. I deduced that you wished my translations to remain in "for the rest of us" mode. The digital nature of Zarathustran speech poses no problems for a digital machine like me, but I assumed you would prefer not to be told about it.

NEEPLPHUT [*diplomatically*]: It seems that perceptual patterns and sensory mechanisms may not be as fundamental as we tend to think, does it?/does it not? (delete whichever is inapplicable). Let me tell you about the seismo-sensitive quasi-antennoids of the . . .

■ MYSTERIOUS OR MYSTICAL?

One of the recurrent problems in explaining how our minds work is that of free will. A machine is deterministic; it follows rigid laws. How can a machine make a free choice? Some scientists, among them Roger Penrose in *The Emperor's New Mind,* trace free will to the indeterminacy of quantum mechanics. There are problems with this suggestion, not the least being that it seems to put free will on the same level as roulette. We don't choose, we just blindly roll the quantum dice. Does that make you feel any less deterministic than having no alternatives at all? René Descartes attempted to answer the question of consciousness by separating the mind from the brain. According to his theory of dualism, the brain is made from matter and behaves like any material object, but the mind is totally different, not made from ordinary matter and not constrained by the laws that govern ordinary matter.

We can understand sensory organs such as eyes and ears by taking them apart and working out what the various bits do. We are helped in this by having developed a good understanding of technological sensors—devices designed to detect various kinds of signal. It's not a trick that works so well for understanding big features of brains, like free will or consciousness, because we don't have anything that we *do* understand to compare them with. Even sensation is a problem, because our brains do much more than merely detect signals emitted by their surroundings. They give us the impression of being *in* those surroundings—which of course we are, but not in quite the manner that our brain makes us believe—and they give us a strong feeling of self. We not only think, but we think we can monitor our own thinking. Microphones do a similar job to the ear, and the chemistry department at Warwick University is currently developing an artificial nose to detect tiny quantities of chemicals. (The artificial nose is very good at distinguishing between different brands of beer, for instance.) But—despite considerable contrary propaganda by some factions of the Artificial Intelligentsia—humanity is nowhere near developing a machine, however sophisticated its electronics, that can think, or that displays any suggestion that it is conscious of its own existence.

The main reason for that failure is that although we can think, we have no idea how we think. Despite the equal but opposite propaganda of some anti-AI factions, we see no reason in principle why a machine couldn't think. A typical argument against "strong AI"—the alleged possibility of intelligent machines—is John Searle's "Chinese room." Imagine yourself locked in a room with various Chinese documents, which you can't understand. In fact some are questions, some are answers. You are given rules in English, written on pieces of paper, telling you how to manipulate those documents so that you match the right answers to the right questions. These rules are so good that your responses are indistinguishable from those of a native speaker of Chinese. You're now in the position of a computer that has been programmed to respond like an intelligent person, but you know that you have no comprehension of what you're doing. Searle deduces that a computer that follows this procedure can't really be intelligent, because intelligence includes understanding. But even if his argument is valid, it doesn't prove that we can't build an intelligent computer some other way.

But what do we mean by "understanding"? It could be argued that even if you don't understand Chinese, the combined system—you plus the rules you're putting into operation—*does*. You, as only one part of that system, won't be aware of this; but if the rules are so good that by using them you always give responses indistinguishable from those of a real Chinese, then the rules must "understand" Chinese.

Searle's answer to this is to suppose that you internalize the rules, too—say, by memorizing them. But this assumes that your brain has a virtually unlimited capacity. You can't memorize every possible question and every possible answer in every possible conversational context: There are just too many combinations. The only kind of rule that you could internalize would be some kind of meta-rule—a rule about rules—that would generate correct responses in a wide variety of situations. But that surely moves one step closer to genuine understanding. To understand something is to grasp it as a whole; and meta-rules compress data, so you're closer to such a grasp than you would otherwise have been. Indeed, to get adequate compression you'd need a very effective meta-meta-meta- . . .-rule, and that would be even more like true understanding. And, incidentally, rather similar to what Chinese people must do when they understand Chinese.

Let's apply Searle's argument to your brain. The conclusion is that you just *think* you're intelligent, when actually all you're doing is putting into operation a huge list of rules, meta-rules, and meta-meta- . . . -rules that you've acquired from your parents, your culture, school, television, and so on. Therefore you're really just a dumb organic machine, going through the motions without understanding anything you're doing. Searle would naturally disagree that this conclusion follows from his arguments, and he would explain his reasons thus: "The idea is that while a person doesn't understand Chinese, somehow the *conjunction* of that person and bits of paper might understand Chinese. It is not easy for me to imagine how someone who was not in the grip of an ideology would find the idea at all plausible." But all that does is restate his disbelief in strong AI. Searle has missed a key point, because he's put a *person* in the room to manipulate the paper, and that leads to potential confusion between the person's intelligence and the system's. In point of fact, the person might equally well be a computer; its sole contribution is to read the rules and act upon them—mechanically. (It must be mechanically, because the person doesn't understand Chinese.) If there's any intelligence in this system, it doesn't lie in the conjunction of person and rules; it's entirely confined to the rules.

Can a system of rules embody intelligence and understanding? Searle would no doubt find this even more obviously ridiculous—but it's what the rules that govern the operation of our brains do. Instead of bits of paper, the brain uses nerve cells and electricity and chemistry to specify and to implement the rules, and we don't actually know what its rules are; but the belief that the brain isn't governed by the laws of physics is just a thinly disguised form of Cartesian duality. Searle is confusing the process that an intelligent entity must carry out—which must obey its structural rules—with what it is like, from the inside, to *be* an intelligent entity. If the neurons in the head of a Chinese person could discuss the "documents" they were processing, then the conversation would go something like this: "Hey, Neuron G-8449, do you understand what all this stuff's about?" "Nope, sure beats me, Neuron K-4436. I just count the nerve impulses and push this button *here* when I get to seven."

The Chinese-speaker understands them, though.

Strong AI isn't an issue. Of course an intelligent machine is possible—in principle: Just copy the organic machinery of a working human brain. But abstract principle isn't worth much: What counts is being able to put the principle into action. The interesting question is, Can we actually do it? And there we run into serious trouble, because we can't copy a brain, we don't know what makes us intelligent, and we don't know how our own consciousness works—it's something of which we are not conscious. We don't even know how our own memories work, though we do have a lot of evidence to suggest that the brain has at least three different memory mechanisms—short-term memory, primary memory, and working memory. Oddly enough, computers also have three main types of memory: short-term memory within the central processing unit; "random access" working memory (RAM); and long-term memory on hard disk. Have we unconsciously modeled our machines on our own brains, or is a triple memory the easiest way to get the job done?

We shall return to these philosophical questions in chapter 12, but for the moment our aim is less ambitious. We wish to assemble some evidence that whatever "mind" may be, it is not some mystic force divorced from the material world. You can dissect individual neurons out of a brain, but you can't dissect tiny bits of mind. But then, you can dissect axles and gears out of a car but you will never dissect out a tiny piece of motion. The ability of a car to move is an emergent property—a process that it can carry out by virtue of its overall organization. Mind seems to be an emergent property of brains, more mysterious than the motion of a car because we can't (yet?) watch the mental wheels going round, but no more mystical. It is emergent monism, not Cartesian dualism, that must hold the key to the understanding of consciousness. That is, mind is a process, not a thing, and it emerges from the collective interactions of appropriately organized bits of ordinary matter.

■ STRUCTURE OF THE BRAIN

The main evidence that mind is a process carried out by brains is that there are clear links between particular physical regions of the brain and particular aspects and functions of the mind. If mind were simply a mystical add-on, there would be no good reason to associate various bits and pieces of it with bits and pieces of the totally different material brain. But in fact the human brain divides quite naturally into pieces with quite different functions and structure (Fig. 26). The piece known as the cerebrum consists of two cerebral hemispheres, highly convoluted structures that look a bit like large walnuts; this is the part of the brain that most of you will instantly recognize *as* brain. The cerebral hemispheres have three layers: an outer gray layer known as the cerebral cortex; a central layer of white matter; and a deep mass of gray matter known as the basal ganglia. The two hemispheres are not entirely separate; thick bands of nerve fibers, rather like telephone cables, link the two halves. The biggest of these is the corpus callosum. Deeper inside the brain are two structures known as the thalamus and hypothalamus.

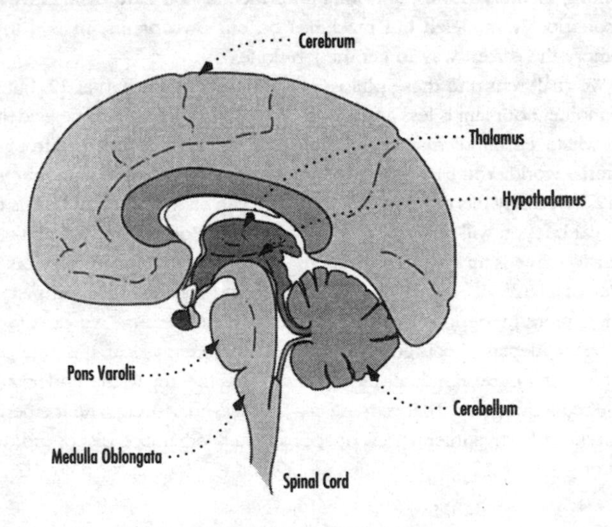

FIGURE 26

Simplified anatomy of the human brain

Each of these components has an intricate structure of its own. Something that complicated could surely have evolved only because every bit of it does something; otherwise why bother to make so many bits and pieces? And we now have some idea what. We can tell what a given bit does by observing what goes wrong when particular regions are damaged in accidents. In the same way, having observed that a car with a missing brake shoe doesn't stop very effectively, we deduce that brake shoes are somehow involved in bringing vehicles to a halt. As we've already explained with regard to the analogous question of genes and alleles, such evidence must be treated with due caution. A car with a broken windshield wiper may finish halfway up a tree because the driver failed to see where he was going in a heavy thunderstorm, but that doesn't imply that the primary function of a windshield wiper is to keep cars out of trees.

The cerebral cortex contains regions—motor areas, which we earlier called the motorium—that control bodily movements. These regions are capable of controlling the activity of muscles so that they exert the forces needed to make the limbs move effectively. Different parts of this region control different muscles, and to a great extent these parts are arranged rather like a map of the entire body, with the legs being controlled by parts near the top of the hemisphere and the arms and face by parts that are lower down. Next to the motor area is a sensory area, which receives input signals from the sense organs. You can't swing through the jungle like Tarzan without some idea of where you're going, and the sensory area is perpetually sending messages to the motor area to tell it where everything is in relation to the appropriate parts of the outside world, so that the motor area can figure out just which signals it should send to the muscles. The cerebellum is heavily involved in coordinating movement with sensory input. Sequences of movements that "work" are somehow retained, probably in the cerebral hemispheres, so that animals can *learn* to run and jump and reach for a banana. When the requisite motion is needed, the motor areas somehow access this mental template and put it into effect.

The hypothalamus controls, among other things, your body temperature. It constantly alters your metabolism so that your overall temperature is very nearly constant, despite changes in your environment. This may make it sound like a supercharged thermostat; but the hypothalamus is also the seat of many of your most deeply felt desires, such as aggression, sexual arousal, and the urge to eat and drink.

The thalamus plays a major role in the sensation of pain. Pain begins as signals transmitted by distant regions of the nervous system, but by the time those signals have reached the thalamus, being considerably processed and transformed on the way, the feeling of pain becomes conscious. The brain's emotional response is usually fairly slow—we take a perceptible time to start to laugh, or to burst into tears—but some responses have to be as quick as possible, for survival reasons. The thalamus contains a fast-track pathway for various emotional responses, such as surprise, watchfulness, or alarm. It also acts as a way station for perceptual signals, such as hearing and vision, en route to their own particular areas of the cortex.

The cortex is the area that controls your "highest" functions, the things that you normally describe as thinking, reasoning, remembering, or deciding. Your language ability resides there. One of the most striking features of the organization of the cortex is that the two cerebral hemispheres, apparently so similar in form, perform very different functions. In about 99 percent of right-handed people, the left hemisphere does most of the work when it comes to language. In people who are left-handed or ambidextrous, about 60 percent also consign their language faculty to the left hemisphere, 30 percent give it to the right, and the rest divide it between both. Right-handers' brains tend to be less symmetric than left-handers'; nobody knows why. Among people who use the left hemisphere for speech, the right hemisphere is typically used for functions such as the perception and recognition of specific objects ("Look, there's a banana!"); spatial orientation ("I am unexpectedly hanging upside down from a tree branch"); and route finding ("Shouldn't we have turned left at the river?").

■ SPEAKING OF LANGUAGE . . .

Language is one of the most unusual features of the human brain, and it's a puzzle. How did we acquire language? Language has a self-referential aspect: We use language to talk about the problem of language, and we use some kind of brain-language, the deep symbolic structure of thought, to think about it.

We really don't know what aspect of language is "primitive." There are theories that language arose from animal calls, imitation of physical sounds, or grunts of exertion—suggestions that the linguist Steven Pinker dismisses as "the infamous 'bow-wow,' 'ding-dong,' and 'heave-ho' theories." Gregory Bateson, in his book *Mind and Nature*, relates how he started with one view and suddenly became converted to its opposite. He began by setting up a reductionist hierarchy of linguistic components, in which nouns and verbs were the "atoms" of language, phrases and sentences were the molecules, chapters of a book were organized particles of matter, and a book was an almost-living entity, a linguistic organism. He then rhapsodized about the highest level of all: fiction. We easily distinguish fiction from truth, but linguistically they are virtually identical. He expressed his amazement that people don't rush to tell the police when a murder is committed—in the theater.

After writing this, he needed some relaxation, so he went to the zoo. Just inside the gate was a cage with two baby monkeys *playing* at fighting. They used no nouns, no paragraphs, no chapters . . . but, he suddenly realized, they understood "fiction" perfectly. His whole model turned upside down. Instead of nouns and verbs being primitive elements and fiction high-level superstructure, he realized that the most fundamental level of language was play versus reality, and nouns, verbs, and chapters were the superstructure.

When language is seen like this, as play interaction between curious primate infants, its symbols look like representations of play actions ("Boo!") and play objects (toys). Infants that grow up in the same family, sharing the special noises that Mother makes, automatically have a shared symbolic structure—a mother tongue indeed. Presumably our early ancestors were those proto-humans whose neural circuitry was especially good at such maneuvers. Later ones were those that had evolved a workable shared grammar, so that the lexicon, the list of shared "words," could expand. In this manner the language faculty became built into our brains.

Nearly all modern languages have features in common; they look as though they all descended from a single ancestral tongue. But because of the lack of written records it seems impossible to find out whether there was one primitive language that emerged from Africa with *Homo erectus*—or early *Homo sapiens*—or whether the similarities are caused by the constraints of the human grammar-machine. Whichever alternative is true, the evolution of language is just as complex as the evolution of organic forms.

We have no idea how, where, or when human language really got started. Most of the "higher" animals have some kind of sound-based signaling system, ranging from the alarm cries of sparrows to the interminable, complex, but exactly repeated songs of whales. We think we know something about what the sparrow "language" means: alarm cries, for example, raise the alarm. We don't have any idea what an alarm cry feels like to a sparrow; to understand that, you presumably have to be a sparrow, and that's not possible for an ape, however smart it thinks it is. Whale songs are plain baffling; but presumably whales have good reasons for singing them. They may be just an elaborate "Here I am," or a recitation of the whale's life story, or the equivalent of gossip, or poetry, or indeed songs. Or they may serve some totally different purpose, or none at all. Maybe you can't evolve something like a whale without it developing a tendency to sing that kind of song.

Language exploits, and may even have arisen from, a trick that our brains seem to find natural in any case. We like to wrap up a bunch of complicated ideas in a single mental package, and label it. We use these packages to structure our world and make it comprehensible. The things we label in this way include the great simplicities, such as "cat," "child," "tree," and so forth, but also more abstract items such as "fractal" or "molecule." The reason the world of our daily lives seems to be simple—as we said right at the start of this book—is that we have become very familiar indeed with a small range of these mental packages, the ones that let us cope with most normal events in our structured environment. We see forty or two hundred kinds of thing in the world around us because we have that many packages and labels; we can transact our business with about five hundred words because those are the most potent and portable packages.

Language takes one further step, and associates to each package an arbitrary symbol: a word. The way that our memory seems to work bears traces of a past in which our minds knew about the packages, but not the symbols. Our memories store concepts as complicated patterns of associated ideas, and then very much attach a word as a label (Fig. 27). You fall foul of this feature of our memories when you suddenly find yourself groping for the name of a person or a place that you've known for years. The link between the package and its label is rather fragile—perhaps because the rich association of related concepts that ties the package together isn't available to reinforce it.

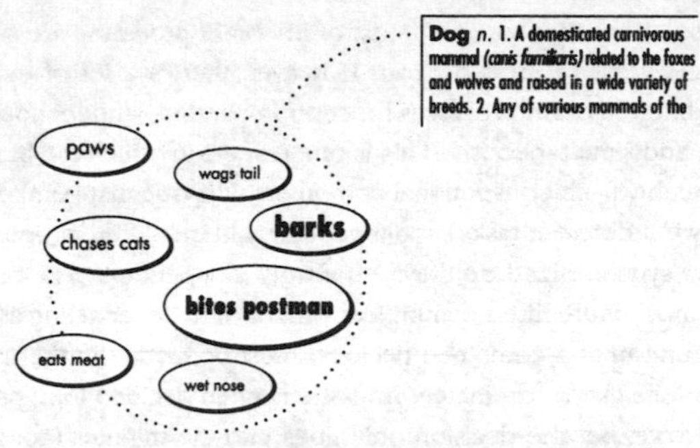

FIGURE 27

Your mental concept of a dog

■ THE MIND'S I

Consciousness is even more of a puzzle than language. Where does the "me" live? How, for that matter, do *you* tell that there is a "me" here at all? It's a difficult problem even to approach. Here's one difficulty that is commonly raised: I know that I am conscious. You *behave* as if you are—but I can't tell if you are for sure. Maybe you're just a clever robot. This problem is a bit of a red herring, and the philosophical principle known as Occam's razor—Don't make unnecessary assumptions—comes to the rescue. You seem to be the same kind of beast as me; your behavior resembles mine rather closely, you go through all the same kinds of motions and emotions—as far as I can tell. So what it feels like to you to be you is probably the same kind of thing that it feels like to me to be me. I can never prove this, but it seems a fair bet. It's not the central difficulty, which is much more enigmatic: How is it that I feel like me at all?

Some recent discoveries shed a little light on the question—or possibly a little darkness. A striking feature of consciousness is that I am utterly convinced that there is only one me. That me is in charge of my thinking, and through that, of my body. *I* decide, not some kind of committee. *I* am the one in charge.

Well, sorry, but probably not. An awful lot goes on inside my brain that's not directly accessible to my consciousness. I don't experience all the intricate neural computations that convert light impulses falling on the retina into "Look, there's a banana!" Most of my brain processes are subconscious, inaccessible to introspection. However, there's a lot of evidence that many different brain processes function in parallel—independently of each other, and simultaneously. This is one respect in which the brain differs considerably from conventional computers. It is true that parallel computers carry out different tasks simultaneously, but the different operations involved are synchronized, so they're not truly independent. The brain, in contrast, is much more like a committee, with each member acting as an independent unit until a combined decision must be made. Furthermore, it makes decisions like a committee: An issue is often decided long before it comes to a vote, but the decision only goes into the minutes (consciousness) after the votes have been counted. For example, if somebody is asked to move a finger voluntarily, and to say when he *first* decides to make the motion, then waves of brain activity in the relevant motor area start up a second or two *before* he says that his conscious mind "makes the decision."

What seems to be going on is that part of the brain is acting as a kind of ringmaster. The lion tamer is in his cage, sticking his head down the lion's throat; the trapeze artists are flinging themselves around up near the roof; and the clowns are running around the periphery causing mayhem and losing their trousers. The ringmaster isn't actually in charge of all this confusion, but he acts as if he is. He acts to bring some semblance of sense and structure into what's going on. And in our brains there is a kind of rationalizer-cum-censor, which shoehorns the separate activities of all the various independent bits into an apparently coherent framework. I can tell that my ringmaster has been at work when I have a dream that ends with an enormous explosion, wake up to the sound of my alarm clock, and recall a vast, coherent set of events that led up to the bang. My brain can't anticipate the alarm going off. And between the time the alarm goes off and the time the noise gets through to my fuddled brain, there isn't time to fit together the complicated tale that leads up to the explosion. So what must happen is that after I hear the alarm, the ringmaster concocts a story that leads up to my dreaming of an explosion, and then convinces me that it happened in a different order. Indeed, it probably convinces itself, for who else can the "me" be but the ringmaster?

I'm not really conscious at all. I just *think* I am. When Descartes argued

"I think, therefore I exist" he might more convincingly have said "I think I exist." I suspect that it's the same for you, but only you can know. And your ringmaster is probably fooling you in just the way that my ringmaster fools me—sorry, himself.

6 SYSTEMS OF INTERACTIVE BEHAVIOR

A man went into church and prayed to God, asking for his assistance in helping the poor.

"Dear Lord, thrice-blessed Creator, if only I should acquire $10,000, then I promise on my mother's grave that I would give $1,000 of it to the poor."

He waited a decent interval, but nothing happened. A thought struck him, and he tried again.

"Lord, if you don't trust me, you can deduct the $1,000 in advance, and just give me the balance."

In the last four chapters we have examined the relation between the apparent complexity of nature and the simple rules that underlie it. We have shown that the reductionist strategy—take it apart, see what the pieces are, understand how they fit together—provides simple explanations for many puzzling complexities. The behavior of atomic nuclei is explained by the properties and interactions of protons and neutrons; and we could have told you how these in turn are explained by the interactions of particles such as quarks, photons, and gluons operating under more exotic rules, but we didn't want to go further in that direction. The numerology of electron shells explains the chemistry of elements as expressed in Mendeleev's periodic table. Complementary behavior of electron shells, where an electron "missing" in one atom can be provided by one that is "spare" in the other, explain how they combine together to form molecules; and the Lego-block flexibility of those particular rules makes it clear that chemistry will be a complex area containing enormous diversity. The long-chain chemistry of carbon is a predictable consequence of special features of its "self-complementary" electronic structure—four pimples and four sockets. The resulting complexity of organic compounds opens up the way to the organized complexity of DNA, proteins, and other important biomolecules. The structure and processes of simpler living things are determined by their DNA genetics and its ability to code for proteins. More complicated life-forms develop through a sequence of stages, each of which is read out from the genetic structure in more or less the same way as happens for simpler organisms; but there is a hierarchical structure of genes that make proteins, genes that regulate genes that make proteins, genes that regulate *those,* and so on. The origin of life itself, or of complex life from simple life, is seen to be the inevitable consequence not just of these processes but also of their inherent imperfections. Mistakes in copying can occur; most of them are disastrous but short-lived, but the occasional improvement flourishes and reproduces. This selection mechanism provides the asymmetry that (usually) drives the evolutionary process toward increasing complexity, both of organisms and of their development.

These explanations provide mechanisms that are so natural, so logically compelling, that once they have been pointed out they appear almost obvious. Atomic theory makes immediate sense of the regular proportions in which chemical elements combine. Subatomic structure makes it "obvious" that sodium, with one spare electron in its outer shell, will team up with chlorine, which lacks one. We can easily see why, when animals are competing for space or food, those progeny that compete better survive better, and thus produce more of their own kind. Such explanations have a simplicity and a directness that is immediately appealing.

Not all scientific explanations are that direct, but we can find more complicated ones of the same kind when, for example, the poisonous metal sodium and the poisonous gas chlorine combine to make salt, without which many creatures cannot live. There are extra bells and whistles, of course, and the occasional snag; for instance, physicists think that salt crystals are cubical because quantum mechanics forces them to be, but nobody yet knows how to deduce this mathematically. By delving ever deeper we can describe in intricate detail how simple laws at one level are responsible not just for complexity but for specific pieces of complexity on some other level.

This scientific worldview, which has been built up over centuries by explaining complicated puzzles in terms of internal simplicities "one level down," is humanity's greatest construction. It reduces complicated questions to internal simplicities—hence the term "reductionism"—and it has given us enormous power to understand and manipulate our world.

REDUCING A BRAIN?

Reductionism equips us with a variety of mental funnels, with complexities at the top, deeper simplicities below. These funnels can be strung together, so that explanations of a given level of complexity delve deeper and deeper into underlying simplicities. As we explained in chapter 1, this structure of nested funnels provides a chain of logical explanation that leads in the reverse direction, "upward" from simple laws to complicated features of the natural world. The resulting insights tend to be presented in a deductive form: "*These* laws imply *this* phenomenon, which explains *that* observation." In contrast, the discovery of the structure tends to be inductive: "*This* observation would make sense if *that* phenomenon was taking place; and that would make sense if nature obeyed *these* laws."

When we look down our reductionist funnels at the the deeper levels of physics and chemistry, what we find is mathematics: wholistic numerology (electron shells), geometry (buckyballs), equations (Einstein's famous "$e = mc^2$" relating energy to mass). The logic of reductionism is most precise in the mathematical depths, and it becomes gradually more fuzzy as we ascend to the more complex levels of biology. By the time we reach Darwinian evolution the model has become verbal rather than mathematical. It is, however, cast in very precise and subtle language, and much of it is supported by mathematical submodels. The explanatory logic is still very precise, but its style has subtly changed.

The reductionist strategy seems far less successful when we think about still higher levels of organization than evolution. The chain of explanation from "lower" to "upper" level becomes more diffuse, and a close look shows that some of its links are missing. We have seen that only the simplest achievements of the visual system are yet understood in this way; we don't really know how the human eye and brain decide that one object is behind another, and we certainly don't know how a particular series of nerve impulses translates into "Hey, there's Uncle Charlie getting off the bus. Gee, he looks tired." There are at least two reasons why we run into trouble with questions like this, why we can't yet understand in detail how even a simple brain works. One is that we don't yet know enough about how nerve cells work, and we don't have accurate, or even rudimentary, circuit diagrams for a brain. Even an expert electronic engineer would have trouble understanding how a circuit worked if he didn't know what its components did or how they were linked together. In short, one reason we've run into trouble is that we haven't yet performed the reduction to the lower level in enough detail.

A second problem, more subtle and potentially more serious, is the sheer complexity of the system that we are trying to reduce. The brain has ten billion neurons, each connected to (roughly) a hundred others—one trillion connecting wires. Even if we had an accurate circuit diagram, it would have to be stored on a very big computer system; we can't, so to speak, hold it in our minds. Despite this, we do have the beginnings of a reductionist picture, and in particular we can see not only how neural networks can possess computational abilities, but how those could have evolved.

To explain something is to state it in terms that are accessible to a human brain, so the problem is self-referential, and that rightly gets everybody worried. Can the internal workings of a brain be accessible to that brain itself? Some people think the answer is an obvious no; for example, a brain doesn't have enough "bits" in it to encode its own connections. This argument is not conclusive, however. Human DNA also contains too little information to describe the brain's connections, so the same argument seems to imply that brains can't develop at all, which we know is false. (There are more than three billion bases in human DNA, but a trillion connections in the brain.) This suggests that when the structure of the brain is represented as a DNA data string, it becomes compressible. There must be a simpler *process,* encoded in DNA, that generates a complicated brain, just as the simple rules for a Mandelbrot set generate an incredibly intricate shape. If so, why can't a human brain comprehend that process, hence "understand" itself? And in any case maybe a group of brains, a team of scientists, could among them understand how one brain works.

Similar difficulties arise when we try to use the reductionist philosophy to explain interactions between or within populations of living things. Even when the individuals seem to be doing very simple things, their interactions add up to more complexity than we can deal with. Indeed, given a system of interacting objects, the number of ways for pairs of them to interact is very close to half the square of the number of objects. As the number of objects increases, the number of interactions increases much faster, like this:

NUMBER OF OBJECTS	NUMBER OF INTERACTIONS
10	45
100	4,950
1,000	499,500
10,000	49,995,000
100,000	4,999,950,000
1,000,000	499,999,500,000

The interactions can have a very tiny effect compared to those of individuals, but if the number of individuals gets big enough, then it is the interactions that matter most. Unfortunately, if the effect of any particular interaction is tiny, we may not be able to work out what it is. We can't study it on its own, in a reductionist manner, because it's too small; but we can't study it as part of the overall system, because we can't separate it from all the other interactions.

This is one of the main reasons why we don't have effective explanations in ecology, epidemiology, or economics. The new area of complexity theory pays a lot of attention to just these areas, searching for a better approach. Despite intense studies of AIDS, we cannot confidently predict the number of people who will be infected in twenty years' time. Nobody knows how to predict stock-market crashes. There are no big areas of reductionist causality in social science or management studies. When we find an explanation that seems convincing, it always turns out that for every expert there is an equal and opposite expert who can convince us of the reverse story. Nobel prizes have been awarded to economists whose theories flatly contradict each other.

However, we can still see enough of the reductionist links, enough of the feasibility of reductionist explanations, to convince ourselves that in principle the chain of logic holds together. Dawkins calls this hierarchical reductionism. All the mental funnels nest together into a treelike structure, but you don't explain things at the top by pushing them all the way down to the bottom; you push them one level lower, and appeal to what is already "known" at that level. Then you explain that level in a similar way, and so on; the complete explanation is implicit, not explicit. We mentioned in chapter 1 that physicists who anticipate the unification of quantum mechanics and general relativity talk of a Theory of Everything. This may sound a grandiose term, but what they mean is that in principle everything that happens in the universe must be a consequence of "true" particle physics. They believe that the tree of funnels has a single root from which all else grows. In such a view the failure of supply-side economics in the United States does not involve phenomena disconnected from particle physics; it just involves consequences of particle physics that are so complex that it is beyond current human understanding to work out the details. The details would, for example, involve completely accurate models, on the atomic level, of the brain structure of every human being in the United States, if not of every human being on earth.

We've seen that Descartes, pondering the intricacies of the human mind and the lumpen properties of dull matter, was led to a philosophy of dualism: Mind is one thing, matter another, and never the twain shall meet. In contrast, the thinking behind reductionist science, made explicit in Theories of Everything, is that of emergent monism. In this view mind is an emergent property of matter, expressing itself when that matter is organized in an appropriate and highly complex way. We don't really understand the link between matter and mind—but then, until recently we didn't understand the link between carbon chemistry and atomic theory. Indeed, we don't properly understand how the crystalline nature of salt derives from quantum mechanics. So a reductionist theory of mother love is some way off.

In this chapter we pursue the evidence for emergent monism. We concentrate on a few simplified mathematical models, which demonstrate that particular features of the material world necessarily have surprising consequences, consequences that resemble various important emergent phenomena. These models are bridges of snow spanning crevasses of ignorance. However, they give us some degree of confidence that the territories at each end can be connected together rationally and without the introduction of totally new influences, outside the normal physical world. Darwin did much the same thing for biology with a verbal model: Random heritable differences plus natural selection imply evolution. Modern versions of Darwinism import mathematical methods and provide rigorous examples of simplicities becoming spontaneously more complex; they replace the snow bridges by rope ones.

■ MODELS AND LAWS

Down the funnels from the lower levels of chemistry and cosmology we see the mathematics of quantum theory. Down the funnel from the higher level of evolution, or economics, we also see mathematics. It's not surprising if some scientists, especially mathematicians, think that you see mathematics down every funnel; and if you do, that's because it *really is* at the bottom of everything. They think that the Theory of Everything will turn out to be mathematical, an equation you could wear on a T-shirt.

However, there are important differences between the funnels that lead to mathematics from the top end of the reductionist story and those that lead to it from lower down. The logic of the lower funnels is precise: They use mathematics as a technique. The mathematics is seen as fundamental: Planets follow elliptical orbits *because* the law of gravity is mathematical and ellipses are a direct consequence; hydrogen atoms possess a particular set of energy levels *because* Schrödinger's equation says they do. The mathematical laws are "real"; they're what nature actually does—and the evidence is impressive. For instance, in quantum electrodynamics the theoretical prediction for a quantity known as Dirac's number is 1.00115965246, whereas the experimental value is 1.00115965221.

The logic of the upper funnels is much more fuzzy. There, mathematics is used as an explanatory device, a way to gain insight from simplified models incorporating key features of the phenomenon being explained but deliberately excluding others. For example, the spread of an epidemic may be explained by a model that assumes that contact between people is random. That's not really true: You don't toss a coin before deciding whether to go in to work. A very complex social structure determines which people you interact with. However, we can get some general insights into epidemics, such as their tendency to break out suddenly at irregular intervals, by making the simplifying assumption of random contact.

The higher levels of the reductionist story use mathematics as a metaphor, not as a precise representation of nature. We're not saying that the mathematics at the bottom of the funnel from evolution is different *as mathematics* from that at the bottom of the chemistry funnel. In fact it's identical, the theory of differential equations. What you see at the bottom of the funnel depends on the direction from which you're looking, as well as on what's down there; it isn't just a thing, but a way of using that thing. There is only one mathematics; the whole subject hangs together as a unity. (Anything that didn't fit in wouldn't be considered mathematics anyway.) But the same piece of mathematics can be interpreted in many different ways. An ellipse can be a planetary orbit or the shadow of a plate on a wall.

If the mathematics at the bottom of the biology funnel had remotely the same kind of interpretation as that at the bottom of the chemistry funnel, then Linnaeus's classification of living creatures would play the same role in biology that Mendeleev's periodic table plays in chemistry. Just as chemical elements have particular numbers associated with them—the number of electrons, protons, and neutrons their atoms contain—so animals would have numbers, or other mathematical features, associated with them. Evolutionary development would resemble nuclear reactions, with a specific and limited choice of available paths. But biology is far more malleable. You can't classify animals in any meaningful way just by the number of legs, ears, nostrils, antennae, and so on. You certainly can't argue that a six-legged species cannot evolve into a four-legged one because that would violate the law of conservation of legs.

This may be because we haven't yet found the right features in biological systems. If such features exist, they must lie at a deeper level of organization than end products like legs and nostrils. Homeotic genes, for instance, *may* explain such features as front end, back end, the dorsal/ventral distinction, and segmentation—and hence, perhaps, the number of legs. But until we find such features we have no idea what kind of mathematics, if any, really lies at the bottom of the biology funnel, or how to interpret it. At the present stage of science, we see mathematical laws down the chemistry funnel, but only mathematical models down the biology one.

Even though mathematical models do not correspond to the whole of reality—indeed, *because* they do not correspond to the whole of reality—they offer definite advantages. Because mathematics is more precise than words, it can handle more delicate distinctions. It can also direct attention to features that are not directly observable, such as average infection rates. And it can be used in thought experiments to show that many of our cherished beliefs—such as that in the impossibility of systems spontaneously becoming more complex—are false.

Mathematical models have a disadvantage, too—a trap that has caught more than one top scientist. For instance, J.B.S. Haldane once said that it was impossible for more than 10 percent of the individuals in a species to carry "bad" mutations because he had calculated that the population would die out if it lost 10 percent of each generation. But some perfectly viable natural populations—cod, plaice, and oysters, for example—lose 99.999 percent of each generation. The quality of a mathematical conclusion is determined by a lot more than just the accuracy of the calculations.

There are three types of mistake. Errors made within the model are the *easy* type to spot. Harder are errors made in the explicit assumptions that lie behind the model. The hardest of all to spot are the implicit assumptions in the worldview that suggested the model. For example, suppose you're setting up a model of biological development based upon the idea of DNA as a message. You would naturally tend to focus upon such quantities as the amount of information in a creature's DNA string and the amount of information needed to describe the animal's physical form. If you then model development as a process of information transfer—lots of messages buzzing to and fro—you will implicitly have built a model in which information cannot be created. You will then be able to "prove" that humans can't develop a brain because the amount of information needed to list every connection in the human brain is a lot more than the total amount of information in human DNA.

But we do have brains. The error is not in the model but in the tacit assumptions that lie behind it. Impeccable mathematics can produce nonsense if it is based on nonsensical assumptions. "Garbage in, garbage out," as the computer scientists say. You might expect a book by a mathematician and a biologist to praise the precision of mathematics as an instrument for digging out surprising biological truths. On the contrary, we both warn you not to take mathematical models too seriously. Surprising consequences are fine, but consequences so surprising that they don't make any sense are almost certainly based on false assumptions. Don't be impressed by mathematics just because you can't understand it.

PREDATOR-PREY MODELS

Here's an example of a historically important mathematical model in the area of population dynamics. Around 1925 the Italian biologist Umberto D'Ancona was studying fish populations, and he came across data for the number of fish caught near the port of Fiume, in the northern Adriatic, for years that included World War I. He was especially intrigued by the percentage of selachians (sharks, skates, rays) in fishermen's catches, because it increased dramatically during the war. He knew the increase must be related to the reduced level of fishing in wartime, but how? Selachians are predators; food fish are their prey. Why should a reduction in fishing provide a disproportionate benefit to predators? He asked the Italian mathematician Vito Volterra for help, and he devised a mathematical model of the interaction between predators and prey—a system of equations based on a few key assumptions about how their numbers vary. If you know what a population is now, and how fast it is currently growing, then you can use these equations to work out how big it will be a small instant into the future ("next"). By repeating this process you can find out what the population will be at any future time.

The model predicts that the two populations will always vary periodically. That is, no matter what the numbers are now, they will eventually return to the same numbers, and repeat the identical cycle over and over again. Next, Volterra calculated the average population sizes. The effect of fishing is to decrease both the supply of food fish and that of selachians. The number of prey then increases more slowly; the number of predators decreases more rapidly.

Fishing obviously changes the population averages, but how do those changes interact with the population dynamics? Volterra's model implies that fishing increases the average food-fish population, but decreases the population of selachians. This may sound paradoxical; it illustrates how mathematical models can be superior to verbal ones. How can more fishing possibly lead to more food fish? Easy! The population of predators is decreased by fishing, so they eat fewer prey—which more than compensates for the extra food fish caught by the fishermen. Conversely, if there is less fishing (as happened during the war) then the average number of selachians goes up, and the number of food fish goes down—just what D'Ancona had noticed.

This effect is called Volterra's paradox. It also applies to the use of insecticide to keep down pests. When the pest known as cottony-cushion scale insect was introduced to the United States from Australia and threatened to wipe out the citrus industry, a natural predator—an Australian ladybug—was introduced as a form of biological control, and the population of scale insects (now cast in the role of prey) dropped. Subsequently it was found that the insecticide DDT kills scale insects, so the citrus industry tried using it to reduce the pest population still further. Perversely, the population of pests increased. Volterra could have told them why: DDT also kills ladybugs.

■ THE IMBALANCE OF NATURE

Some people have a very oversimplified view of ecologies. They talk of the balance of nature, as if this is somehow static. Think of a forest that contains, say, rabbits and foxes. If there are too many foxes then the rabbits will not sustain them, so the fox population drops; but when the fox population drops there is then plenty of food hopping around the woods, and up the population goes again. It does look as though there is some optimum value for the fox population, and so there is, *given a fixed number of rabbits*. However, the rabbit population isn't fixed; it's hopping around too. All populations vary; creatures die and new ones are born. What occurs is not a static balance but a dynamic. The interactions of predators and prey—or the far more complex food-and-parasite webs that occur in a real ecology—drive the population numbers in well-defined, mutually related ways, just as the interactions of gravity, propulsive forces, and aerodynamic lift define a dynamic that determines how an aircraft will behave. Volterra's analysis shows that, even in a very simplified model, you usually don't get static populations. His model is actually much too simple, but more sophisticated variants that come a lot closer to what happens in the real world exhibit similarly "unbalanced" behavior. Oscillations are the norm.

This kind of thinking focuses attention not so much on a particular mathematical model (like Newton's law of motion), but on a particular *class* of models: dynamical systems. A dynamical system, in nontechnical terms, is one whose state at any instant determines the state a short time into the future without any ambiguity. You "know" what dynamical system you're talking about if you "know" the rule for passing from the current state to the "next" state. A dynamical system, in these terms, is deterministic: The present state determines all the future behavior, and only one future is possible. There is no choice, no chance, just a predetermined series of changes. The proof is simple: The present state unambiguously determines the next state, which determines the next state, which. . . . By stringing together a series of short-term predictions, you can foresee all the future behavior.

This method is much more powerful than you might think. For instance, suppose you knew a rule that, given the weather now, would tell you what the weather will be one second into the future. At first sight this might not seem to be much use: Would the TV company really be interested in such a short-term forecast? But by applying your rule again to your one-second prediction, you get a two-second prediction. Then a three-second prediction, four, five. . . . In principle you could predict the weather a million years ahead. Astronomers use just such a method to predict where all the bodies in the solar system will be a million years from now.

CHAOS

What we have just described is the perfect paradigm of law and order. The laws—the rules of the dynamical system—are simple and precise. Once you know what is happening now—the "initial conditions"—then you know the future. Therefore all future behavior must be simple and predictable. There is a direct link from patterns in what is observed—the elliptical orbit of Mars, say—and patterns in the underlying dynamical laws. Simple laws lead to simple behavior.

Actually, it's not like that. For example, the equations used to predict weather are indeed dynamical systems, just as deterministic as those for eclipses, and just as simple. But in practice nobody can make accurate forecasts of what the weather will be next week, let alone a million years from now. Why can't we predict the weather? Surely a deterministic system must be totally predictable? It is, in the following strict sense: If you know the current state *exactly,* then it determines the entire future behavior. There are two loopholes here, and both are important. One is that the calculation needed to work out the future state may be too difficult. The other is the word "exact." Real measurements always involve small errors. You can't use a ruler to measure a thousandth of an inch. The most precise measurements yet made in physics are accurate to about nine decimal places. Is that exact enough to make the future totally predictable?

No, it's not. It *is* for the simplest kinds of dynamical behavior, such as a steady state, or periodic change though a repetitive cycle. Inaccurate data on what the state is now produce equally inaccurate data on the future state, but not noticeably *more* inaccurate data. This is especially obvious for a steady state. If my measurement is in error by a thousandth of an inch today, and nothing ever changes, then it will still be in error by the same thousandth of an inch in a million years' time. But for some types of dynamic behavior, accuracy to nine decimal places is of no use for prediction. Neither is accuracy to a hundred decimal places, or to a million, or to any finite number. In these systems, "exact" has to mean "accurate to infinitely many decimal places."

The common feature of such unpredictable but deterministic systems is the process "stretch-and-fold." If the dynamics kneads the system like a lump of dough, stretching it out and folding it back on itself, then states that are close together always get pulled apart. On the other hand, states that are far apart may suddenly be folded together. The system can't settle down to anything simple, because simple structures are pulled to bits, but it can't escape altogether, because it's perpetually folded back into the same space. Like a ball in a pinball machine, it is pushed away from all the pins—the simple types of behavior—but it can't escape from the table. What do you do if you're not allowed to behave simply, but you can't get away? You are forced to do something complicated. The pinball bounces from pin to pin, never doing the same thing twice. This kind of complex behavior, produced by simple, deterministic rules, is called chaos. Before computers became powerful enough, hardly anybody noticed that it could occur; whenever they ran into it, the problem got too hard, so they gave up. They didn't ask why it had gotten too hard; they just went off and worked on a different problem. Now that our computers are up to the task, the dreadful truth has become inescapable: Chaos is everywhere. It is just as common as the nice, simple behavior so valued by traditional physics.

Suppose you're playing pinball, and you manage to start the ball in a way that gives a really high score. It's a deterministic machine: All you need do is start the next ball in precisely the same way, and you're on your way to a huge score. But in practice, however careful you are to reproduce the initial conditions, the next ball does something quite different from the first. What's happening is that the "stretch" in the dynamics is repeatedly magnifying any tiny differences in the initial conditions—the speed of the ball, its precise starting position. Most of the stretching actually occurs when the ball hits a pin: Tiny changes in the initial angle at which it hits can lead to big changes in the outcome. But the important thing is that stretching does occur. Suppose, for simplicity, that every time there is a stretch, differences are multiplied by ten. Then a difference in the hundredth decimal place will show up after a hundred stretches, and a difference in the millionth decimal place will show up after a million stretches. *Any* difference will show up if you wait long enough.

This phenomenon is called sensitive dependence on initial conditions, or more colloquially, the butterfly effect. ("If a butterfly flaps its wings in Tokyo, then a month later it may cause a hurricane in Brazil.") The butterfly effect is a characteristic feature of chaos. No matter how small the initial error is, it will eventually become comparable to the predicted values—and from that moment on, your prediction will bear no useful relation to the actual behavior. Chaotic dynamical systems amplify tiny differences hidden far along the decimal tail, well below any error threshold you may care to set.

This is an enormous problem for reductionism. If you look down a funnel and see chaos, then your explanation will suffer from the butterfly effect. In principle the funnel down from economics leads (via the molecules of shoppers in the supermarket) to quantum mechanics. But if an error in the trillionth decimal place of one molecule of the checkout clerk will swamp the actual behavior, you're in trouble. That's one reason why the weathermen make so many mistakes; it's not their fault.

COMPLEXITY FROM SIMPLICITY

Another striking feature of chaos is that "butterfly effect" behavior is very complex. Let's look at one particular deterministic dynamic rule which applies to numbers: "Chop off everything before the decimal point and multiply by 10." For example, the rule turns 3.14159 into 1.4159. That determines the behavior one time step into the future; to go further ahead you crank the handle and repeat the rule over and over again: 4.159, 1.59,

5.9, 9. You can *see* the decimal tail in the initial conditions shifting up one place at each stage. "Multiply by 10" provides the stretch, and "chop off everything before the decimal point" folds everything back (numbers bigger than 10 never show up). The butterfly effect is also highly visible. We stated the initial value 3.14159 to five decimal places, and predicted the resulting motion five steps into the future. Where does the system go after reaching the number 9? Well, if the initial condition was really 3.141597, we'd actually be at 9.7, and go to 7 next. But if it was 3.141598, we'd be at 9.8 and go to 8. The sixth step depends on the sixth digit of the initial condition, and we never specified what that was. So if we ask where a given initial value goes in the long run, we need to know its entire decimal tail. This is impossible in the real world, so in practical terms the system is unpredictable.

It is nevertheless informative to put your mathematician's hat on for a moment and ask what would happen if you *could* predict it. There are several possibilities. If the initial number is, say, ⅓, which in decimals is 0.3333..., then successive values generated by the rule will be 3.333..., 3.333..., and the pattern is rather obvious: It's a steady state. If the initial number is 1.21212... then successive values are 2.1212..., 1.21212..., and the behavior is periodic, repeating every two steps. But maybe the initial number is more complicated—say, π. The dynamic rule is very simple; but the successive digits of π, which show up when you apply that rule with π as initial state, are extraordinarily complicated. There is *no* recognizable pattern to them, except various coded mathematical versions of "These are the digits of π." The behavior, in short, is chaotic, even for this very simple rule. Moreover, different initial conditions can lead to totally different kinds of dynamics: steady, periodic, or chaotic. They can all be mixed up together in the same dynamical sytem.

That was an example of mathematical chaos, but many real physical systems are equally unpredictable. Perhaps the simplest is the forced pendulum. It's like an ordinary pendulum, a heavy bob swinging from a pivot on the end of a rod, and it has a tiny bit of friction that will eventually slow it down and cause it to stop swinging altogether. To prevent this, you "force" it, agitate its support from side to side in a perfectly regular, periodic manner (Fig. 28a). This should drive its oscillations and compensate for the friction, or so you imagine. It turns out that the detailed motion can be ordered or chaotic, depending on the initial position of the pendulum and the frequency of the imposed wobbles. If it's chaotic, then the fine detail will be unpredictable, but we're after a different, much coarser kind of prediction.

On a coarse level there are three basically different things that the pendulum can do. It can hang downward and wobble to and fro (Fig. 28b); it can whiz around like a propeller in a clockwise direction; or it can whiz around in a counterclockwise direction (Fig. 28c). It can also balance on end (Fig. 28d); we'll come back to that unusual but crucial state later. For the moment let's focus on the two spinning motions, which happen when the energy is high enough. Given the initial position of the pendulum, and the forcing frequency, can we predict just the general kind of motion, clockwise or counterclockwise?

What happens is best shown on a diagram. Plot the initial position horizontally and the forcing frequency vertically. Choose a point in the diagram, and read off its coordinates to get a position and a forcing frequency. Run the pendulum using those values. If it settles down to clockwise motion, paint the chosen point black; if counterclockwise, leave it white. Do this for all points of the diagram. What you would expect to find is a black region (depicting those initial conditions that lead to clockwise motion) and a white one (counterclockwise) with a well-defined boundary, something like the top left diagram in Figure 29.

And so you do. However, you may start wondering about how to draw the line between black and white, clockwise and counterclockwise, and take a closer look at it. Magnify a small region near this boundary, as in the top right diagram, and you'll find that the boundary isn't a simple curve at all: it breaks up into roughly parallel stripes of black and white, some thick, some thin. If you magnify a boundary region still more, you get the same kind of behavior. However closely you look at the boundary, it never looks like a sharp edge. It fuzzes out into puff-pastry layers of black and white.

If you try to predict the motion of the pendulum for initial conditions close to this boundary, you'll have terrible trouble. The tiniest error will swap you from a black region to a white one, from clockwise motion to counterclockwise motion. The tiniest change produces an enormous difference—another kind of butterfly effect. The structure is intricate because the boundary between clockwise and counterclockwise motion corresponds to a pendulum that is balanced vertically upward on a pivot that is agitating from side to side. We haven't talked much about this possibility

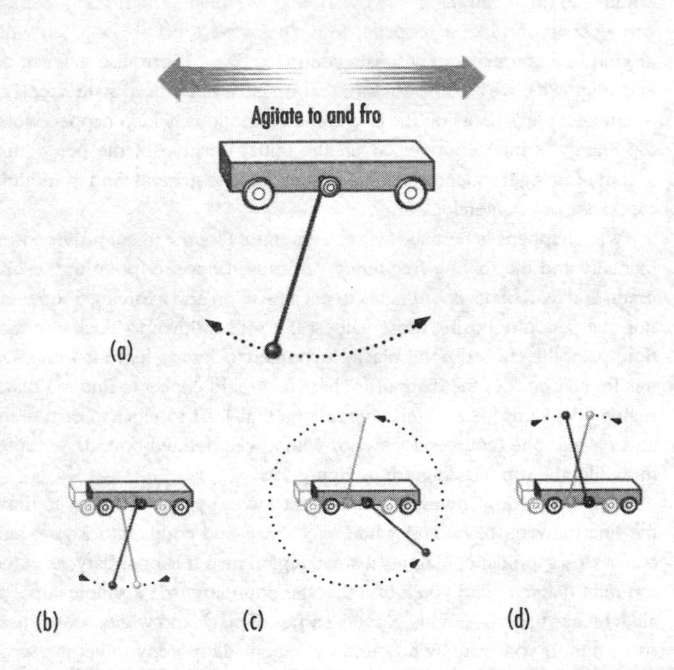

FIGURE 28

(a) *A forced pendulum;* (b) *the pendulum hanging vertically downward;* (c) *the pendulum whizzing around like a propeller;* (d) *the pendulum balanced vertically upward*

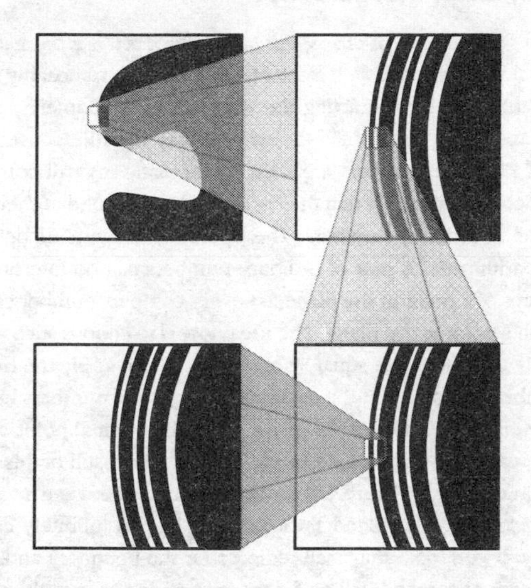

FIGURE 29

The boundary between clockwise and counterclockwise motion in a forced pendulum. No matter how much you magnify the picture, you never see a well-defined edge. (Pictures run clockwise from top left.)

because it's a rather delicate state, and it depends in a very delicate way on initial position and forcing frequency.

Mathematicians call this kind of chaos fractal basin boundaries. There are two competing types of motion; each has a "basin" of initial states that produce that particular motion. But boundaries between basins are not tidy curves, drawing the line between different possibilities. There *is* no line, just a grayish fractal fuzz. Boundaries in real life are often similar; it may not be possible to draw a precise line that distinguishes two obviously different extremes—not even legal/illegal, alive/dead, or male/female.

DATA COMPRESSION

The most famous icon of chaos, the Mandelbrot set, is a dramatic instance of a fractal basin boundary. It results from a rule not noticeably more complicated that that for extracting the digits from π—namely, "Square the number and add a constant"—though the way the rule is used is slightly different. First, the numbers are what mathematicians call complex numbers, which doesn't mean that they're complicated, but that they're formed from pairs of ordinary numbers. The operation "square" is defined using complex arithmetic. A pair of ordinary numbers can be interpreted as the coordinates of a point in the plane, so every complex number can be identified with a point in the plane. The idea now is to choose such a point; call it c. Apply the rule once: square c and add c. Then apply the rule again to *that* number, and so on. If the resulting sequence of numbers heads off toward the infinite reaches of the plane, color the original point c white. If it doesn't escape like that, color it black. You saw the result of this process on page 21 in chapter 1, Figure 4: It looks like a cross between a cat, a cactus, and a cockroach. The Mandelbrot set also has an infinitely complicated boundary. If you look at a small region near the boundary and magnify it, you see ever-new detail, taking far more complex shapes than the stripes that occur in the forced pendulum. There are spirals, blobs, sea horses, fans, curlicues, trees, crystals, tracery. . . . A wave of the butterfly's wing (a very tiny change in the initial number c) can convert sea horse to crystal and curlicue to tree as c's decimal tail shifts up a few places. This small-scale intricacy—always beautiful, but unpredictable and in a very rigorous sense uncomputable—goes on *forever*. The set is a fractal: It has distinctive structure on all scales of magnification.

There are two ways to tell a computer how to draw a Mandelbrot set—or, more correctly, a good approximation to a Mandelbrot set, which is the best you can ever do. One is to copy what fax machines do: Scan the picture along a series of parallel lines, sending out a signal whenever you encounter the set and no signal when you don't. This produces a huge quantity of electronic bleeps or nonbleeps, and the computer receiving this message can reconstitute the set, dot by dot. Alternatively, you can just send the rule that generates the Mandelbrot set; it's short and simple, a few hundred bleeps at most. If you measure complexity by the number of bleeps—the traditional measure of information—then a description of the picture, the thing you actually see when you look at a Mandelbrot set in a book, requires huge amounts of information. The picture is complicated. The rule that prescribes the set's shape, however, requires very few bleeps; it is simple. We have compressed the huge quantity of data required to describe the Mandelbrot set into the much smaller quantity of data needed to define the rule that generates it.

It's a bit like trying to give directions to a friend who is coming to visit. You can fax her the entire map of Philadelphia; or you can send a much shorter message: "First road on the left after Burger King and second on the right; park under the third streetlight and it's the house with the stupid gnome in the front garden." But for the Mandelbrot set, there's a twist: Given the second message *alone* (the rule) she can reconstruct the entire map of the town (the Mandelbrot set). There is no equally simple rule to reconstruct Philadelphia.

As we briefly mentioned in chapter 1, there are two ways to interpret this. The first is that the Mandelbrot set isn't really complicated at all. It's just as simple as the rule that generates it. It only looks complicated because you don't know what the rule is. One cliché describes it as "the most complicated object in the whole of mathematics." That's not really true, but you can see why the cliché gained currency. (A map of Philadelphia is actually far more complicated than the Mandelbrot set, because the roads and their names are arbitrary and "random" data cannot be compressed. Note how this all depends upon a key distinction between truly random data and the pseudo-random data produced by deterministic chaos.)

The second interpretation is that simple rules can produce complex results. It may look like a map of Philadelphia but it might be the consequence of simple rules. Indeed, according to the Theory of Everything, it *is*. So are Philadelphia, Burger King, and the stupid gnome in the front garden. To transmit the map of Philadelphia by telephone, send the equations of the Theory of Everything and the initial conditions for the universe. After that it's purely a matter of calculation.

In interpretation one, complexity is conserved between cause and effect. The snag is that now you can't work out the complexity of anything without considering all the things that might be causing it. You don't know how complex a map of Philadelphia is. In interpretation two, complexity is easy to calculate, but it's not conserved. It may be easier to send the map by phone than you think—but it won't be easy to find out how, because you have to contemplate *all possible rules* that might generate it.

The business of science is to infer the nature of rules from observations of their consequences. Either interpretation carries a very similar message for science. Interpretation one: Your observations may look complicated, but perhaps they're really simple. You just don't know the simple rule that lies behind them. Interpretation two: Just because what you observe is complicated, that doesn't mean it has to arise from a complicated set of rules. There is little real difference in the two interpretations, as they affect science, because science doesn't have the luxury of knowing the rules in advance. Both encourage us to seek simplicity within apparently complex data.

SPACES OF THE POSSIBLE

The human brain's strengths do not include several attributes that would be very useful to a mathematical scientist, such as the ability to perform rapid calculations flawlessly or to memorize huge chunks of data perfectly. (A few very unusual individuals can do these things, but more often than not they can do nothing else of interest.) This is why scientists and mathematicians are making increasing use of computers. Computers don't render the human brain obsolete, though; the brain supplies abilities that today's computers lack, such as judgment, creativity, and the ability to recognize patterns. The human visual system is an incredibly powerful information-processing device. Somewhere along the evolutionary track we acquired the ability to imagine, and from that point on we were able to use our visual system for quite different purposes, operating with internal representations generated by the brain instead of sensory inputs from the outside world. Think of a hippopotamus with an ear trumpet. Now think of something that we haven't told you to imagine.

Imagination lets us give shapes to things that don't really have shapes, things that don't exist. Our language is full of revealing phrases. We talk of "circular logic"—"if only it weren't so crowded, a lot more people would come." A circle is a visual image for any process that chases its own tail. That phrase is itself a visual image, and one that would have made sense to a caveman; perhaps a stepping-stone to the use of geometric images, internal abstractions of general forms, was the use of metaphorical images

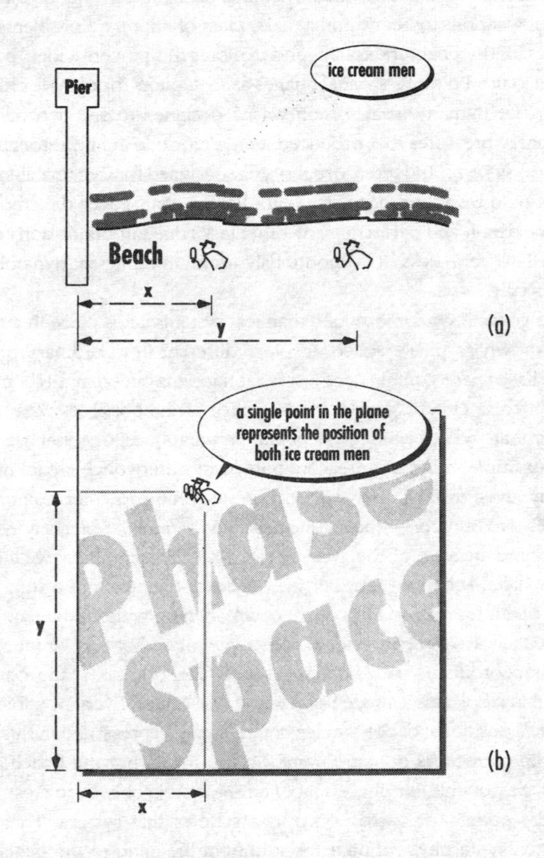

FIGURE 30

Two ice cream men: (a) physical space; (b) phase space

based on the natural world of animals and plants. The great French mathematician Henri Poincaré invented a way for the human brain to "see" dynamics in the mind's eye, thus enabling us to bring one of our most powerful weapons to bear upon a major class of important problems. Most of the rest of this chapter explains and explores this powerful idea, because we want to use Poincaré's visual images as metaphors for complex kinds of change. The human visual system wasn't designed to see dynamics; the evolutionary pressures that produced it didn't include the representation of dynamical systems. Indeed a visual system designed for dynamical-systems theory would be able to visualize high-dimensional spaces directly, zoom in for fine detail, and perform many other tasks that our brains don't do terribly well. Nevertheless, it is enormously useful to represent dynamic concepts visually.

Here goes. The geometry of dynamical systems takes place in a mental space, known as phase space. It's very different from ordinary physical space. Phase space contains not just what happens but what might happen under different circumstances. It's the space of the possible. What do we mean by that? What's possible depends on what questions you ask.

For example, think of a beach a mile long with two ice cream men on it. At any given instant, they will each be at some particular point on the beach. So two numbers—their distances x and y from the pier, say—suffice to determine the state of the system (Fig. 30a). (For simplicity we'll ignore their distance from the water, but we've shown the beach as a strip with a definite width in order to fit both ice cream men on without overlap.) Now we introduce a different picture, Poincaré's phase space. We interpret x and y as coordinates in a fictitious space (Fig. 30b), a plane one mile square, that associates a single point to the two values. Every possible combination of positions for the two ice cream men is represented in this space and is represented as a single point. So the plane, not the beach, is the space of the possible (or phase space) for the system of two ice cream men, and every point in it corresponds to a state of that system. This plane doesn't exist as a physical object, but only in the mind of the mathematician. Yet it exerts a very real influence, as we now demonstrate.

It would be most convenient for the customers if the two ice cream men were to station themselves in two different places, one-quarter and three-quarters of a mile along the beach; then nobody would have to walk more than a quarter of a mile. But that never happens! In practice the ice cream

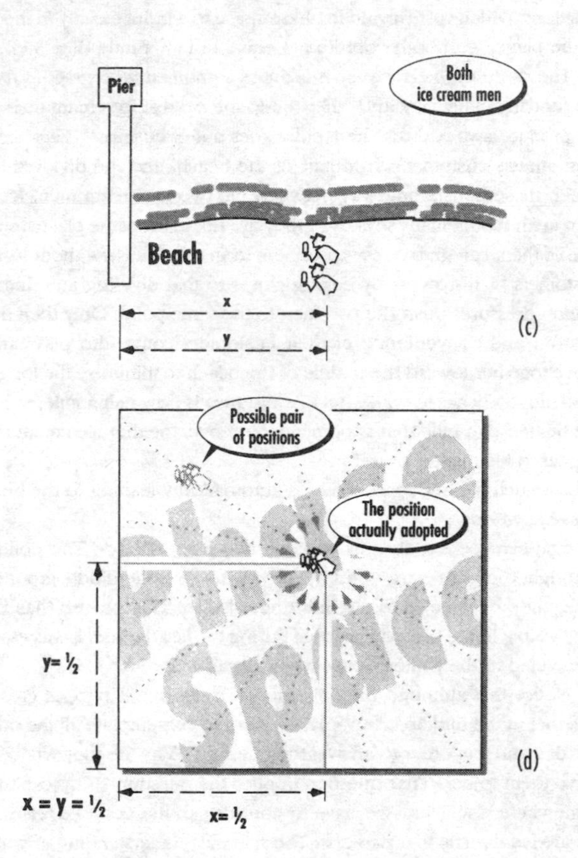

FIGURE 30, CONTINUED

Why do they both end up in the middle (c)? Because the dynamics on phase space (d) drive them there.

men insist on being in the *same* place, in the middle (Fig. 30c). Why? The answer is that there is a dynamic in phase space that drives them to those positions, which correspond in phase space to a point exactly in the middle of the plane, with both coordinates equal to half a mile (Fig. 30d).

The cause of the dynamic is laziness combined with greed. Customers are fundamentally lazy and will walk to the nearest ice cream man (unless his product is so bad that he rapidly goes out of business). Each ice cream man attracts customers from part of the beach, and the dividing line between those parts is midway between the two ice cream men. Ice cream men are fundamentally greedy; if they are not in the same place then either one of them can grab a few customers from the other—without losing any customers of his own—by edging closer to that dividing line. Inexorable market pressures push the two men to the same point. Only then does the comfort and convenience of their customers come into play, and they move together toward the middle of the beach to minimize the longest distance anybody has to walk. But that distance is now half a mile, rather than the quarter of a mile that is possible. Moreover, neither ice cream man has increased his market share!

So much for free market forces automatically leading to the best of all possible worlds.

You can see exactly this effect in two-party politics. The political left and right can grab extra votes by moving toward the middle ground. Over time, they become virtually indistinguishable. This process has become very visible in the United Kingdom during the last decade; it was essentially completed in the United States many decades ago.

Notice that although the only thing we observe is two ice-cream men together in the middle, $x = y = \frac{1}{2}$, we have to contemplate all the other values of x and y in order to answer the question "Why are they where we observe them to be?" That question implies the question "Why can't they be somewhere else?" and we have to consider all the possible "elses." Thus we are led directly to a paradox: The scientific understanding of what actually occurs is based upon the contemplation of possibilities that don't. Phase space must contain all possible questions, not just the final answers.

Precisely because the space of the possible includes things that don't actually occur, it must be chosen; and it's not always obvious what the best choice is. For example, what if the ice cream men can move toward or away from the sea, as well as along it? Or float along on a boat? Using a bigger space may tell you more because you're comparing the actual with more possibilities; on the other hand, too big a space is unwieldy.

What is the space of the possible for Volterra's model? His fish lived in the Mediterranean, but that's not the space we have in mind, any more than the beach was the right space for studying the dynamics of the ice cream men. The space of the possible for a dynamical system is the space of its states—the things that are happening to it as a whole. And it isn't just the states that *do* occur; it's also those that *might* have occurred instead. The important features of Volterra's predator-prey system are the number of predators and the number of prey, and in principle it could be set up with any values for those two numbers. We can think of the two variables as coordinates on a plane, just as we interpreted the two distances of the ice cream men as coordinates. A point in this plane does not represent a shark, nor even the position of a shark in the Mediterranean; it represents a certain number of food fish (the horizontal coordinate) *and* a number of predators (the vertical coordinate). For the ice cream men, a single point represented two distances; but here a single point represents two population values, both those of predator and prey, simultaneously. It's a population plane, not a geographical one.

All possible combinations of population values occur in phase space, but once those values are decided, the dynamic joins them up with a specific set of lines of change that link any particular pair of values to the pair that it will become an instant into the future. Unlike the points of phase space, these lines can't be chosen arbitrarily. They are determined by the dynamical process whereby predators and prey interact—by Volterra's chosen rules. As time passes, representative points move along the lines of change, so the dynamic causes points to flow through phase space. As model predators and prey interact according to Volterra's dynamic, the representative point wanders around in the population plane. The entire plane "flows" along its dynamic, just as an ocean flows, driven by the dynamic of its currents.

What is the flow pattern? Where do the currents run in the population plane? Volterra's analysis shows that the population changes periodically, in cycles. After a certain time, the path traced out in the population plane must return to its original position, and thereafter repeat. That is, the path must close up into a loop. The dynamic phenomenon of periodic motion has been geometrized as a closed loop. Indeed, we can visualize all possi-

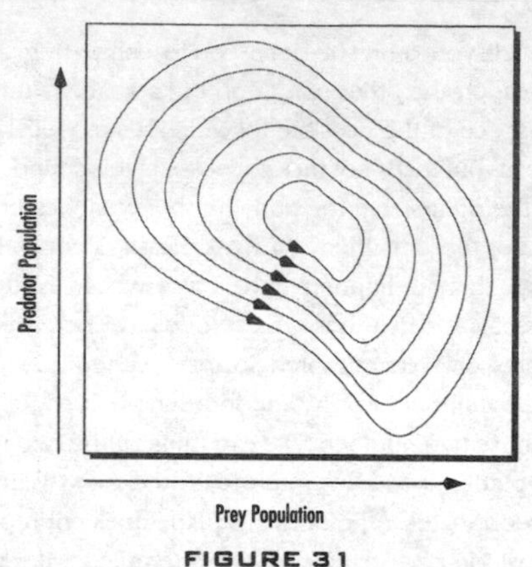

FIGURE 31

Phase portrait for Volterra's model

ble dynamics in the Volterra model, starting from all possible initial conditions; it is a nested set of closed loops and it looks like an archery target painted by Salvador Dali (Fig. 31). As time passes, the points in phase space that represent the state of any given system flow along these curves. This kind of "universal" picture of a dynamical system is called a phase portrait. Figure 30d was a phase portrait, in fact.

ATTRACTORS

The Volterra model predicts that any initial numbers for the population lead to a periodic cycle that returns to those same numbers. There are lots of possible cycles (the nested loops). Real populations that vary periodically tend not to be like that; instead, they settle down to one particular cycle of values. What we need is a better model, with a phase portrait like Figure 32. This picture has only one closed loop, drawn as a heavier line. Everything else spirals toward it, except for a single point inside it.

The closed loop represents stable dynamics. If some small disturbance from outside the dynamical system drives the point off the loop, the spiral motion pushes it back on again. In contrast, the point in the middle represents an unstable steady state. Anything that starts exactly on it will stay fixed forever; but the slightest disturbance will cause it to spiral out and

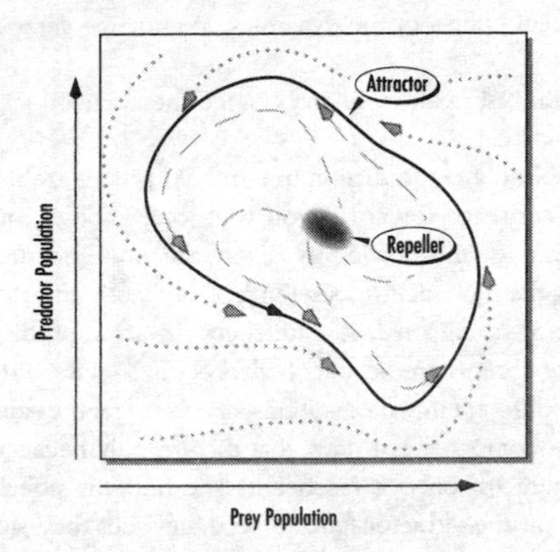

FIGURE 32

A more realistic phase portrait for predator-prey dynamics

away. In topological language, the loop is called an attractor and the point a repeller. An attractor, in general, is a region of phase space that "attracts" all nearby points as time passes. For the system of ice cream men, the attractor is the state in which they are both in the middle of the beach. Attractors are of crucial importance because they capture long-term dynamic behavior. If you watch a typical dynamical system and wait for a while, it ends up on an attractor.

Stable steady states—like that of the ice cream men—are point attractors; stable periodic cycles—like the one described by the modified Volterra model—are closed-loop attractors. The attractors of chaotic dynamical systems are far more complicated. Usually they are fractal, with fine structure on all scales. That fine structure is geometrical evidence for the butterfly effect: Arbitrarily small changes matter. Classical, predictable dynamical systems progress elegantly and regularly along attractors with simple geometric shapes—smooth curves and surfaces. Chaotic systems wander irregularly over fractal attractors. Their path is still elegant, but much less predictable. Classical dynamics is like battery-farmed chickens; chaotic dynamics is free-range. In Volterra's classical system, a fish or two added or subtracted from the equation makes little real difference, but in a chaotic predator-prey system, a single fish scale extra would change the entire subsequent course of the dynamics, in a marine version of the butterfly effect.

However, the fish scale wouldn't change the attractor, just the precise motion on it.

This may sound a contradiction in terms. Attractors are the things that the dynamics converge toward if you wait long enough; but once they reach the attractors, they promptly diverge again—and drastically. But that's precisely the right picture. Anything off a chaotic attractor is "folded" toward it; but anything on it is "stretched" in an unpredictable way—except that one thing *is* predictable: It always stays on the attractor. Figure 33 shows how different initial conditions all end up on the same chaotic attractor, but the computer can draw that picture only because the system stays on the attractor once it reaches it. The butterfly effect means that nearby points on the attractor tend to separate—but they stay on the attractor. Think of a Ping-Pong ball in an ocean, with complicated currents at the surface. If you release the ball from below the surface, it floats upward. If you drop it from above, it falls downward. It is attracted to the surface, no matter where it starts from. But once *on* the surface, it is buffeted to and

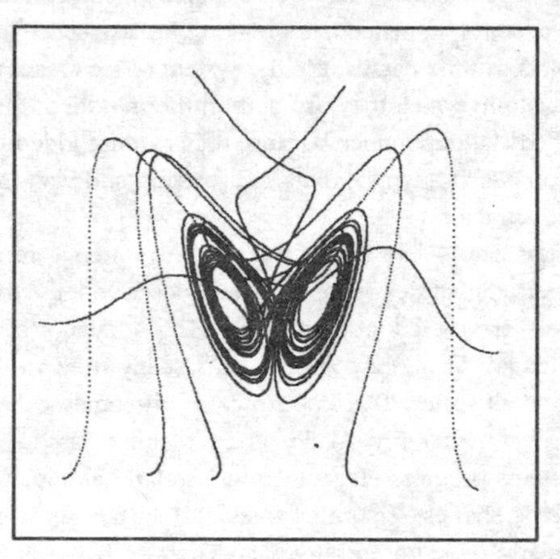

FIGURE 33

How different initial states all end up on the same attractor. This attractor, the mask-shaped object in the middle of the picture, was discovered by Edward Lorenz in 1963 while he was investigating weather prediction.

fro by the currents. The ocean surface is the attractor, and the ball always ends up there; but the dynamics on the surface can be very complex and unpredictable.

Attractors are emergent phenomena in dynamical systems. It's very difficult indeed to tell what kind of attractor is present by looking at the equations for the dynamics, especially if the attractor is chaotic. Instead, you have to solve those equations (usually by computer, and thus approximately) and see where the system goes. The name "attractor" appeals to mathematicians, but it carries the unfortunate suggestion that dynamical systems are goal-oriented—that states end up on the attractor because they know in advance that they have to go there. On the contrary, we only find out what the attractor looks like by watching where initial states go. With good reason, biologists have developed a deep distrust of goal-oriented explanations, such as Lamarck's theory that acquired characters can be passed on to succeeding generations. This makes them suspicious of an attractor as an explanation or a model, which in turn makes them suspicious of dynamical systems as explanations. However, they really shouldn't let themselves be put off like that. In Aristotle's time people thought that objects fell to the ground because that was their "natural place"—a goal-oriented theory if ever there was one. Newton replaced it with the theory that the earth's gravity attracted them toward the ground; they were not seeking a predetermined goal, but merely obeying the forces that acted on them. In our new terminology we can say that the ground is an attractor. This sounds like Aristotle's theory, but mathematically it's the same as Newton's. So when we use attractors as images for, say, evolution, we're not trying to suggest that the system "knows in advance where it's going." All we're saying is that the dynamical equations push it around according to certain rules.

THE LEGEND OF THE HAGGIS

We want to move your thoughts toward a geometric image that puts dynamics into phase space more evocatively. Instead of painting phase space with flow lines, we want to deform it into a landscape, with the "flow downhill" dynamic of gravity. There is an immediate obstacle: You can't do this in any system that has periodic cycles. Only in Maurits Escher's disturbing etchings can you flow downhill and get back to where you start. So we may have to throw in a certain amount of motion that isn't gravity-powered, and when we have to, we will.

Visitors to Scotland, when served a certain traditional native delicacy, often ask what it is. They are then told the legend of the haggis. The haggis, it is said, has its right legs shorter than its left. This forces it to run around and around the sides of hills, clockwise (as viewed from above). When the haggis is in season, hunters climb the hillsides and confront the creatures, forcing them to reverse direction so that they try to run counterclockwise. Because of the differences in length of their legs, they overbalance and tumble to the bottom of the hill, where they are collected by helpers with bags.

If you were to hover over the haggis hunt in a helicopter, you would observe herds of haggis running around and around the hill. If the slope of the hill were uniform, they would all stay on the same level, and what you'd see would be just like the closed curves of Volterra's predator-prey model. The flow lines of Figure 31 are like the contour lines of a landscape. What about Figure 32, the improved model? That corresponds to haggises on a hill whose (concave) slopes change with height, like Figure 34. A haggis that starts high up the hill will stagger downward as it wanders around the peak; one that starts too low will stagger upward. In between, there is one level at which the slope is just right for a haggis to stand stably. Look-

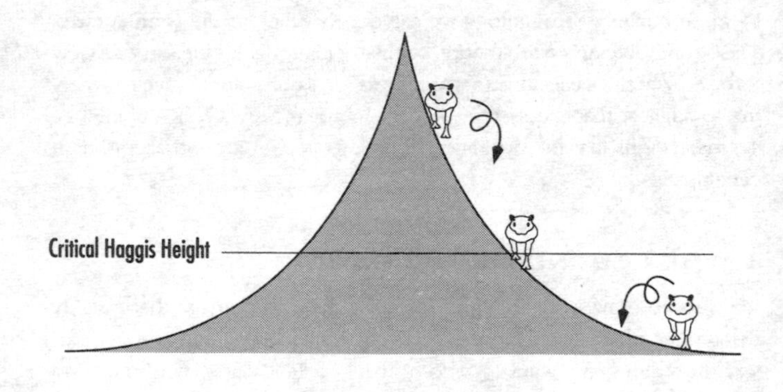

FIGURE 34

An attractor in haggis dynamics

ing down on the motion from above, you would see spirals leading outward from the peak and inward from the base, converging on a closed curve at the critical haggis height. That's exactly the structure of Figure 32.

In future, whenever we use a landscape as an image for dynamics, think of the haggis.

CATASTROPHE THEORY

What other curiosities are there in the grab bag of dynamical systems? One of them is catastrophe theory, which in the early 1970s caused the same kind of stir that chaos theory did in the late 1980s. It sounds as if they ought to be the same thing, and popular accounts do sometimes give that impression, but actually they address two very different ideas. Catastrophe theory is about steady states; chaos theory is about complicated ones. Catastrophe theory is about how states change when you alter the system a little; but in chaos theory you don't alter the system, you just leave it running. Obviously you can combine the two ideas, and think about what happens to chaotic states when you alter the system a little; just such a synthesis is now being developed. It ought to have a really splendid name, such as calamity theory, but it's usually called dynamical systems theory or bifurcation theory.

Let's explore what catastrophe theory has to offer, but in the language of today instead of that of the 1970s. A dynamical system can have just one attractor, or several. If it has several, then depending on what conditions you start from, you may end up with different long-term behavior. Dynamical systems can also have adjustable parameters—features that are fixed in any particular instance but can change from one instance to another. The growth rates in Volterra's model are adjustable parameters. In each predator-prey system, these growth rates take particular numerical values, but different systems have different growth rates. What makes catastrophe theory interesting is that the combination of adjustable parameters with multiple attractors can sometimes be explosive. Adjusting the values of parameters by a tiny amount can cause the entire geography of phase space to change, with remarkable results.

Here's an example, which we approach tangentially by way of beetles. The fossil record of beetles is very strange: Over periods of millions of years, at a given site, particular types of beetle seem to come and go. They can disappear for vast periods and then suddenly reappear, perfectly formed, exactly as before. Do they somehow unevolve, returning to a form that had been replaced by a fitter one? No, the answer is much simpler. Beetles like particular temperatures. As the climate changes, they move around, following the zones that afford the temperatures they like. Those zones can wander away from a given geographical location and then wander back. The phase space for beetles is the landscape in which they live—for once phase space and real space turn out to be the same—and the zones with the right temperature form attractors for the beetle dynamic.

The climate parameters cause the attractors to move, and sometimes more. Let's consider a rather well-behaved climatic system: a mountain range in which temperature decreases steadily and uniformly with height. We'll also take a long-term view and ignore seasonal variations; what interests us is ice ages. Suppose the beetles like a fairly cold environment. During an ice age, they will be found spread all over the plains. As the ice melts, they begin to retreat up the mountains, forming a number of rings lying between the altitudes at which the temperature becomes intolerably hot and intolerably cold. As the ice retreats, the rings move higher. Some beetles have chosen the wrong mountains; their rings reach the peak of the mountain and then shrink and disappear entirely as the temperate zone "lifts off" from the mountain top. (The beetles don't lift off; you don't get a cloud of them hovering above the mountaintop.) In the short term they may fly to another mountain, but that mountain will already have its own beetle population anyway, and you don't see long-term effects of that migration. On a geological time scale, the attractor that was at the top of the mountain has simply disappeared. Other rings can split up as the temperate zone moves up across a pass; what was a single ring below the pass divides into two, one going up the peak to the left of the pass, and one to the right (Fig. 35). So attractors can split up. As the ice begins to advance again, the reverse processes happen: Attractors can merge, or suddenly reappear. Notice that the temperature changes are continuous but the changes to the attractors are not.

This is not an isolated example. We could tell a similar story for water beetles in lakes. As the lakes dry up, they can split into smaller pieces, or disappear; when the rains come, they can appear in a dry depression or merge as they grow. Much of the time the number of attractors (lakes) and their general pattern stays the same, but every so often it changes with a

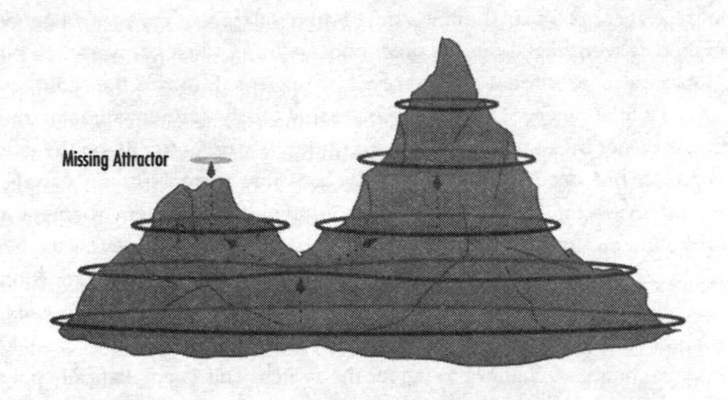

FIGURE 35

Qualitative changes in attractors. Continuous variations can cause attractors to split into pieces or disappear. Reversing the process can cause attractors to merge or appear from nowhere.

disappearance or a merger. Such a "qualitative" change in a dynamical system, brought about by a continuous change in a "control" variable such as temperature or rainfall, is called a catastrophe or a bifurcation.

Catastrophes possess many subtle features. One is the surprising production of big changes in behavior from small changes in circumstances. The dramatic change from having beetles on top of a mountain to having no beetles need not be brought about by equally dramatic changes in temperature; it's more a case of the last straw. Just make the temperature that tiny bit hotter and *wham!* This is a remarkable answer to a question we didn't even ask at the beginning—namely, How can large changes in behavior result from small changes in circumstances? We just followed our mathematical noses, and up the answer popped. Another subtle feature is that after the catastrophe you may end up on an attractor that you never knew existed. It was busy developing, out in the mathematical space of the possible, but you didn't notice it because the new attractor wasn't being physically expressed. Suddenly it *is* expressed, and the mathematical fiction gobbles up your own reality and lands you in totally unexpected circumstances.

The earth's climate is a huge dynamical system, and it seems to have at least two attractors: the mild kind of climate we are currently experiencing and ice ages. You may think it a trifle odd to talk about ice ages when everybody's worrying about the greenhouse effect and global warming; but global warming is just a change of a few degrees. It moves the "mild" attractor a little. Ice ages take our climate to a totally different attractor, one that is colder by a hundred or more degrees. We are currently on the mild attractor, but out in the space of the possible there lurks an ice age. Disturbances—and we're throwing enough junk into the environment to create them ourselves—could switch our climate to the cold attractor. We could be in the position of the hilltop beetle, just on the verge of extinction, blissfully unaware that a mathematical fiction in the space of the possible is about to become reality. And the really nasty feature is that it may take only the tiniest of changes to trigger the switch. This is emphatically not a fantasy. It is a respectable way of expressing some of the things we know about climate change. We could tell a similar story about earthquakes, hurricanes, avalanches, and heart failure. The universe is always ready to realize an unexpressed potential. Mathematical fictions can bite.

It is particularly important to understand that the apparent paradoxes of chaos and catastrophe do not stem from the oversimplifications of the models. This mistake is made repeatedly by biologists, meteorologists, ecologists, and especially economists. If a simple model throws up a paradoxical phenomenon, some people blame it on that particular model instead of asking how general the phenomenon (not the model) might be. Complicated models usually do things that are more paradoxical than simple ones.

■ LANGTON'S ANT

We will generally use dynamical systems and the associated paraphernalia of attractors, chaos, and catastrophes as our images for systems that change over time. But we also want to discuss one further type of mathematical system known as a cellular automaton. Basically it's a grid of switches that can be set to a number of different positions, but what matters is the rules for working the switches. One of the simplest cellular automata is known as Langton's ant, after its inventor, Christopher Langton. The ant moves either north, south, east, or west on a square grid of black and white cells, following three simple rules:

- If it is on a black cell it makes a 90° turn to the left.
- If it is on a white cell it makes a 90° turn to the right.
- As it moves to the next square, the one that it is on changes color from white to black, or the reverse.

You may think that Langton's ant must be a remarkably simple animal, but no. In fact, it poses a problem that is currently baffling mathematicians.

As a warm-up, suppose you start the ant in an eastward direction on a completely white grid. Its first move takes it to a white square, and the square it started from turns black. Because it is now on a white square, the ant's next move is a right turn, so that it is then facing south. That takes it to a new white square, and again the square it has just vacated turns black. After a few more moves it starts to revisit earlier squares that have turned black. If you try out the rules you'll find that the ant's motion gets quite complicated—and so does the ever-changing pattern of black and white squares that trails behind it. Every so often during the first few hundred moves, the ant produces a nice, symmetrical pattern. Then things get rather chaotic for about ten thousand moves. After that, the ant gets locked into a cycle that repeats the same sequence of 104 moves, whose overall result is to move it two squares diagonally. It continues like this forever, systematically building a broad diagonal "highway."

This behavior is curious enough, but computer experiments suggest something even more striking: If you scatter any number of black squares around before the ant sets off, then it still ends up building a highway. For example, Figure 36 shows the pattern that forms when the ant starts inside a particular solid rectangle. Before it builds the highway, it builds a "castle" with straight walls and complicated crenellations at the corners. It keeps unbuilding and rebuilding these structures in a curiously "purposeful" way until it gets distracted and wanders off . . . building a highway. The problem that is baffling mathematicians is that nobody can *prove* that the ant always ends up building a highway, for every initial configuration of (finitely many) black squares. But it certainly seems to.

That's an example to give you the flavor, and we'll say more about it in chapter 13.

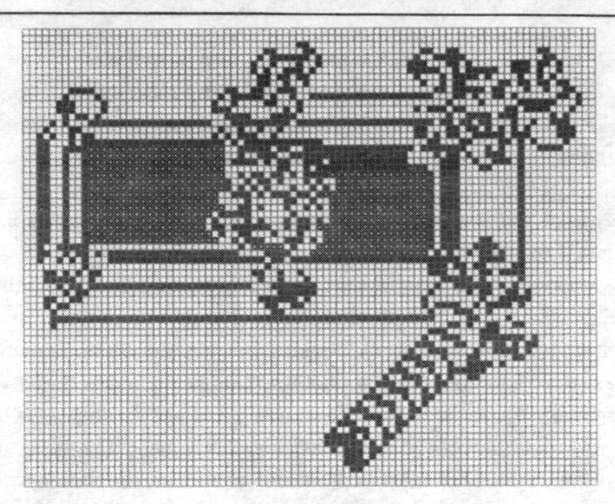

FIGURE 36

The pattern created by Langton's ant when it starts inside a black rectangle. Small white dots mark those squares of the original rectangle that have not been visited. The highway is at lower right.

■ THE GAME OF LIFE

Cellular automata in general were invented by John von Neumann, for a particular purpose. He wanted to make a self-replicating machine. He was thinking of abstract mathematical "machines," and so are we, but the development of nanotechnology is opening up a serious prospect of real von Neumann machines. A von Neumann machine, real or mathematical, is a kind of general-purpose programmable factory that, when provided with a blueprint, can build anything. Then all you have to do is provide it with its *own* blueprint . . .

This is strikingly similar to the way living cells do the job: cell as chemical factory, DNA as blueprint. Von Neumann wanted not just a verbal argument about hypothetical machines, but some structure (albeit one existing only in the mathematician's imagination) that actually did reproduce itself by his method. And that's where the cellular automaton came in. Think of a huge chessboard, each of whose cells can contain one of a fixed collection of markers—colors, numbers, carved wooden pieces, it doesn't matter. Let's pick colors to be specific. At time 0 the game is set up by assigning colors to the cells in some manner. At each subsequent instant of time, the color of each cell is changed, depending upon the colors of its neighbors according to some fixed system of rules. A typical rule is "A red cell that is adjacent to two greens, one blue, and one pink must turn purple." Such a structure can contain well-defined "objects," arrangements of cells in a given shape and with given colors. A simple object might be nine pink cells arranged in a 3 × 3 square. A more complicated one might involve a million cells, forming a vast, complicated shape, with particular colors in particular places.

Von Neumann devised a set of rules, and an object that automatically reproduced itself just by following the rules for changing the colors of cells. If you started with one such object and waited, eventually you'd get two. His approach used a lot of colors and a lot of rules, but it worked. Many years later John Horton Conway invented an automaton with much simpler rules that turned out to be able to do the same kind of thing. Conway's automaton used only two colors: black and white. And there were only three rules:

- A cell that is white at one instant becomes black at the next if it has precisely three black neighbors.
- A cell that is black at one instant becomes white at the next if it has four or more black neighbors.
- A cell that is black at one instant becomes white at the next if it has one or no black neighbors.

In all other cases, cells maintain their color. The neighbors of a given square are the eight cells adjacent to it vertically, horizontally, and diagonally. All changes are deemed to be made simultaneously.

Conway called his automaton the Game of Life, or Life for short. The idea is to start with an object made up of black cells, and the rest of the board white. Then you follow the rules and watch how that object changes. For example, a 2 × 1 block dies out at the first move. A 2 × 2 block doesn't do anything, so it survives indefinitely. More interesting is a simple shape called a glider: It moves. It changes shape in a four-step cycle, after which it has moved one cell diagonally. More complicated shapes, spaceships, move horizontally or vertically. A "glider gun," which changes through a fixed cycle of thirty shapes, fires an endless stream of gliders.

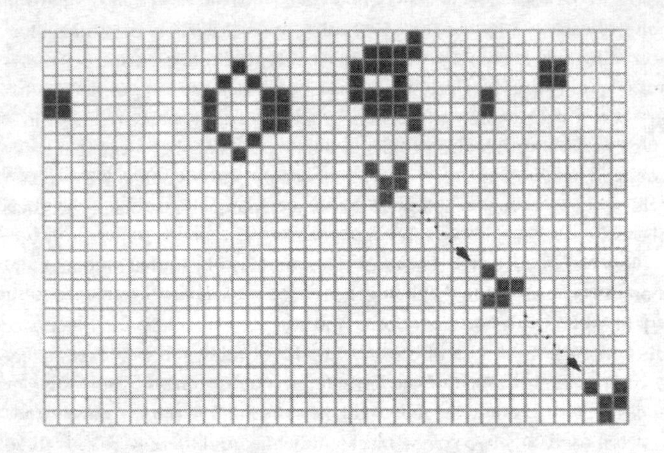

FIGURE 37

Emergent phenomena in the Game of Life. Glider gun (black) fires a stream of gliders (gray).

These are very simple examples of emergent phenomena, and they emerge from Conway's rules (Fig. 37).

Conway's Life runs in a reductionist universe, which really does have a Theory of Everything, namely Conway's rules. Given a starting shape, its future is completely determined by those rules. But in practice it may be very hard to predict what will happen, even though it's all implicit in the rules. To make it explicit, you have to follow the rules and see what happens; there don't seem to be any short cuts. Big shapes can collapse, small ones can grow, and there are always surprises. Three of the things that emerge from Life are programmability, undecidability, and replication.

First, Conway proved the existence of an initial configuration that acts like a programmable computer, using pulses of gliders instead of electrical impulses to carry and manipulate information.

From this he deduced that the outcome of the game is inherently unpredictable, in the following sense: There is no way to decide in advance whether a given object will survive indefinitely, or disappear entirely. He did this by appealing to a theorem in mathematical logic which says the following: There cannot exist a computer program that can decide in advance whether any given program, when run on a given machine, will go on forever, or will stop. The only way to find out is to run the program and watch. If it stops, you know; if it keeps going, you have no idea whether it will continue to keep going, or whether it's just about to stop as soon as you give up and go away. The theorem is called the undecidability of the halting problem, and the proof, discovered by Alan Turing, is simple and clever. The basic idea is that if there were such a computer program—a universal decider, so to speak—then you could set it up to monitor its own progress, and to stop if and only if it didn't stop. This is a paradox, so such a program can't exist.

Once you have programmable computers, it's not hard to pinch von Neumann's trick and design self-reproducing machines. So Life—a two-state cellular automaton with only three rules—has self-reproducing computers. Given that, you could set up self-reproducing "animals" with "genetic programs" that interact with each other (just program the Life computer to simulate such a system). You could "irradiate" the board with gliders to cause random mutations. Then you could sit back and watch evolution at work.

COMPLEX SYSTEMS

Reductionism holds that everything in the universe, however complicated, is the outcome of simple laws. We find those laws when we look down our mental funnels; having found them, we can make them work for us. However, reductionism is not as complete as it may seem to be. Imagine an intelligent configuration in Life, one with a kind of cellular mind, which has received—perhaps from an oracle—the message that everything in its universe is governed by those three basic rules. It knows the Theory of Everything for the Game of Life. Would it be satisfied with them as an explanation of its intelligence?

No. It would want to know exactly how those rules lead to intelligence; it wouldn't be content with the rules alone, no matter how many experiments the Life physicists did to verify them. The same goes for us. We want more than just the rules; we want to know what they imply about our place in the universe, and exactly how the possibility of being human follows from them.

One of the things that today's science lacks is an effective theory of complexity. We can't make any great sense of economics, for example; we don't know exactly what causes global recessions, and we certainly don't know how to predict them. We don't know exactly how hurricanes form, or where they will go. So it's a bit ambitious to tackle human existence or intelligence. However, as the twentieth century draws toward a close, attention is increasingly focusing on the mechanisms of complexity. For example, Stephen Wolfram has set up a Center for Complex Systems at the University of Illinois; and George Cowan recently founded the Santa Fe Institute in New Mexico to tackle just such questions and to bring scientists from many different traditional disciplines together in the common cause of understanding complexity.

There is a new scientific discipline called complexity theory. It's exciting and important stuff. Doors seem to be opening that lead into realms previously undreamed of. It has become possible for psychologists to wonder about consciousness without being ridiculed by their peers. We are beginning to discover that complexity has its own laws—but we don't understand them terribly well, not yet. It's important to realize that those laws are not just the statistical regularities that can be found in any random system. The key to complexity is not statistical "gas laws" for economics or weather. We want details.

The key isn't things like the Mandelbrot set either, though that's certainly a step in the right direction. What we really want is an understanding of broad questions like "Why did vision evolve?" "What causes stock-market crashes?" "Why do people get depressed?" Or, going for the jugular, "What collapses the underlying chaos to allow the emergence of simple, robust structures?"

Reductionism has led us to those questions—but can it answer them? Could complexity theory answer them instead? And if not, what other method of explanation should we look for?

7 COMPLEXITY AND SIMPLICITY

A duke was hunting in the forest with his retinue of men-at-arms and servants; he came across a tree. Upon it, archery targets were painted, and smack in the middle of each was an arrow. "Who is this incredibly fine archer?" cried the duke. "I must find him!"

After continuing through the forest for a few miles he came upon a small boy carrying a bow and arrow. Eventually the boy admitted that it was he who shot the arrows plumb in the center of all of the targets.

"You didn't just walk up to the targets and hammer the arrows into the middle, did you?" asked the duke worriedly.

"No, my lord. I shot them from a hundred paces. I swear it by all that I hold holy."

"That is truly astonishing," said the duke. "I hereby admit you into my service." The boy thanked him profusely. "But I must ask one favor in return," the duke continued. "You must tell me how you came to be such an outstanding shot."

"Well," said the boy, "first I fire the arrow at the tree, and then I paint the target around it."

In chapter 1 we asked where the organized complexity of the universe comes from. Now we have an answer, the reductionist answer, which focuses upon relations between levels of description. Complexity at any given level is a consequence of the operation of relatively simple rules one level lower down. Simplicity breeds complexity through sheer multiplication of possibilities; organization derives from the tidy simplicity of the laws. Thus the interactions between a small range of fundamental particles account for all of the chemical elements and the manner in which they bond to other elements. The complexities of chemistry are those of a Lego set: myriad copies of a limited selection of atoms, assembled at will, subject only to the rules for chemical bonding. The complex phenomenon of life is at base chemical, relying on the unusual properties of long-chain carbon molecules, and in particular the computational and informational properties of DNA that lend it the ability to replicate. This unique feature of carbon is (simply) a consequence of its atomic structure. The complexity of biological development is the working out of a DNA "program" that, among other things, organizes the production of proteins. The complexity of a living organism is the result of an evolutionary game played over huge periods of time according to a rule so simple that it is really just a logical tautology: Winners win. The complexity of intelligence is the result of a similar evolutionary game, played by networks of neurons, interacting through sensory input and output. The complexity of ecosystems and societies is the result of a network of transactions and communications between individual organisms.

The term "reductionism" carries the connotation that complexity on one level is *reduced* to simplicity on another level. Mental funnels lead down from complexity to simplicity (Fig. 38), and the deeper they go the more mathematical they become. A coherent chain of explanation runs upward, from subatomic particles to sacred cows. Rules are revealed by reductionism, and reductionism rules.

We hope that you have enjoyed, understood, and been at least to some extent convinced by this story as we have told it so far. We think that it is true, insofar as anything known to human beings is true. Not, to be sure, in

FIGURE 38
Reductionism summarized as a funnel

every detail, but in its broad sweep. It is certainly the way in which nearly all scientists think about explanations. It works, not just as a philosophical viewpoint but as a practical proposition. Its results directly or indirectly affect the daily lives of billions of people: television, jumbo jets, satellites, silicon chips, aluminum cans, plastic spoons. It is an enormously useful way of looking at the relation between simplicity and complexity.

But . . .

Neither of us believes that reductionism alone is enough, nor that it is the most useful way of looking at simplicity and complexity. And in the remainder of this book we hope to convince you of that. The last six chapters haven't been a waste of time, though; we had to establish an agreed background of conventional science before we could start on anything less conventional. That background is valuable in its own right, but now we want to take you beyond the reductionist story. We want to extend it, not overturn it.

Because the reductionist story is a very tidy one, we were able to give you a road map before setting out; we could tell you that we would begin with atoms, and work our way up through chemistry, DNA, organisms, evolution, intelligence, and ecosystems to human society. Now we want to explore the same general ground again, but in an off-the-road vehicle. This time we can't offer you a road map, because there are no roads. We want to show you the gaps in the reductionist story, and how we think science might begin to fill them in. That leads us into uncharted territory, where the best we can do is put up the occasional warning sign.

We came to our perception of the limits of reductionism from very different directions, for the entrenched thought patterns of mathematicians are very different from those of biologists. Despite this, our thinking on these issues is amazingly similar, as we discovered early in our acquaintance. It was this discovery that led to the present book. Over the last few years, we have evolved a common viewpoint which turns the reductionist worldview upside down—or, perhaps more accurately, inside out. As we explained in the preface, we claim no originality for many of the ideas. But we have done our best to put the ideas together in a new way. The upshot is that a mathematician and a developmental biologist agree that the best way to think about the complexity of the universe is not just to examine how it can be generated by the interaction of simple components, obeying simple rules, and producing complexity one level further up.

Scientists have been asking the wrong question.

They have focused upon complexity as the thing that requires explanation, and they have taken simplicity for granted. The answer to complexity turns out to be fairly obvious and not, in itself, especially interesting: If you have a lot of simple interactors, and let them interact, then the result can be rather complicated.

The interesting question is precisely the opposite, the question that most scientists never thought to ask because they didn't see that there was a question to ask. Where does the *simplicity* come from?

RATIONAL AND IRRATIONAL

Let's begin by making it clear what sort of simplicities we're thinking of. Simplicities like the numerology of atoms—whole numbers of electrons, protons, neutrons. Magic numbers like 8 and 18 that, by way of electron shells, determine the atoms' chemical properties. Simplicities like the gas laws, which relate temperature, pressure, and volume in gases, which we know are really extraordinarily complicated systems of vast numbers of atoms, bouncing off one another at tremendous speed. Simplicities like Newton's law of gravitation, asserting that the attractive force of gravity falls off according to the inverse square of the distance. The deep biological simplicities, such as hemoglobin, chlorophyll, the double helix of DNA, and homeotic genes. And purely mathematical laws, such as the Pythagorean theorem and the fact that the circumference of any circle is π times its diameter.

Integer ratios in chemistry led to the acceptance of atomic theory long before anybody could see an atom. Does God love whole numbers? We humans seem to think so. "God made the integers," said the mathematician Leopold Kronecker; "all else is the work of man." Kronecker was a wholistic fundamentalist. We were baffled by chlorine's atomic weight of 35.46, and sighed with relief when it turned out to be a mixture of three-and-a-bit parts of 35 to one of 37. We've gotten used to π, mostly because it shows up in so many places, but still find it peculiar that the universe should employ such a curious number when plain 3, or maybe $^{22}/_{7}$, would be so much tidier. We are always impressed by integers, whole numbers. We would be less impressed by Newton's law of gravity if it involved the inverse 2.167th power. An inverse 2.167th law looks like an empirical fudge; an inverse square law looks like a universal truth.

We like "clean" geometry too: circles, ellipses, the regular solids. Buckminsterfullerene attracts scientists' attention because it is a new form of carbon, but even more because it is a truncated icosahedron. What is it that attracts us to whole numbers and regular geometric shapes?

A mathematical image captures what we're up against. A so-called real number is anything that can be represented by a decimal, possibly nonterminating. Buried in the real numbers, like diamonds in clay, are the integers: the whole numbers 0, 1, 2, 3 . . . and their negatives. From the integers you can build so-called rational numbers; these are fractions such as $\frac{2}{3}$ or $\frac{22}{7}$. The ancient Pythagoreans discovered that some real numbers are irrational: They cannot be represented as exact fractions, however big the numerator and denominator may be. Their first example was $\sqrt{2}$, the square root of two. Later on it turned out that π is irrational. There are infinitely many irrational numbers, and infinitely many rational numbers, but Georg Cantor showed that, in a reasonable sense, there are more irrationals than rationals. Overwhelmingly more.

Despite the prevalence of irrationals, they and the rationals are intimately mixed together. Given any irrational number, you can find a rational number—indeed, infinitely many—as close to it as you wish; correct, say, to a billion decimal places. It's not hard to do this: Write the irrational as a decimal, and throw away all decimal places after the billionth. Terminating decimals are always rational. It is also true (and only marginally harder to prove) that given any rational number, you can find an irrational number—indeed, infinitely many—as close to it as you wish. Rationals and irrationals are mixed together like sugar and flour in a cake and remain mixed even if you look at them through a microscope, however powerful it might be.

Let's phrase this in terms of simplicity and complexity. Rational numbers are "simple" in comparison to irrationals: They are built from simple ingredients (integers) by simple rules ("Make a fraction"). Irrationals are more complicated; their description is generally more involved. For example, you can't talk about π without first sorting out circles and lengths and things like that. So, in the world of real numbers, the simple ones and the complicated ones are mixed together so finely that if all you do is look at the first billion decimal places, you can't tell which kind you've got. The mathematics at the bottom of the reductionist funnel may seem to offer infinite precision, but every simple nugget of mathematics is surrounded by clouds of far more complicated—but almost indistinguishable—pieces.

■ CLOUDS OF IMITATORS

The world of possible scientific theories (not just the true ones, but everything you could imagine and a lot more besides) is just the same. When thinking about the dynamics of a system, we found it useful to consider a mathematical fiction: the space of all possible states. Even though the system does not occupy all of those states, but selects them according to the dynamic, we cannot understand the selection mechanism unless we contemplate the possibility that the system might have done something different. In the same way, when considering scientific theories, we have to work in theory space, the space of all possible theories. Not just the theories that have been agreed upon as a result of lengthy series of experiments, but the alternative theories that might have been adopted had the evidence been different, or indeed had the evidence been the same. The reasons for accepting a particular theory tacitly involve the reasons for rejecting its competitors; so we have to consider the competitors, even if they eventually turn out to be "wrong."

In this mental space of possible theories, every theory is surrounded by a cloud of different but almost indistinguishable others. As close as you like to a simple theory—such as Newton's law of gravity—there is a halo of nearby theories that are as complicated as you wish, theories like "Actually it should be the inverse 2.000000000000000000000000001st power law." The halo is not that straightforward, though; it includes theories such as "The inverse-square law holds universally, except for one atom in the Andromeda galaxy at 7:15 P.M. on Christmas Day, 1842." That's not a very appealing theory, we admit; but if it *is* true, and Newton's law is false, it's too late to find out. (It is, however, falsifiable, which according to philosophers is what makes a theory scientific. It could have been tested experimentally by observing all the atoms in Andromeda at the specified time and noticing that one of them didn't do what it should if it obeyed Newton's law.)

Newton's halo also includes general relativity, a geometric description of gravitation. Relativity says that mass is really the curvature of space-time, which means that the kinds of masses we encounter in everyday life behave pretty much as Newton said (a slightly curved space-time looks nearly flat), but enormous masses don't (a highly curved space-time doesn't look flat). This is why Newton's law looked pretty good for a long time and why we still use it to put communication satellites into orbit. It also means that you could state the theory of relativity as "The inverse square law—or something pretty close—holds, provided masses aren't too big." But we wouldn't be impressed by a theory that was stated that way, because it's uncomfortably close to "The inverse-square law holds universally, except for one atom in the Andromeda galaxy at 7:15 P.M. on Christmas Day, 1842." What impresses us about relativity is its mathematical simplicity and elegance.

Simple and complex theories surround Newton's law, and all other theories of nature, in much the same cake-mix manner that rational and irrational numbers surround the number 2. Given these dense clouds of imitator and competitor theories, how do we choose the "right" ones? Over the centuries, scientists have devised a working philosophy that places the emphasis upon simplicity. The principle known as Occam's razor asserts that assumptions should not be made unnecessarily complicated. Occam's razor is a way of sorting the simple theories from the complicated ones. So when scientists select theories, they do not use just the criterion of agreement or disagreement with observations. They also have aesthetic principles in mind. They want the theory to be universal, not peculiar to some particular place and time. They want it to be elegant, not held together with chewing-gum and string. They use these aesthetic principles to remove the cloud of "trivially" competing theories that necessarily surrounds every theory. Paul Dirac took a rather extreme view, saying that he would prefer a false but beautiful theory to a correct but ugly one.

Occam's razor isn't a scientific theory; it's a philosophical principle, a meta-theory, a theory about theories. And it has its problems. At any given instant in the development of science, Occam's razor is great for chopping away unnecessary detail and concentrating your mind on what currently seems to matter. However, as science develops, theories that started simple tend to get more complicated. For example, we've already mentioned that when DNA was first discovered people thought that genes were just connected segments of DNA. They might have been scattered all over the genome, of course, but by Occam's razor, they shouldn't have been. The trouble is, many genes are *not* connected segments of DNA, and some genes *are* scattered all over the genome. Occam's razor is a working rule of thumb, not an ultimate answer.

Another problem with Occam's razor is that over time we revise our views of what is or is not simple. The number π started out looking pretty complicated, but by now we've become so used to it that it seems almost as simple as the number 2. Below we will tell the story of two new simplicities that have crystallized out from chaos theory: our old friend the Mandelbrot set, and a new one called the Feigenbaum number. Without chaos theory to explain why they really are simple, both would have seemed fiendishly complicated.

■ OCTAM'S TROWEL

Dirac's criterion of mathematical beauty suffers from the same problem: What appears beautiful at one period of history may seem rather trite, or even nonsensical, at another. Ptolemy's theory of epicycles (see page 25, Figure 5 in chapter 1), which for centuries dominated scientific thought about the motion of the solar system, is an instructive example of the human tendency to invent brain puns. The ancients expected planetary orbits to be circles because circles have an air of elegance and perfection. When circles didn't work, they built more complicated orbits by fitting circles together—huge, elaborate mathematical constructions, circles whose centers moved along circles whose centers moved along circles. It worked, in the sense that epicycles were entirely sufficient for predicting whereabouts in the zodiac Jupiter would be at some future date, or when the next eclipse of the sun was going to happen.

NEEPLPHUT: It is amusing/only to be anticipated (delete whichever is inapplicable) that you still use Pnurflpeef's outmoded law to navigate through space.

CAPTAIN ARTHUR: Pnurfl— Oh, you mean Newton's law of gravity. We use it because it's simple.

NEEPLPHUT: And wrong.

STANLEY: Yes, but the errors are so small that they really don't matter. It's not worth the extra computations to use relativistic gravitational theories.

NEEPLPHUT: You have not yet discovered octacycle theory, I gather.

CAPTAIN ARTHUR: Not to my knowledge. What is it?

NEEPLPHUT: A simple, accurate, and exceedingly octimal way to calculate planetary motion. It is based on the underlying principle that the ideal form for a planetary orbit is a regular octagon.

CAPTAIN ARTHUR: What? But octagons have corners! You don't see planets turning corners! What's wrong with ellipses?

NEEPLPHUT: No planetary orbit is an ellipse.

CAPTAIN ARTHUR: That's because perturbations from other planets distort the orbit from its ideal form.

NEEPLPHUT: That is just a convoluted way of agreeing with me. Perturbations are a futile invention to save the false ideal of ellipses. Octacycles are the true ideal.

I say "ideal" not, as you do, to save a falsehood, but because no planet actually travels in a simple octagon. Nature abhors naked simplicities, and while a single octagon may be octimal, it is far too simple to be credible. We have a philosophical principle that we call Octam's trowel, because it is generally laid on very thick. It says that theories should not be rendered unreasonably simple. A true orbit is therefore the combination of many octagonal motions. Centered upon a point pursuing the primary octagon is a smaller secondary octagon, and centered upon a point pursuing the secondary octagon is a yet smaller tertiary octagon, and so on. You can get very accurate orbits by choosing the sizes and speeds correctly.

CAPTAIN ARTHUR: But that's just the old epicyclic theory!

STANLEY: Using octagons instead of circles.

NEEPLPHUT: Pardon?

STANLEY: An ancient and totally outmoded terrestrial theory of the planets represents their orbits as superimposed circles in just the same way. It worked—but for the trivial reason that *any* orbit can be so represented, as accurately as you wish. Even an octagonal one, I guess. It was discarded because it was purely empirical: It gave no real insight into the causes of planetary orbits.

NEEPLPHUT: Of course not. Since the true ideal is the octagon, a circle is a poor approximation to reality.

CAPTAIN ARTHUR: But don't you see that your theory of octacycles has the same faults? It too can "explain" any orbit whatsoever. Even an elliptical one.

NEEPLPHUT: But Captain, that is exactly my point. Octacycles explain your slightly flawed ellipses, *and* without any of this "perturbation" nonsense. Moreover, they are fully in accordance with the principle of octimality, and with Octam's trowel. I rest my case.

We couldn't agree with Neeplphut, however well his method fit observations, for the same reason that we now dislike Ptolemy's theory. The trouble with epicycles is that they can always be made to work, no matter what shape planetary orbits may be. You can fit any orbit to a system of epicycles. It may have to be a complicated system, but there is as much flexibility in it as the supply of epicycles permits—and that supply is up to you. The more epicycles you're willing to use, the better fit with observation you can get. If you want an octagonal orbit, for example, you can do it with epicycles. It will have lots of tiny curves everywhere, but not exceeding the level of error that you're willing to allow. You can also produce circular motion using octacycles, so anything you can do with circle-based epicycles can be done with octagon-based ones—or heptagonal ones, as any septimist would hasten to point out.

To the ancients, the success of epicyclic theories confirmed their belief in the circle as the perfect figure. We now see that this is nonsense: Any shape would do the job. (It is true, however, that some shapes lead to easier calculations than others, and so have greater practical value—which is one reason why Ptolemy's circles survived for such a long time.) Neeplphut, stuck in his species' octimist paradigm, has problems seeing the point, but it's a powerful objection nonetheless. If a theory is so flexible that it can "explain" anything whatsoever, it's probably nonsense.

■ FEIGENBAUM AND MANDELBROT

The old simplicity of epicycles was entirely practical. It worked. We threw it out anyway in favor of a more subtle, but somehow "deeper," more satisfying, simplicity. The story of science is that of repeated revolutions in our conception of the simple.

One of the great surprises of chaos theory is the discovery of totally new simplicities, deep universal patterns concealed within the erratic behavior of chaotic dynamical systems. The first of these unexpected simplicities was discovered by the physicist Mitchell Feigenbaum, and it's known as the Feigenbaum number. As is the case with most great discoveries, he had the devil of a job getting it into print.

We'll explain it in a physical interpretation. If a small quantity of liquid helium, cooled very close to absolute zero, is heated from below, it forms tiny convection cells, in which the helium circulates and carries heat upward. If the temperature at the bottom is increased a little, then the cells begin to wobble periodically: wobble-wobble-wobble-wobble over and over again. At a higher temperature, pairs of consecutive wobbles become slightly different: wobble-*wobble*-wobble-*wobble*. The period doubles in length; you now have to wait for two wobbles before everything repeats. As the temperature rises still further, the wobbles group into fours—wobble-*wobble*-WOBBLE-*WOBBLE* repeated indefinitely. This doubling of the period by the creation of ever finer differences between consecutive sequences of wobbles is called a period-doubling cascade. Each successive step in the cascade occurs as the result of an ever smaller rise in the temperature. In mathematical models there is a particular critical temperature, and when this is reached, the period has doubled infinitely often, resulting in chaos. The period-doubling cascade is a route from order to chaos. It's important because it is one of the commonest such routes.

Liquid helium is not the only system to behave in this way. Water dripping from a tap goes through a similar period-doubling cascade, from drip-drip-drip-drip to drip-*drip*-drip-*drip* to drip-*drip*-DRIP-*DRIP*, as the tap is turned on further; again the cascade culminates in chaos. Electronic oscillations, models of the interactions between predators and prey, and models of blood-cell production also behave in this curious manner. The period-doubling cascade is a common feature of dynamical systems—although it is only in the last twenty years that we have known about it.

Feigenbaum was experimenting with one of the simplest models of such a cascade using a pocket calculator, and he discovered a strange pattern. If we were to interpret his pattern in terms of the liquid helium, it would go like this: The amount by which the temperature must rise, in order to double the period, decreases geometrically as the period gets longer. Each such increment is roughly $1/4.669$ times as long as the previous one. For instance, if the temperature has to go up by 1 degree to increase the period from 1 to 2, then it must go up by only $1/4.669$ of a degree to increase the period to 4, by $1/4.669^2$ to increase it to 8, and soon (Fig.39). The precise ratios

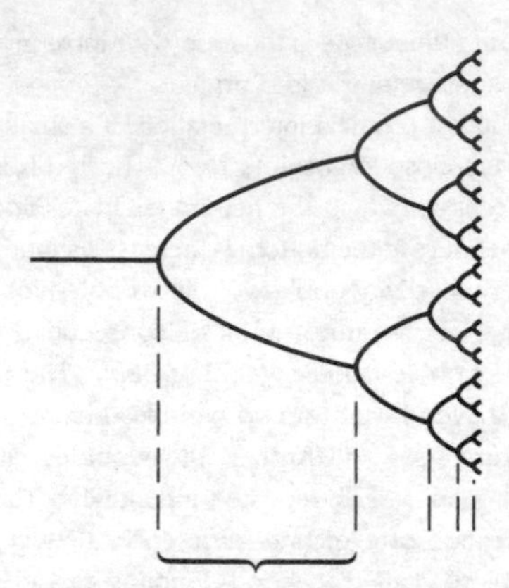

FIGURE 39

The Feigenbaum number lurks within every period-doubling cascade.

are not exactly equal to 1/4.669, but as the period gets larger, that value becomes a better and better approximation.

This observation alone wouldn't have been very important—a decreasing geometric series is an easy way to cram infinitely many events into a finite space, and any such series must have *some* common ratio. But when Feigenbaum tried a different mathematical equation with a period-doubling cascade, he got the *same* ratio, 1/4.669. Indeed, the two ratios agreed to twenty decimal places. That was a great surprise and a great mystery; it obviously wasn't coincidence. More experiments showed that virtually any mathematical equation with a period-doubling cascade produced the same universal ratio. There was a new, special number in mathematics, an emergent phenomenon rising from the chaotic ocean like Botticelli's Venus. A new, totally unexpected simplicity, emerging from some of the most complex behaviors known to mathematicians. Its value was 4.669 and a bit.

Another icon of chaos also provides a lot of insight into the relation between simplicity and complexity. This is the infamous Mandelbrot set, which we described in chapters 1 and 6. Its complexity is not just baroque elaboration, but "natural." It comes from a simple dynamical rule. The multicolored Mandelbrot sets that paper the walls of the planet (literally, on posters) are really just ordinary black-and-white ones like Figure 4 of chapter 1 that have been decorated using "painting by numbers." The idea is to count how many dynamical steps it takes for the numbers to reach some threshold value, say 100, and then to color-code pixels (the tiny dots on the computer screen that combine to make pictures) according to that number. The amazing thing about the Mandelbrot set is that such a simple process produces such an incredibly complicated object. In fact, it is so complicated that it is uncomputable, even though everybody uses computers to draw it. The drawings are mere approximations; the exact object is, in a very important sense, unknowable.

Despite this, there are all sorts of patterns in the Mandelbrot set. Here's one. The set can be described as a roughly heart-shaped "body," on which sits a smaller "head," on which sits a yet smaller "head," and so on. There's a lot of other stuff too: a long spike on top, blobs around the edges, blobs on the blobs, thin tendrils winding every which way, miniature Mandelbrot sets complete in every detail. But let's concentrate on the sequence of ever-decreasing heads. How fast do they decrease? It turns out that each head is approximately $1/4.669$ times the size of the previous one.

The mathematics of liquid helium has no obvious or direct connection with that of the Mandelbrot set, yet the same universal constant turns up. We'll explain the reasons for the Feigenbaum number's universality in chapter 12, but you don't have to understand the mathematical details to see the point. The Mandelbrot set is a new simplicity, albeit an apparently rather complicated one. It contains another new simplicity, the Feigenbaum number. Both simplicities are "natural laws" that hold throughout the realm of chaos.

EMERGENCE

The sudden discovery of new simplicities amid some highly complex, apparently unstructured muddle is a fairly rare event—a lot rarer than the discovery of some new detail of the complexity. At least 999 out of a thousand scientific papers are about complex details, but the one that we treasure and for which we award a Nobel Prize is the one that reveals a new simplicity. It is as if simplicities are all around us, but scattered rather thinly. Some scientists are rather good at laying hands on them; they must have the right kind of mind, seeing the world with unusual clarity. Albert Einstein specialized in big simplicities, and so did Paul Dirac, Gregor Mendel, and Dmitri Mendeleev.

Why do such simplicities exist, and how do we discover them?

In chapter 6 we talked of emergent phenomena, as philosophers call regularities of behavior that somehow seem to transcend their own ingredients. For instance, chemicals possess characteristic colors. It's not that sulfur atoms possess a rudimentary yellowness or carbon atoms a rudimentary blackness. Indeed, some forms of sulfur are orange, red, or purple, as Jupiter's moon Io dramatically demonstrates: It looks like a pizza, but most of its surface is just various forms of sulfur. And carbon atoms can form transparent diamonds as well as black soot. The colors are not present, not even in cryptic or rudimentary form, in the atoms from which the chemical is made. The color of bulk matter depends upon how its atoms interact collectively with light. The collective structure of bulk matter reflects light at certain preferred wavelengths; those determine the color. Color is an emergent phenomenon; it only makes sense for bulk matter.

Life is an emergent phenomenon, too—emerging from chemistry by way of DNA. The Feigenbaum number is an emergent phenomenon, emerging from chaos by a route we have yet to describe. Emergence is the source of new simplicities, but since we understand the process of emergence rather poorly, that's not a terribly helpful observation. What it does is help make respectable the idea that a collection of interacting components can "spontaneously" develop collective properties that seem not to be implicit in any way in the individual pieces.

Emergent simplicities "collapse chaos"; they bring order to a system that appears to be wallowing hopelessly in a sea of random fluctuation. Sometimes we can get glimpses of the mechanism of emergence. For example, people used to think that a gas was an infinitely divisible fluid. Before we discovered atomic theory, the underlying simplicities down the mental funnel from gases were thought to be the gas laws—"Pressure is proportional to temperature," and so on. Nowadays when we look down the funnel from gas dynamics we see a more complicated kind of mathematical pattern, known as statistical mechanics. A gas is not a uniform continuum characterized by simple quantities like pressure, temperature, and volume. It is a seething confusion of atoms, and the laws they obey are not the gas laws, but the laws of classical (or even quantum) mechanics. However, the gas laws emerge from the mechanical rules when we focus not on individual atoms but on their statistical features—in particular, their average values. The pressure of a gas is the average number of molecules inside a given region, and temperature is the average kinetic energy of molecules. These statistical properties of averages are large-scale features that emerge from the microdynamics, and they are features that other large-scale systems can use. For instance, your continued existence depends upon the atmosphere maintaining its normal range of pressures and its normal oxygen content. Because you are very big, on atomic scales, and your movements are very slow, your reaction to gas molecules is insensitive to tiny details. You "smooth out" the fluctuations by not noticing them, so you can build reliable structures upon the statistical features. Your lungs trawl the atmosphere for shoals of oxygen molecules, and it doesn't matter how they get into the net as long as the total catch is its usual size. A very tiny, fast-moving creature would have to hunt down individual gas molecules like a hunter-gatherer on the savannahs, and would be terribly vulnerable to random shortages.

Is *all* emergence just averaging, gas laws, statistics?

Statistical regularities are certainly one important and widespread mechanism for emergence. Curiously, the existence of statistical regularities probably traces back to the deterministic nature of chaos. Determinism implies underlying laws, and the regularities of statistics are traces of those laws on the macroscopic level. It may sound a very strange thing to say, but *really* random systems would not possess statistical regularities. No averages, no standard deviations, no correlations. Years ago Lewis Fry Richardson wrote a paper entitled "Does the Wind Possess a Velocity?" He obtained the answer no. Turbulent flow involves ever tinier eddies, so that the instantaneous speed and direction of wind is meaningless on very fine scales. If something doesn't exist, you can't average it.

However, many emergent features do not come from statistics. There is nothing statistical about π, the Feigenbaum number, the Mandelbrot set—or chlorophyll, DNA, or homeotic genes, for that matter. Any large-scale simplicity can be exploited by a system that is suitably attuned to it. This is especially clear in the workings of the stock market. If any trader starts to act in a manner that has any kind of predictable structure—statistical or not—then other traders can notice it and use it to their own advantage. In the stock market, this leads to disruption and instability, but in physics or biology it leads to recognizable and exploitable features that can be used as building blocks for higher-level functions and structures. For example, flowers can exploit accidental differences of the emergent feature "color" of chemicals by erecting upon them a sophisticated advertising campaign to attract bees. The success of the campaign depends on the bees' ability to see the colors, though—so as far as a flower's public-relations department is concerned, "color" just means "what bees can see."

Statistics is just one way for a system to collapse the chaos of its fine structure and develop a reliable large-scale feature. Other kinds of feature can crystallize out from underlying chaos—numbers, shapes, patterns of repetitive behavior. Many of those features have their own intricate internal structure (for example, the Mandelbrot set) which is quite different from the underlying rules that generated the feature in the first place. The intricacy of the Mandelbrot set bears no obvious relation to the simplicity of the process that produces it. The rules for making a Mandelbrot set are dynamic; the internal intricacies are geometric. That's why we described it a few pages back as a complicated simplicity; there's no contradiction.

THE INFORMATION IN MATHEMATICS

Mathematics wallows in emergent phenomena. It also came to terms, long ago, with something that often puzzles nonmathematicians. By definition, all mathematical statements are tautologies. Their conclusions are logical consequences of their hypotheses. The hypotheses already "contain" the information in the conclusions. The conclusions add nothing to what was implicitly known already. Mathematics tells you nothing new.

Except, of course, that it makes things explicit rather than implicit.

Before Feigenbaum discovered his number, everything we now know about it was already true. The number couldn't have turned out any other way; it was implicit in the definition of a dynamical system. It couldn't have been 7.332 instead of 4.669. But we couldn't *use* the number for anything until Feigenbaum noticed it. Indeed, if mathematicians try to use something that's true but unproved, their colleagues jump on them, screaming "How do you know that's true?" and won't let them publish their paper.

It's not enough for something to be true; you have to know it's true, and be able to explain why. Otherwise, you don't know whether it's safe to use it. Whether or not a particular berry is safe to eat is implicit in its nature; depending on its chemical constituents, the berry either is, or is not, poisonous to humans. But a parent, faced with a berry for which the implicit has not been made explicit, will sensibly discourage any childish alimentary experiments—and that decision is a sensible one, even if it later transpires that the berry was indeed safe all along.

We could show you scientific articles in which, for example, an engineer argues that a particular piece of new mathematics is useless *because* it is tautologous, saying that "mathematics never conveys new information." Unfortunately, the same remark holds true for all the treasured mathematical techniques of engineering—differential equations, tensor analysis, complex variable theory, or whatever.

All higher-level simplicities are somehow implicit in the lower-level structure that gives rise to them. However, that doesn't mean that the higher simplicities aren't useful, that they're "nothing new." Physicists generally believe that the laws of quantum mechanics imply that salt crystals are cubical, but nobody knows how to make this explicit. They don't know how to make the existence of *any* crystalline structure explicit. It was a huge triumph recently when somebody made the existence of gases an explicit consequence of quantum mechanics. Far from being unimportant or tautologous, higher-level simplicities are the bread and butter of science—the simple, recognizable features of otherwise complicated theories that we use to understand the natural world.

The representation of a higher-level simplicity as an explicit consequence of lower-level complexities tells us intellectually interesting things about the connection between the levels. But often it adds no useful gadgets to a working scientist's tool kit, because a high-level simplicity is much easier to think about than some chain of complexities that causes it. When you use a hammer you don't want to worry about its molecular structure. Mathematicians know this well. "A theorem," said Christopher Zeeman, "is an intellectual resting point"—something you can stand on to proceed further. Something you *know*, can encapsulate, grasp as a whole. Key scientific concepts have this same quality.

■ IMMUNITY TO CHAOS

Philosophers of science distinguish between two aspects of a theory. One is its laws, which are expected to hold in considerable generality; the other is the initial conditions to which those laws should be applied, which are pretty much arbitrary. For example, Newton's theory of gravity assumes the inverse-square law of gravitational attraction, but if you want to calculate where the moon will be in a million years' time you need more—namely, where it and the rest of the solar system are now. Initial conditions are so arbitrary that it's not reasonable to expect a theory to provide laws for them too.

However, it can be important to ask how sensitively the system depends upon its initial conditions. One of the ways in which post-chaotic mathematical thinking has changed deeply is that it places considerably more emphasis on this question. Its answer is the butterfly effect: The subsequent behavior may eventually be totally different. The same law applied with indistinguishably different initial conditions can predict totally different events. When chaos is present—as it very often is—"negligible" differences are no longer negligible.

This affects the way we test theories. It is no longer always appropriate to start with today's data as initial conditions and then predict tomorrow's. Instead, we must test the process encapsulated in the "laws" of the theory. We have to get our hands on at least some aspects of the mechanism itself, not just its input-output characteristics. Chaotic theories are extremely sensitive to small errors in initial data, which can totally change their predictions. This poses problems for the reductionist methodology. The crystalline structure of salt, for instance, cannot depend in any important way upon how the atoms vibrate as the crystal is forming, or salt crystals would not form at all. In a reductionist derivation from lower-level theories, be they simple or complex, this insensitivity has to "come out in the wash." Since the motion of atoms is chaotic, their precise behavior *is* sensitive to initial conditions. So the process of crystallization must "ignore" the details of the initial conditions.

Since we are talking of atoms, we ought to think about quantum mechanics too. Quantum systems don't exhibit chaos in the conventional sense, but any classical (that is, nonquantum) theory of large numbers of particles certainly does. Quantum systems aren't chaotic because the infinitely fine structures that are important for chaos are forbidden in quantum mechanics, thanks to the uncertainty principle. In compensation, quantum mechanics as currently interpreted has an even stronger kind of unpredictability: chance. If this interpretation is correct, then quantum mechanics is a barrier that reductionism cannot completely cross. Quantum mechanics has its own form of small-scale chaos—genuinely random fluctuations, rather than the deterministic but effectively random fluctuations of conventional chaos.

The big simplicities of crystallography have to be immune to this underlying chaos, be it classical or quantum. Crystals do exist. Our contention is that this kind of stability must lead to immunity not just to the underlying chaos but to many of the other mechanisms that characterize the lower-level theory. Crystal lattices are not just immune to small-scale chaos; they are immune to most of quantum mechanics. If atoms behaved like little balls of Newtonian matter, you'd still get crystal lattices. If atoms were octahedrons (as Neeplphut would no doubt prefer) rather than spheres or probability clouds, you'd *still* get crystal lattices.

We can even make a good guess why crystal structure is immune to most fine details of atomic theory. The main thing we need to know is that physical systems tend to minimize their energy. If a crystal is made from a lot of similar particles, under appropriate conditions of temperature, pressure, and so on, then the way to minimize energy is to pack them together as densely as possible. Lattice packings are widely conjectured to be the densest. This hasn't yet been proved—a proof claimed in 1991 remains controversial—but the evidence in its favor is compelling. This argument in favor of an atomic lattice is independent of the shape of the atoms or their detailed properties; energy minimization is enough. So this particular feature of any underlying reductionist description is all that we really need to know.

There is thus a very strong, compelling aspect to large-scale simplicities. It is wrong to think of them as just complicated consequences of the underlying laws that happen to possess some striking, simple description. Crystal lattices are not just phenomena that emerge from quantum mechanics. They have a universal aspect; they are phenomena that will emerge from any theory sufficiently close to quantum mechanics that involves identical roughly spherical atoms and energy minimization. This kind of universality is common to many, perhaps all, emergent phenomena; it is why they are able to emerge at all.

This is an absolutely crucial point for the rest of this book. *Universality implies replaceability*. If quantum mechanics suddenly turned into subatomic octacycle theory, our explanations of crystal structure would hardly change. The same goes for our explanations of life. It is here that the reductionist story loses much of its force. Physicists take pride in the hoped-for Theory of Everything, on the ground that it will be truly fundamental, that in principle it will explain *everything* that we observe on a human scale. But in order to exist at all, the phenomena that we observe on a human scale must be independent of the detailed organization at the lower level.

We hasten to add that by "independent of the detailed organization" we do *not* mean that the phenomena are not consequences of the detailed substructure. We are talking about explanations and understanding, not about the specific physical mechanisms employed by nature. We're not trying to tell you that atoms do not exist. We're saying that if somebody gave an explanation of why cats drink milk that started from atoms, then the same general kind of explanation would work for many different assumptions about the detailed behavior of atoms. By implication, if somebody did a huge, impenetrable quantum-mechanical calculation whose end result was "The computer says that the cat has its nose in a saucer," *we* wouldn't consider it a meaningful explanation.

Phenomena that are independent of detailed substructure are *universal*—"nature's theorems." They arise for general, widely applicable reasons, not because of special details. So, if you changed the Theory of Everything, it would make no difference at all to almost everything that it allegedly explains. It wouldn't change the explanations, either, anywhere above the subatomic level, because those explanations are immune to subatomic physics. Look in a biology text and see how many times the explanations involve quarks.

■ IS IT ALL IN THE MIND?

Ptolemy fooled many generations into accepting a theory that relied on an invented pattern, an allegedly deep truth that was neither true nor deep. To what extent is current science an equally misleading delusion of the human mind? The main feature of Ptolemy's epicyclic theory that reveals its chewing-gum-and-string essence is its wealth of adjustable parameters—all of those epicycles, as many as you want. Too much flexibility is always suspect. Kepler's improved theory, that orbits are ellipses, isn't like that. Even less so is Newton's inverse-square law. Kepler's theory has only two adjustable parameters, the two axes of the ellipse; Newton's has none. Those theories are rigid and elegant.

Yes and no. They are certainly elegant, in the sense that they appeal to our aesthetic ideas of simplicity. And they appear rigid, don't seem to have many adjustable parameters. However, in fact they do—not in themselves, but in the cloud of competing theories that surrounds them. Kepler, for example, tried a whole series of curves as possible orbits, and only reluctantly accepted the ellipse (which to him seemed too neat to be probable). If the inverse-square law hadn't worked, Newton would have tried the inverse cube, or whatever else came to mind. Their theories passed observational tests by conforming to planetary motion, but they had been selected to do exactly that. Kepler's theory passes *only* the test that was used to construct it to begin with: It has no predictive value (except perhaps that the elliptical nature of the orbit was established using Mars, so that transferring it to the other planets amounts to a prediction). Newton's law does much better: It explains not just the observations against which it was selected, but many others—the precession (motion of the axis) of the moon, the way that Jupiter and Saturn disturb each other's orbits, and so on. We are rightly impressed by Newton's inverse-square law: it explains an awful lot with such simple ingredients.

Nonetheless, human beings are irredeemable wholists, and their judgment is swayed considerably by tidy numerology. That Newton's law comes out with an exact power of 2 is impressive, even though relativistic corrections actually cause the number to vary away from that value, which would be precise only if space contained no matter at all. General relativity explains the integer 2 as being characteristic of "flat" space-time. Here's another tale of (wrongheaded) wholism. In quantum physics there is an important number known as the fine structure constant, whose value is very close indeed to $1/137$. The formula for the fine structure constant is a fraction, and it makes just as much sense upside down, so there is a fundamental constant very close to 137. That's a whole number. Many physicists tried for a long time to derive the number 137 from "deeper" principles, but it now looks as if the physical constant isn't an integer at all. Physics is littered with numbers; *some* of them have to be close to integers.

The moral is clear. When we find a simple numerical or geometric pattern in nature, then as a species we have a habit of immediately thinking, "That's it!" We often seek no further, and more than once this habit has led us astray. Our minds are imperfect instruments, with odd preferences for simple geometry; presumably this is because of the architecture of our brain structure, especially its visual system. Are we just selecting patterns, such as epicycles, that appeal to us and imagining them to be deep truths about nature? Does nature do things quite differently? Is it "all in the mind"? Our patterns are, after all, ideals. No real crystal has a perfectly regular atomic lattice. There are impurities (the wrong atom in the right place) and dislocations (the right atom in the wrong place). Nature seems happy enough to make crystals that way. Is it sensible to give such prominence in our thoughts to ideals that are never actually attained?

SHERLOCK HOLMES STORIES

Are the big simplicities of the universe just illusions, born out of human prejudice? Does nature possess patterns, or are we inventing them? Our answers will occupy the rest of this book. Along the way we will encounter all kinds of attempts to explain the big simplicities of nature. Our immediate concern is to dispose of one argument of this sort, which runs along these lines: "This theory is so finely tuned that even the tiniest change would wreak havoc; therefore it must be true." An example of this kind of argument is the so-called anthropic principle in cosmology. Roughly speaking, the story is this. If the values of certain quantum parameters were even slightly different from their observed values, then the long sequence of events that is needed to produce an intelligent being would have broken down. We've already mentioned that if Planck's constant were even slightly different, then the nuclear resonance that lets stars create carbon wouldn't exist, and without carbon, where would human life be? And we've explained how that leads to the anthropic principle: Given all this, is it sensible for us carbon-based humans to ask why Planck's constant takes that particular value?

Let's be clear what we're discussing here. We're not discussing whether this universe has a different value of Planck's constant from the usual one. We know full well that it doesn't. What we are doing is a thought experiment in universe-space, the space of all conceivable universes: Can we conceive of a universe with a different value of Planck's constant, in which an intelligent carbon-based creature like us could ask the same question? People who believe in the anthropic principle argue that the answer is no, add that all life must be based on carbon because of carbon's unique ability to provide a skeleton for complex molecules, and conclude that our universe is the only one that could produce intelligent life. Imagine yourself to be an intelligent carbon-based life-form (come on, try, it's not that hard really) and suppose that you don't know whether your universe is the one that we currently live in. Now ask yourself the question, "What is the value of Planck's constant?" You should be able to intuit the answer 6.6262×10^{-34}, because that value is a necessary condition for your own existence.

It sounds like a powerful argument, but it falls down in one crucial respect. It assumes that while exploring universe-space by changing Planck's constant, you don't change anything else. Sherlock Holmes stories have the same fragility. If you change just one event, the story falls apart. In "The Red-headed League," if Jabez Wilson had not possessed red hair, then the trick of luring him off the premises of his pawnbroker's shop would have failed. All of which implies, of course, that there is only one possible Sherlock Holmes Story . . . and now you see the fallacy, for there are hundreds of Sherlock Holmes stories. Each one is so carefully crafted that changing a single feature will destroy the logic. But you don't do that. You change several features in a consistent way. Jabez Wilson could have been bald, in a story called "The Bald-headed League."

It is certainly interesting to know that if we change Planck's constant while keeping everything else the same then humanity would become impossible. Or that if we change the mass of the electron, while keeping everything else the same, ditto. But these two facts together do not imply that, in order for us to exist, Planck's constant *and* the mass of the electron must both have their observed values. Perhaps if both were changed at the same time. . . . The anthropic principle's argument is not exploring universe-space, as it tries to imply. The argument is exploring only a few very specific directions within an enormously multidimensional universe-space—the directions along the Planck's-constant axis or the electron-mass axis. This kind of argument for the anthropic principle is like claiming to have explored the whole of the universe when you've only gone "down the road to the chemist," as it says in *The Hitchhiker's Guide to the Galaxy.*

CONTENT AND CONTEXT

Reductionism is fine for technology, because technology (sensibly) chooses to work in the simplified worlds in which reductionism provides the desired answers. But if our intention is to understand the universe rather than to manipulate tiny pieces of it, then we must recognize that most of reductionism is heading relentlessly in the wrong direction. It's making us look down through the funnels when we should be looking outward to everything around us. And anthropic principles are no substitute, being full of logical holes. We can't get away with thinking that simplicity is tautologous—as anthropic principles wrongly do—or that it requires no explanation at all, as reductionism tacitly assumes by not asking that kind of question.

We might get away with such thinking at the level of atoms—after all, you've got to start somewhere. If *everything*'s complicated, we're beaten before we start. So for the moment let's buy the reductionist assumption that there really are simple laws at the heart of the natural world, and track through the underlying story of the first half of this book: the idea that reductionism explains complexity. The complexity of chemistry arises because simple atoms can interact in an enormous variety of ways. Fine. Then the complexity of DNA arises because simple molecules can interact in an enormous variety of ways. No, wait a minute, it's not like that at all. First, where does the bit about the simplicity of molecules come from? Not two sentences ago we were arguing that they are complicated consequences of simple atoms. Anyway, there aren't millions and millions of molecules, all doing things as complicated as DNA; there's just one of it. (RNA does the same job in some viruses, so if you prefer, there's two of it. That's still not millions.)

If you take sugar and fat and add energy—say, by heating them in a pan—you get toffee. Toffee is made from staggeringly complicated carbon compounds. There are as many atoms in a typical molecule in a lump of toffee as there are in human DNA. To describe just one of the molecules in toffee, in full detail, is a truly mind-boggling task. Also a pointless one. The details of toffee molecules don't matter, only the general organization. Most of the details might as well be random, and in a sense are. No two molecules in toffee are exactly alike. But you don't see biologists asking for huge research grants to sequence the toffeenome. What comes out of the toffee molecule is a pleasant treat for the sweet-toothed and a decent living for dentists. What comes out of DNA is of a totally different order.

There are several kinds of complexity. One is the Lego-block explosion of the possible. But that is emphatically not what is interesting about DNA. Yes, the Lego-block complexity has to be available to build it, but that kind of complication is downhill to carbon. The problem is to keep carbon atoms from linking up all over the place, not to make them link up. No, the interesting complexity of DNA is not that it contains millions of atoms. It's a more subtle kind of complexity: complexity of organization. Nature has shaken up a sack of Lego blocks, and out comes a perfectly formed fire engine. Yes, it's one of the possible models you could make—but why *that* one?

If you analyze the internal structure of DNA, you can find out in infinite detail how it is organized. But no matter how intensely you analyze the internal structure, you will never find out why it is organized that way. The answer is not inside the DNA. It is not something that can be uncovered by a reductionist approach. The answer lies outside the DNA molecule. There is a lot of DNA around because DNA is the sort of stuff that, once there's some of it, soon there's an awful lot of it. It replicates. Richard Dawkins devised the image of "selfish DNA" as well as that of "selfish genes." (Indeed, by the time he came up with it, "gene" and "DNA" had effectively become synonymous to him.) If you see a lot of copies of some particular segment of DNA all over the place, it's because that segment happens to be unusually effective at getting itself copied, in the overall context in which DNA replication takes place—namely, living creatures.

We're not saying that DNA is organized as it is *in order to* produce life. There is no sense of purpose in DNA. It has no way of knowing that creating life would be a good way to get replicated. Its ability to replicate is indeed just very complicated chemistry, so complicated that we have no idea how a lot of it is done. A hugely complicated list of chemical reactions calculated by supercomputers doesn't offer any kind of explanation of the prevalence of DNA. However, there is a very compelling explanation indeed, in terms of a particular simplicity of the otherwise incredibly complicated workings-out of DNA chemistry among living organisms. It is so simple that it is tautologous. *Replicators replicate.* Things good at getting replicated will appear in great quantity. And that simplicity *outside* DNA is a far more powerful explanation than endless computer calculations of its internal structure could ever be.

However, there is a price to pay. The external explanation in terms of higher-level simplicities may tell us why there is a lot of DNA around, but it doesn't give us any idea of how it replicates. Externals are for whys, internals for hows. As we have said, we want to go beyond reductionism, not replace it.

Here's another "why" question. Why do herbivores on the plains of the savannah have eyes on the sides of their heads, whereas carnivores have them at the front? (They do: Think about zebras, giraffes, gazelles, lions, and hyenas.)

A genuinely reductionist explanation would look deep inside the animals' cells, sequence their DNA, and describe the chemical changes that lead to eyes at the side of their heads. It would tell us how the eyes end up where they do, but it wouldn't tell us why. Many different DNA sequences produce similarly placed eyes in herbivores, and many different DNA sequences produce similarly placed eyes in carnivores (though not in the same place as herbivore eyes, of course). The biochemistry is rather complicated, and you wouldn't expect to find much in the way of common patterns.

However, there is a very compelling "why" answer; one so general that it would, for instance, also lead us to expect all Zarathustran mobile insectivorous plants to have sensors at the front, while their deep-rooted and immobile insects would have sensors at the side. (The plants come to the insects to be pollinated, you understand. Then they eat them.) We mentioned it in chapter 4. Animals live in ecological systems, and forces from outside cause them to evolve. Carnivores need eyes at the front in order to focus their attention on their prey as it tries to elude their clutches. Herbivores need to keep an all-around watch for predators, and this is most easily achieved with eyes on the sides.

If we knew what DNA sequences meant for development, then the first approach would tell us a lot about how the eyes get where they do. It is an explanation in tems of content, of internal structure at lower levels. The second explanation tells us why the eyes are positioned where they are. Not a "why" in the sense of "purpose"; we're not trying to say that the eyes "know" where they are supposed to be, that God, Octimism, or a disembodied evolutionary compulsion placed them there. All we are saying is that the eyes are part of an external system with its own logic, its own dynamic, just as a reductionist explanation says that they are built up from internal components that have their own logic or dynamic. The two types of explanation parallel each other; they are equally "scientific." They just look at the problem from two different directions.

TWO TYPES OF TRANSPARENCY

The same two explanations can be distinguished in a completely different manner. The explanation of eye placement in terms of the evolutionary effects of predator-prey interaction is simple, direct, almost obvious. The explanation in terms of DNA and biological development, in contrast, is incredibly complicated and really rather arbitrary. It doesn't explain why the DNA sequence is what it is. But then, the evolutionary explanation doesn't explain how the DNA sequence affects the development of eyes.

The distinction this time is a little deeper, and it is not that between simple explanations and complicated ones, though it tends to work that way. We want to use the word "transparent," and we have to explain that the word is used in two exactly opposite ways. When people say that the workings of some piece of machinery are transparent, they mean that they know what's going on inside. The image is that of a watch inside a transparent case. So when they observe one cat keeping a close eye on another cat's food bowl, they accuse the poor beast of transparency; they can "see" how its mental cogs are ticking.

However, there's a second, exactly opposite meaning. Computer scientists say that a piece of software is transparent if you don't need to know how it works in order to use it. This time the image is that of a piece of machinery made entirely out of glass; you see right through it to whatever lies beyond. It is this second sense of "transparency" that we want. So from now on, that's what "transparent" will mean: The whole process is see-through, not just its casing.

Incidentally, if it makes you feel more comfortable, it's not unusual for a word to mean two opposite things. The word "balance" has three meanings, all opposite to each other (so to speak). One meaning has to do with equilibrium; we say somebody has a well-balanced personality, for instance. Another invokes the idea "delicately poised"; we say a situation is balanced on a knife edge. In the first image, "balanced" means "stable"; in the second it means "unstable." The third? "Balance of payments" means the difference between imports and exports—the amount that must be added to produce equilibrium.

Back to explanations. The evolutionary explanation is transparent, in our sense that it functions somewhat independently of the precise internal workings of biochemistry. The biochemical/developmental one is opaque; it depends crucially on the finest chemical details. Transparent explanations are often contextual, and they express a worldview that is more common in the arts than in science. In science we dissect a tree into its components, section its cells, and sequence its DNA. In the arts, we look at a tree and see *without* analysis where it stands in God's creation, what its place is in the universe. The whole idea that the mind can perceive the complex totality of the world comes from art. For instance, all major movements in painting after about 1800 (if not earlier) have philosophical aspects; the viewer is changed and sees everything "in a new light." If we could somehow combine the significant aspects of the scientific and artistic viewpoints, we could see the entire universe in a new light.

The reductionist explanation of herbivores' eyes is internal, opaque, and complicated. The contextual one is external, transparent, and simple. However, these features need not go together in this way. An external explanation could be opaque, and a transparent one could be complicated—though usually it won't be. All eight combinations of internal/external, opaque/transparent, and simple/complicated are possible. (The Zarathustrans would love this eightfold classification.)

Mechanisms, as well as explanations, can be transparent; and transparent mechanisms can collapse chaos and create emergent simplicities. This is an important clue about the emergence of simple patterns, and it leads to two general ideas. First, complexity at a lower level can generate new simplicities at a higher one. Second, instead of looking inside things to find out how their complexity arises, you can look outside them to find out why their simplicity arises, and seek explanations in terms of context as well as content. What may seem to be inexplicable emergence may well be selection from outside.

The rest of this book explores those ideas, retracing the ground of the first half but emphasizing the contextual view. This is the sense in which we turn the book's first half inside out. We will be looking for hidden simplicities of a nonreductionist kind and trying to elucidate the contextual mechanisms that produce them. In the final chapter, we'll try to sum up what we think we've learned, and where we think it might lead.

8 THE NATURE OF LAWS

Ivan and Pyotr were watching the installation of the Omsk telephone exchange. "These modern gadgets," sighed Pyotr. "I can never understand how they work."

"But it is easy," replied Ivan. "It is like a very long dachshund. You twist his tail, and he barks in the middle of the city."

"Ah," said Pyotr. "That explains telephones. But what about radio?"

"Radio is exactly the same," said Ivan, "but without the dachshund."

Our exploration of the influence of context upon the behavior of the universe begins with physics and chemistry. At first it might seem strange that such things as contextual physics and chemistry could have any meaning; but that's because reductionism attains its fullest development in those areas, where we are so accustomed to the reductionist view that we tend to ask only questions that have reductionist answers. In fact, most of the "big picture" questions in physics and chemistry—the arrow of time, the interpretation of quantum mechanics, the nature of the physical universe—depend at least as much upon context as upon content. So do many questions about common patterns that arise within systems whose insides are extremely different (for example, both galaxies and bathwater follow spirals).

Scientists get themselves enormously confused about big questions. The reason is that they try to tackle them in the same way that they tackle the small questions. When trying to understand the gas laws, they come across principles such as the second law of thermodynamics, whose usual interpretation is that the amount of disorder in the universe steadily increases. This principle works brilliantly when applied to the behavior of gases, but it sits uneasily with the rich complexities of life on our planet. In a similar manner, the linearity of quantum mechanics—the fact that any two quantum states can be superposed to yield another—works brilliantly for electrons but leads to the paradoxical plight of Schrödinger's cat, which seems to make no sense whatsoever.

Both cases have a lot in common. Each rests on a single, rather simple mathematical law—irreversibility in one, linearity in the other. Each law is exceedingly general in scope: *Every* mechanical system is time-reversible; *every* set of quantum states can be superposed. Each runs into conflict with our senses when it is applied to the complex systems that make up the world on the human scale. To a physicist, an atom is simple, but a cat is an enormously complicated system of around 10^{26} atoms. To a child pouring out a saucer of milk, a cat is simple but an atom is incomprehensibly complicated.

We shall argue that the apparent paradoxes inherent in these problems rest upon an inadequate understanding of the role of context. The simple laws of physics answer particular, simple questions in their own appropriate contexts: gases; electrons. They seem to answer similar questions in much wider contexts, such as cats. So why do they also seem to give *wrong* answers, answers in conflict with everyday experience? Because in the context of everyday experience they are addressing quite different questions from those they appear to be addressing.

And that puts the whole question "What is a law of nature?" in a very different light. Laws of nature are not eternal, abstract truths. They are patterns that prevail in some chosen context: The laws that you find depend upon the questions that you ask. And not just in physics and chemistry.

THE SECOND LAW

The second law of thermodynamics—"Disorder always increases"—seems to rule out the increasing complexity and organization exhibited by living systems. It is often used as the basis of an attack on Darwinism for just this reason. Thermodynamics is one of the most pervasive, but also one of the most subtle, branches of science; and it seems to be a horrible trap for the unwary. Thermodynamics works beautifully in its original context, heat engines. In most other areas it is usually no more than a metaphor, one that has often been stretched far beyond its breaking point. But it is an amazingly seductive metaphor, because it seems to rule out a lot of things that people find uncomfortable. Many people, scientists among them, have turned their backs on radical new discoveries because they thought thermodynamics ruled them out.

A good example is the production of diamond by artificial means. In the late eighteenth century it was discovered that diamond is a form of pure carbon, made from the same ingredients as soot or graphite. The difference is that the atoms in a crystal of diamond are arranged in a very different way from those in a crystal of graphite. It is this arrangement that gives diamond its desirable properties of hardness and sparkle. Ever since the discovery of their chemical identity, people have tried to convert cheap graphite into valuable diamond by artificial means. At normal temperatures and pressures, graphite is the thermodynamically most stable form of carbon. That is, it retains its molecular structure despite the random vibrations of molecules that we call "heat." The reason is that the chemical bonds in a molecule are a bit like springs. It takes a certain amount of energy to compress or stretch a spring away from its natural shape and size, so every chemical molecule has a definite "energy level" that prescribes the amount of energy that is stored in its bonds. Graphite is at the bottom of an "energy well"—if its molecular structure is distorted in any manner, then the energy becomes higher. So it takes extra energy to change graphite's molecular structure—quite a lot of extra energy, in fact.

Diamond, in contrast, becomes thermodynamically stable only at pressures of around 60,000 atmospheres and temperatures of 2,700°F; and it is generally assumed that natural diamonds have formed under such extreme conditions. The obvious route from graphite to diamond is to create similar conditions in the laboratory, and in 1953 a Swedish company managed to follow this route. A few mavericks thought about low-pressure methods, but the majority of workers in the field considered such ideas to be akin to dreams of the perpetual-motion machine. Their reasoning was simple: Since graphite is thermodynamically stable at low pressures, and diamond is not, then diamond cannot form at low pressures. If it did form, it would immediately convert to graphite.

In fact, this reasoning is nonsense. Diamond is "metastable," which means that although it is not the configuration of carbon atoms that has the lowest possible energy, it has a lower energy than any "nearby" arrangement of atoms. The difference between the energy levels of diamond and graphite is not so great, roughly that between solid and melted butter. Despite this, the energy needed to change graphite into diamond is huge; you have to boil it, breaking every bond in its crystal lattice. You need almost as much energy to make the change in the opposite direction, and for the same reason: The problem is not the difference in energy levels, but the height of the barrier in between. Common soot is another metastable form of carbon, and its energy is actually *higher* than that of diamond; the energy hierarchy is graphite, diamond, soot. So the scientists who thought diamond would spontaneously transform into graphite at ordinary pressures should also have argued that soot would transform spontaneously into diamond.

They didn't, of course. They *knew* that the laws of thermodynamics ruled out the low-pressure formation of diamond, and they weren't interested in soot. As a curious coda to the story: Low-pressure synthesis of diamond is now a reliable piece of technology, becoming important in electronics, where diamond-based devices offer potential advantages to traditional silicon chips.

The diamond story is not so much one of important ideas being suppressed because of misguided reliance on misunderstood thermodynamics; it's more one of ideas never getting discovered at all because nobody was willing to risk a career on them, or because nobody could get the funds to carry out the work even if he was willing to take that risk. Later in this chapter we shall describe in detail a case—the existence of oscillatory chemical reactions—where an important discovery was simply ignored on the grounds that thermodynamics implied that it couldn't happen. We mention these two examples of scientific ostriches hiding their heads in the sand of thermodynamics in order to demonstrate that even very basic aspects of that branch of science have been widely misunderstood in the past, and without doubt still are in the present.

CHICKEN FEED

The underlying principle is clear enough: If the laws of thermodynamics seem to conflict with the evidence of your senses, believe your senses and take a long hard look at the thermodynamics. Got that? Good; let's think about the conflict between the second law and the phenomenon of life. You're alive, so it's the appeal to thermodynamics that must be shaky. Where? It's obvious that any resolution of the conflict must depend upon a careful analysis of what thermodynamicists mean by "order" and "disorder." It's much less obvious that the central issue is quite different: not what they mean by disorder increasing, but in what context they can demonstrate that it will.

The second law is usually stated in terms of a quantity called entropy, which measures the amount of disorder in a system; the law states that entropy must continually increase. This finds dramatic expression in the scenario of the "heat death of the universe," in which the entire universe becomes a lukewarm gas with no interesting structure whatsoever. Entropy increase is often explained using the image of a pack of cards being shuffled. Suppose, for example, that the pack is arranged so that all the red cards occur together on top and all the black ones below—an "ordered" state. Now shuffle them repeatedly. You'd expect them to end up mixed together randomly, in a "disordered" state, so shuffling increases the amount of disorder in the pack. Analogously, suppose that at some moment all the oxygen molecules in a room are concentrated at one end and all the nitrogen molecules are at the other. This is an ordered thermodynamic state. After a very short period, however, random collisions will mix all the molecules together, more or less uniformly throughout the room. This is the orthodox picture of the relentless increase of entropy, and it is the standard interpretation of the second law.

Time seems to flow in one particular direction; it has a well-defined "arrow." However, it seems logically and mathematically possible for time to flow backward instead—a possibility exploited by such novels as Martin Amis's *Time's Arrow* and the much earlier *Counter-Clock World* by Philip K. Dick. So why doesn't it? Thermodynamics offers a simple explanation for the arrow of time: it is the direction of entropy increase. Thermodynamic processes are irreversible: Oxygen and nitrogen will spontaneously mix, but not spontaneously unmix. There is a puzzle here, however, because any classical mechanical system, such as a room full of molecules, is time-reversible. In the mathematical equation for classical (that is, nonquantum) dynamics, if at some instant the velocities of all particles are simultaneously reversed, then the system will retrace its steps, back to front in time. The entire universe can "bounce" in time. For example, since boiling an egg is a dynamical process consistent with the laws of mechanics, so is unboiling an egg. Why, therefore, do we never see an egg unboiling?

The thermodynamicist's answer is that a boiled egg is more disordered than an unboiled one, entropy increases, and that's the way time flows. But in that case, how did the chicken ever create the ordered egg from the disordered chicken feed? A common explanation is that living systems somehow "borrow" order from their environment by making it even more disordered than it would otherwise have been. Then they use their extra "negative entropy"—order—to build an egg. There's a certain amount of truth to this, but it seems that chickenkind has been borrowing an awful lot of negative entropy over the millennia.

How does life create order, despite the second law? One answer is that the law applies only to closed systems in thermodynamic equilibrium—systems that are isolated from external influences and have settled down into a balanced state. (In the dynamical language of chapter 6, they are systems that have reached their attractors, instead of still being in a transient state, heading toward the attractor but not yet there.) There are two loopholes that allow life to be consistent with the second law. Either it can be viewed on its own as an open system, one that is subject to outside influences, driven by energy from the sun. Or it can be viewed as a transient state of a closed system—life plus its environment—that is ultimately heading toward equilibrium. But life looks too persistent to be a transient state, and mere openness of the life-plus-environment system does not explain the richness of life. There ought to be a better explanation.

We contend that the entire question is bogus, even for closed systems that are in thermodynamic equilibrium. The card-shuffle image gives the wrong impression of entropy; the entropy of the universe does *not* increase over time; and there is thus no special direction with which to identify the arrow of time. Life contradicts nothing, since there's nothing to contradict. On the other hand, the arrow of time then requires a nonentropic explanation.

STANLEY: Neeplphut, you seem to be having problems with this idea of entropy increase.

NEEPLPHUT: No, I think *you* are confused about what order and disorder are.

STANLEY: Let me demonstrate. Do you have a pack of cards?

NEEPLPHUT: Certainly. I regret/must inform you (delete whichever is inapplicable) that they are not entirely like your terrestrial packs. In particular, there are only eight cards. We call them things like the Squirm of Weirds, but it may be easier if I number them from one to eight with a felt-tip pen. [*Scribbles on cards.*]

STANLEY: Okay, Neeplphut my old buddy, let's have them. Thanks. I'll arrange them in an "ordered" state. [*Lays cards so that their faces read 1 2 3 4 5 6 7 8 in numerical order.*] You agree this is ordered? It has a definite pattern?

NEEPLPHUT: I wish to reserve judgment at this time.

STANLEY: You won't give an inch, will you? Oh, very well. After a shuffle or two, the order becomes . . . hmmm, 7 5 3 1 8 6 4 2. That looks pretty random to me. Order to disorder, yes?

NEEPLPHUT: I hate to contradict, but what *I* see is the initial state 4 8 3 7 2 6 1 5 becoming 1 2 3 4 5 6 7 8. Disorder becoming order.

STANLEY: What?

CAPTAIN ARTHUR: I think Neeplphut is looking at the other side of the cards. Didn't you see him writing numbers on *both* sides?

NEEPLPHUT: Remind me not to play poker with you, Captain. Yes. I wrote the numbers 4, 8, 3, 7, 2, 6, 1, 5 on the back of the cards that *you* see as 1, 2, 3, 4, 5, 6, 7, 8. So *my* final order is 1 2 3 4 5 6 7 8.

STANLEY: But that's cheating. If you'd chosen any other way to number the cards on the back, my particular shuffle wouldn't have led to 1 2 3 4 5 6 7 8 when looked at from your side.

NEEPLPHUT: True; but you have now learned that your intuitive notion of order is sensitive to coding of the data. Incidentally, your alleged random ordering 7 5 3 1 8 6 4 2 is not. It consists of the odd numbers in descending order, followed by the even numbers in descending order. But my point about coding order into apparent disorder stands, whatever your final arrangement may be.

RESURRECTION SHUFFLE

Let's take Neeplphut's analysis a stage further. We must make it clear at the outset that the central issue is a purely mathematical one: the relation between the reversibility of the laws of classical mechanics and the irreversibility of the second law of thermodynamics. We will use physical language to discuss this issue, but please don't start worrying about whether the laws of classical mechanics or thermodynamics actually apply to real systems. That's a separate issue. For the problem we are trying to resolve, an apparent contradiction *within mathematics*, we are entitled to assume that the laws do apply. It's easier to follow the mathematics by using physical imagery, but when we talk about a physical system, we will actually be referring to some mathematical model that represents that system. The same remark holds with regard to our later discussion of Schrödinger's cat: We take quantum and classical mechanics for granted, and consider the relation between them, using physical images for vividness.

In the dynamics of gas molecules, the transition from one state to the "next" state—the state a very short time into the future—is not arbitrary.

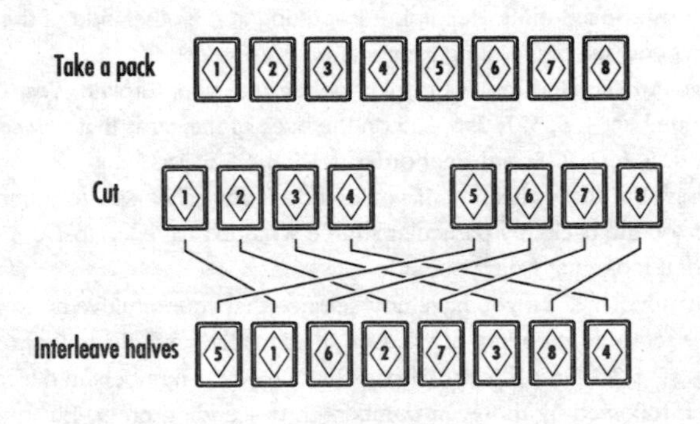

FIGURE 40

Effect of a riffle shuffle on a pack of cards

Despite the rhetoric, molecules don't move at random. They just follow very complicated paths that appear random to our feeble senses. The motion is determined by a single fixed rule, given by the laws of mechanics. The analog for cards is to use a fixed shuffling rule. In fact, the sequence 7 5 3 1 8 6 4 2 was obtained using a riffle shuffle in which the pack is split into two pieces and cards from each piece are arranged alternately (Fig. 40).

We are here using shuffles as analogs of molecular motion. The free motion of molecules in a gas involves no change of energy. Therefore our mathematical shuffles should also be thought of as involving no change of energy—the system of cards is a closed one. Shuffling real cards does take energy, of course, but these aren't real cards. Two successive riffles give Stanley's sequence 7 5 3 1 8 6 4 2. This still has a pattern, as Neeplphut remarked; but with only eight cards and two shuffles, that's not entirely surprising. Nonetheless, the degree of disorder does seem to increase with each shuffle.

Let's shuffle a third time:

SHUFFLE	RESULT
0	12345678
1	51627384
2	75318642
3	87654321

After three shuffles, the pack has been *reversed* in order. It seems difficult to maintain that this is "less ordered" than the state 1 2 3 4 5 6 7 8. What our brains perceive as the degree of disorder begins by increasing, but then starts to decrease again. Not only that: A further three shuffles will obviously reverse the pack again, and get us back to our starting point:

SHUFFLE	RESULT
3	87654321
4	48372615
5	24681357
6	12345678

Disorder can scarcely increase at each stage, if we get back to where we started from.

You may object that we're always using the same shuffling rule. We repeat: *Dynamics is like that.* However, since you insist, let's think about "random" shuffles at each stage. Surely those increase the disorder.

If the initial state is ordered, then random shuffles do indeed disorder it. After a few shuffles, however, it becomes hard to maintain that the disorder is still increasing. In an intuitive sense, disorder reaches a "saturation point" and hardly changes. Moreover, an occasional shuffle may restore the original order. At any given stage, the probability that this will happen is 1 in 40,320, or roughly 0.000025, because there are precisely 40,320 ways to arrange a pack of eight cards. The probability is low, but a restoration of the original order *could* occur. If you shuffled the eight-card pack a hundred times a second, then on average it would take just under seven minutes to return to the original configuration.

There is no inexorable trend toward increasing disorder.

Moreover, our definition of "order" has loaded the dice in favor of disordered states. "Ordered" sequences are relatively rare. How many there are depends upon what you consider order to be, but we can make the point by agreeing that any sequence is ordered in which cards 1 through 4 occur first (in any order) and are followed by 5 through 8 (in any order). Think of a Zarathustran pack of cards, in which numbers 1 through 4 are red and 5 through 8 are black. We will assume any other arrangement to be disordered. The number of ordered states is then 576, so that leaves the number of disordered states at 39,744. There are 69 times as many disordered states as there are ordered ones. So when you shuffle randomly, you are 69 times as likely to produce a disordered state as you are an ordered one. This is a contextual point: When we select a meaning for the word "order," we are choosing a context for the whole of the subsequent discussion. If the context is loaded, so will the conclusions be, but we may not notice because contexts are usually not made explicit.

As the number of cards increases, this ratio becomes larger and larger. With a standard pack of fifty-two cards, for example, and considering "order" to be all the reds followed by all the blacks, there are 495 trillion times as many disordered states as ordered ones. With a normal-sized room containing equal numbers of molecules of oxygen and nitrogen, and taking "ordered" to mean that all oxygen molecules are in the left half of the room and all nitrogen molecules in the right half, there are roughly $10^{10^{60}}$ times as many disordered states as ordered ones. It's absolutely clear why the molecules of oxygen and nitrogen in a room do not spontaneously segregate themselves into the two halves. They could, in principle, but the probability is incredibly small. However, it is not ruled out by our calculation, and thus there is no inexorable irreversibility in the way molecules get mixed up.

■ THE TRUE MEANING OF ENTROPY

Back to cards. We've now seen that we can't attribute to each card sequence a numerical measure, entropy, that always increases when the cards are shuffled. Let's try to do the best we can: Entropy either increases or stays the same. Since all transitions between sequences are possible (even if many are rare) it follows that entropy must remain constant. The same argument goes for gas molecules. If a collection of molecules, moving freely under the laws of dynamics, makes a transition from one state to another, then the two states must have the same entropy. This is less silly than it may seem. Start with our usual room with oxygen and nitrogen segregated; this, reasonably enough, we decide to call an ordered state. Let the molecules follow the dynamical laws for precisely ten seconds. You observe what appears to be a disordered state. However, it is not as disordered as it seems. If you could reach in and simultaneously reverse the velocities of all the molecules, you'd produce an equally disordered state. Now wait ten seconds . . . *voilà,* the molecules have segregated themselves toward the two ends. That's not what a truly disordered state should do. Conclusion: There can be cryptic order, manifesting itself to human eyes only at some future (or, time-reversing, past) time.

Does all this imply that the second law of thermodynamics is wrong? Should we dispose of all the textbooks? We should probably dispose of their authors for choosing such a poor image as card-shuffling and getting everybody confused about irreversible changes in time-reversible systems, but compassion is a virtue. Moreover, there's a perfectly good sense in which the second law of thermodynamics is true, and it's what the textbooks are really trying to explain. It depends upon a particular thought experiment, an idealization of a physical process. That process is the interaction of two previously independent systems. For example, put a partition in a room and pump oxygen into one half, nitrogen into the other. While the partition is there, the two halves do not affect each other and the states remain segregated (ordered). Remove the partition, coupling the gases together, and very rapidly they mix (disorder). Entropy increases. In a similar manner, imagine a pack of red cards being shuffled, and a second pack of black cards being shuffled independently. Each retains its distinct uniform color (order). Now put the two packs together and shuffle the lot.

There is a rigorous mathematical treatment of the concept of entropy, and it is entirely consistent with the interpretation of entropy as disorder. It assigns, in a well-defined way, a constant entropy to any isolated system. While this system remains isolated, its entropy remains the same, as we argued above. However, when two previously isolated systems are permitted to interact, the entropy of the combination can be proved to be greater than the total entropy of the two separate systems. It is in this sense that entropy increases. That makes entropy increase fine as a model for experiments in which gases are allowed to mix by removing partitions in boxes, or for the workings of heat engines and all the classical thermodynamic paraphernalia. But the relentless increase of entropy does not apply to a single "free-running" system—such as the entire universe.

■ KAC'S CATS

Mark Kac—pronounced "Cats," oddly enough, in view of what follows—was a probability theorist, and he had a vivid way to explain entropy increase. Imagine two cats, one black and one white, each with fleas. The white cat has its own set of white fleas, the black cat has black fleas. (The fleas are of the same species; the colors just assign them to their respective cats. Imagine that some experimentalist has painted each flea in its appropriate livery, so that we can keep track of them.) While the cats do not meet, their respective fleas hop around at random on their own personal cats, exhibiting a specific degree of entropy. This remains constant, until the cats meet up. Fleas then begin to hop from one to the other. Soon you get one white cat and one black cat, each with a mixture of black and white fleas. The amount of disorder among the mixed fleas is more than that among the black fleas alone plus that among the white alone. This is clear: The variety of possible interactions has become greater. White fleas can introduce disorder by interacting with black ones as well as amongst themselves. A children's party with ten children is far more chaotic than two parties each with five. There are far more ways to shuffle a pack of fifty-two cards than there are to shuffle two separate twenty-six-card packs.

Now we can see why this notion of entropy, the entropy that does increase, does not conflict with the time reversibility of dynamics. Recall that we must start with two isolated systems and then allow them to interact. This entire process is obviously different from its time-reversal. In particular, the number of isolated systems involved goes 2 → 1; the reversed scenario would go 1 → 2. It would require us to start with a single system and separate it into two subsystems by forbidding them to interact. Imagine two flea-infested cats, which walk apart, each taking its own share of the fleas. The reversed scenario is simply not the same; moreover, similar mathematical arguments show that total entropy decreases when previously combined systems are separated in this manner.

When the laws of thermodynamics were being formulated, people were thinking about systems like steam engines. They concentrated on features that were natural to such systems, such as rapid mixing to a homogeneous state, and they formulated laws based upon models that captured those features. The features that are of interest when studying steam engines, however, are not particularly appropriate to the study of life. The expansion of the universe away from the Big Bang seems to be driving it toward ever more heterogeneous states. Instead of all the atoms spreading themselves about uniformly, they have clumped together into planets, stars, galaxies, and supergalactic clusters. For systems such as these, the thermodynamic model of independent subsystems whose interactions switch on but not off is simply irrelevant. The features of thermodynamics either don't apply, or are so long-term that they don't model anything interesting. Take Cairns-Smith's scenario of clay as a scaffolding for life. The system consisting of clay alone is *less* ordered than that of clay plus organic molecules: Order is increasing over time. Why? Because one type of cat (clay) has gone off with the fleas (organic molecules), declaring that henceforth those are *its* fleas; the other cat has been left with none. A single system has separated itself into two effectively independent parts. The same happens again when the fleas (organic molecules) declare their independence from their host cats (clay) and conventional RNA-and-DNA-based life gets going in its own right.

The tendency for systems to segregate into subsystems is just as common as the tendency for different systems to get mixed together. Molecules of fat, suspended in water, link together to form membranes that separate an "inside" from an "outside." Fatty cell membranes do the same thing. The process of cell differentiation, whereby originally identical cells develop into different kinds—muscle, skin, nerve, kidney, liver—is just like a bunch of cats going off with fleas from the common pool and declaring that henceforth those are their fleas and nobody else gets to lay a paw on them.

TIME'S ARROW

But what of the laws of dynamics? Aren't those time-reversible?

Yes. But the thought-experiment scenarios that demonstrate the irreversible increase of entropy do *not* obey the laws of dynamics. They involve outside interference that "switches on" or "switches off" interactions between the two subsystems.

The relentless increase of entropy, then, is a consequence of considering only scenarios in which previously separate subsystems combine. This scenario has its own directionality: The number of systems must decrease. Dynamics does not possess that kind of directionality, and the scenario, while an excellent approximation to some dynamical phenomena over reasonably short time periods, is not fully consistent with the dynamical laws. It is therefore hardly surprising if it conflicts with global features of those laws, such as time reversibility. If you embed time-reversible laws in a time-irreversible context—which is precisely what we've done—then the result will be irreversible. The context wins.

In the real universe, there are no truly isolated subsystems. It is only the entire universe that is time-reversible. However, could we perhaps time-reverse some almost isolated subsystem? Could there be a planet somewhere in our universe on which eggs unboil and raindrops head skyward? The dynamics on that planet alone would not conflict with the laws of nature. But consider what would happen if you time-reversed some (apparently isolated) subsystem of the universe. For a short period it really would seem to undo its previous behavior, proceeding backward. However, all the time it would be interacting, albeit very weakly, with the unreversed portion of the universe—the gravity of far galaxies, say. That interaction would be chaotic, and the butterfly effect would come into play. Those tiny disturbances would become amplified, and shortly the "reversed" system would be doing something totally different from the time reversal of its past behavior. This argument doesn't tell us what, but it wouldn't be unboiling eggs.

We hasten to add that we are not identifying chaos with entropy, and we are not saying that chaotic systems are time-irreversible. On the contrary, mathematics offers many examples of time-reversible chaotic systems, in which the butterfly effect works both ways: Neither the distant future nor the distant past can be deduced from an imperfect observation of the present state, however close to perfect the observation may be. What we are saying is that chaos prevents reversibility of some part of the system, however close it is to being isolated from the rest.

So it's not entropy that is responsible for the arrow of time. How, then, do we explain the apparently fixed directionality of time's arrow? Here's one suggestion.

You can't run time in two directions in the same universe. You have to choose one or the other. We have a very strong sense of time moving in a particular direction, *relative to our consciousness*. Perhaps that directionality is built into the universe itself, and time really does flow one way, or perhaps we just observe things directionally. In either case, our universe is stuck with one direction out of the two that are in principle possible.

In our world, rocks fall off cliffs and end up at the bottom. If we happen to be creatures living in a universe that had chosen to run its time in the other possible direction rocks would suddenly leap into the air, landing on top of cliffs. If our consciousness also ran backward, we would interpret this as rocks falling off cliffs. But if our consciousness could run the other way—and unless it can, there's no problem to time's arrow in any case—we wouldn't be the least bit surprised. Before a rock made such a leap, we would be able to detect vibrations of the ground, converging from the distant horizon, focusing themselves upon that particular rock, and combining their efforts in one final shove to hurl it into the air. Clearly rocks on the ground attract such waves. We would offer a similar explanation for a tree levering itself upright, after which the lumberjack uses his ax to repair the final wedge-shaped cut in its trunk, picking up wood shavings from the ground and slotting them into the tree. We would explain how rivers are formed by an excess of water in the sea, deposited by "devaporation," creating a pressure that forces the water up into the mountains, in ever-narrowing courses, until it forms pools that spurt raindrops up into waiting clouds. We would marvel at how mature salmon absorb eggs and sperm, fight their way tail-first downstream, past all the water jumps, to the sea, where they become young enough to make the journey upstream to turn themselves into eggs and sperm, ready to be absorbed by the next generation. And we would go around asking each other why time doesn't run the other way in our world, and wouldn't it be strange to live in a universe where rocks fall off cliffs, salmon lay eggs, and rivers run into the sea.

We are the context in which we observe the world.

GAS LAWS ARE CONTEXTUAL

Nowhere could this be more obvious, once you think about it, than in the gas laws. Recall that these are relations between macroscopic (that is, large-scale) quantities such as temperature, pressure, and volume. But on a microscopic scale gases consist of vast numbers of incredibly tiny molecules whizzing around and bouncing off each other.

Gases don't know about gas laws. They don't even know they're gases. All they do is play handball with their molecules. Does a gas have a volume, for example? Well, let's think how we measure volume. We build a box, fill it with gas, and then measure the size of the box, multiply the numbers together, and declare the result to be the volume of the gas.

Volume of what gas?

The gas that we decided to enclose in the box. It's *our* choice; *we* set up the context in which "volume" has a meaning. We observe that molecules don't get out of the box. They don't even know they're in a box; they just keep playing handball. To be fair, occasionally a molecule of gas bounces off a molecule of box, and for a moment it "knows" the box is there, but it is only the collective motion of *all* the molecules that reacts to the global box. The system of molecules in the box "knows" the box is there, all right; otherwise it would escape. But we define that system; the gas doesn't.

Just as reasonably, we could define the "polume" of a country to be the square root of its population. If you're reading this in the United States, you are one member of a system of people whose polume is 15,880 You "population molecules" live your life quite happily without being aware of these fundamental quantities. But in fact your lives are governed by laws comparable to the gas laws. For example, the average number of you per square kilometer is the square of your polume divided by the total area of your country. Wasn't it clever of you to arrange that, when you didn't even know what polume was and, indeed, couldn't affect where the other population molecules decided to live?

Yes, of course—it's a tautology.

But all mathematics is a tautology. Gas laws are (widely conjectured to be) tautologous consequences of quantum mechanics. So the law that relates polume to population density is just as legitimate as that which relates the pressure of a gas to its volume.

We're not saying that the volume of a gas and the polume of a nation are meaningless concepts. We're saying that they are given their meaning by a choice of context. The tautologies of mathematics are important; they state that specific features necessarily follow from particular assumptions. We then get to choose which features are interesting. Statisticians are interested in averages, so averages litter the gas laws. If poets had set up the gas laws, they'd extract words from molecular configurations and look for rhymes.

The same goes for pressure. We choose to set up an experimental system in which the collective behavior that we call pressure can exhibit its effects: We confine the gas in a chamber and then try to squash it. Observing regularities, we invent the feature "pressure" and codify its regularities in laws. Does a herd of elephants have a pressure? Out on the plains, no. But if you enclose them with a fence and then start to shrink that fence, at some point the pressure of increasingly distressed elephants will become great enough to smash the fence down.

ANOTHER THERMODYNAMIC OSTRICH

That's a rather vague general point about context. Here's a more specific one. Although Darwin proposed a mechanism that, he argued, can lead to the increasing complexity of life, it's not feasible to grow dinosaurs from chemical soup in the laboratory. We can, however, experiment with a much simpler system in which order arises spontaneously from chaos. The system is a chemical one, called the Belousov-Zhabotinskii (B-Z) reaction. It oscillates. That is, it goes through the same sequence of changes over and over again.

This is another example of ostriches head down in the thermodynamic sands, as promised several pages back. Until recently, most chemists thought that oscillatory reactions were ruled out on thermodynamic grounds. The reaction should work its way downhill through the energy levels, precluding any return to its previous state. So when William Bray reported an oscillatory reaction in 1921—hydrogen peroxide turning into water and oxygen and back again in the presence of an iodine catalyst—nobody believed him. By the late fifties, Ilya Prigogine had explained to anyone who cared to listen that thermodynamics in nonequilibrium systems is different from what everyone had been taught by textbooks, which dealt only with equilibrium systems but hadn't explained how crucial that assumption is. When in 1958 the Russian chemist B. P. Belousov observed periodic oscillations in a mixture of citric acid, sulfuric acid, and potassium bromate, he had difficulty getting his paper accepted for publication; but a few people did notice, and followed the idea up. Shortly afterward, A. M. Zhabotinskii improved the recipe so that the oscillation showed up as a color change from blue to red, and he discovered that the B-Z reaction is far stranger than anyone had imagined. Not only did it oscillate, as was previously assumed to be impossible; it spontaneously formed regular patterns, which everybody would have assumed even more impossible had they not, by now, begun to believe the evidence of their eyes instead of their thermodynamics textbooks.

If you place the B-Z mixture in the bottom of a shallow dish, and stir it up thoroughly, you get a dull reddish liquid. Bright blue dots appear here

FIGURE 41

Patterns that form spontaneously from chaos in the B-Z reaction

and there, and spread. Then red dots appear in the center of the blue regions, and you see an expanding blue ring with a red center. New blue dots appear inside the red regions, and so on. Soon you have expanding target patterns of red and blue rings. When these run into each other they don't superpose like water waves; they just join up into intricate, convoluted waveforms (Fig. 41).

The B-Z reaction is an extremely impressive example of the spontaneous formation of order, of patterns, from a sea of chaos. If you ever get the opportunity, *do* try it yourself. Personal experience carries more weight than a thousand academic tomes. The patterns are a surprise because in the uniformly mixed chemicals, no particular place is any different, in principle, from any other. We say "a sea of chaos" because the chemical molecules are moving randomly, like those in a gas; the apparent uniformity is just an average feature, like pressure. Yet strong spatial patterns appear, without any external influence. The reason is that the uniform distribution is unstable. Small concentrations of chemicals tend to grow, for a time, giving patches; then they become too large and the reaction snuffs itself out, to start again. And these initial concentrations are set up by chaotic fluctuations on very tiny scales and by random influences such as dust particles. The instability explains why the uniform state isn't seen; but more subtle analysis is needed to explain why the result is a coherent system of waves—a large-scale simplicity emerging from an underlying chaos.

In a reductionist spirit, you can analyze the ways in which the B-Z chemicals react, study their reaction kinetics, set up huge computer models, and reproduce the "target pattern" waves. You have now "explained" the spontaneous formation of patterns in this system. However, many other chemical systems produce similar patterns. An example is the reaction of carbon monoxide with oxygen on the surface of a platinum catalyst—such as takes place in a car's catalytic convertor, to reduce pollution. Colonies of slime mold—a biological system, albeit one with a lot of chemistry going on—also behave in this way. Each requires its own reductionist explanation, and what's missing from that point of view is any insight into why all these disparate systems exhibit similar patterns.

Such insight requires a contextual view, just as was needed to understand the positioning of herbivores' eyes. The circular target patterns are a new "law of nature," in the sense that they are a regularity in the natural world with its own definite structure and implications. But this is a law about laws, not a law about things. It doesn't say that particular atoms behave in particular ways. What it says is that such patterns are likely to arise in any dynamical system with suitable features. Those features include spatial extension (spread across a flat region, say), a particular type of oscillatory local dynamic, and coupling rules for how one region influences its neighbors. The B-Z patterns are consequences of a meta-law, a law about laws, a common pattern shared by an entire class of rules. They are only a representative of much more universal phenomena, of a common pattern that applies equally to chemicals and to slime mold. Those phenomena do not depend on detailed properties of the underlying setup, such as which chemicals you use, or even whether you use chemicals at all. They depend upon certain universals—a planar spatial distribution; local oscillations; certain kinds of short-range coupling. They are independent of any underlying chaos; indeed, they are independent of a great many regularities as well.

■ SCHRÖDINGER'S CAT

The B-Z reaction is a fairly down-to-earth example of the importance of context in physics and chemistry. Thus fortified, we try stronger meat: the deep question of the nature of quantum mechanics. Niels Bohr said that "if you're not confused by quantum mechanics then you haven't understood it." Daniel Greenberger added that "Einstein said that if quantum mechanics is right, then the world is crazy. Well, Einstein was right. The world *is* crazy."

Quantum mechanics involves irreducible randomness. Radioactive atoms decay according to chance. In principle there is no difference between an atom that is about to emit a gamma ray and decay and one that is not. Read that again: *No* difference. No internal structure that triggers the decay. Just pure chance. Randomness alone is not such a problem; we may not feel comfortable with "God playing dice," but we're used to living in a world that may do nasty things unexpectedly. However, there's a further feature of quantum systems that is much more disturbing, one that bears no relation to anything that we encounter during our daily lives. A quantum system does not exist in a single, well-defined state, but in some kind of superposition of all possible states, determined by its wave function. This raises a fundamental question: What is an experimental observation? Physicists measure properties of quantum systems—the spin of an electron, say. They get well-defined numbers, not quantum superstitions. How does this happen?

For many years the most popular explanation among physicists has been the Copenhagen interpretation, which we mentioned in chapter 2. It holds that when you make an observation—and only then—the wave function collapses into a well-defined state. We also mentioned a famous thought experiment to dramatize the difficulties with the Copenhagen interpretation: the sad fate of Schrödinger's cat. The cat is in a box with a radioactive atom, whose decay affects a sensor that releases poison gas to kill the cat. Until you open the box, the cat is allegedly hovering in quantum limbo, superposed between life and death. You collapse the cat's wave function to one or the other by opening its box. This sounds rather unlikely, but several generations of physicists have convinced themselves that it is a correct statement of the facts.

However, as T. S. Eliot puts it in "The Naming of Cats," "THE CAT HIMSELF KNOWS." The cat is an observer, too. Why can't the cat collapse his *own* wave function? Does the collapse occur only when a human being looks into the box? When a scientific instrument on the approved list probes the box? When a human being looks at the records made by a scientific instrument on the approved list? If you study the literature on Schrödinger's cat, you will find all kinds of arguments about the role of consciousness, Cartesian mind/matter duality, heaven knows what else. Actually, the whole philosophical problem rests on a failure to consider the context in which an observation can be taken. An observation is seen as a simple act performed—it is seldom said how—by a detached external observer.

The crucial phenomenon is not how the measurement gets into the mind of the observer, but how the instrument that makes the measurement converts the wave function of a quantum system into a well-defined number. You point your "spinometer" at the wave function of the electron and out comes a spin of either +½ or −½. This number *subsequently* is processed by the human observer's brain, but wasn't the damage already done when the pointer on the spinometer swung to one or other of its two positions?

A spinometer is a large-scale system. If we choose to think about it within a quantum paradigm it is composed of huge numbers of particles, and possesses its own incredibly complicated wave function. But it's more convenient to think of it classically, as a coil with a pointer and a scale marked +½ / −½. From this point of view an observation is some quantum-mechanical process that begins with an electron and somehow amplifies its properties as it interacts with all the particles in the spinometer. Eventually the classical phenomenon "pointer on +½" or "pointer on −½" occurs. This is interpreted as an observation of the quantum spin of the electron. What physicists actually mean by "observation" when they perform one is thus a highly complex process, an emergent phenomenon that involves the interaction of a huge system of particles.

A cat is conventionally viewed as a classical system; in particular, the properties "alive" and "dead" are classical properties. But a cat can also be seen as a very complex quantum system, just like the spinometer. Consider the sequence of events. The radioactive atom decays, the poison gas is released, the cat dies, the observer opens the box. At what stage is the "measurement" made? The usual Schrödinger's cat story leads us to imagine it is at the final stage, that of opening the box. But surely a measurement must be made earlier? How does the machinery know whether to release the poison gas or not? There is a sensor that decides whether the atom has decayed. The operation of the sensor is *already* an observation, in the normal sense of pointing a macroscopic instrument at a quantum system and getting a definite yes or no response. Unless the sensor "knows" that the atom has decayed, nothing will happen to the cat. The role of the human observer's consciousness is irrelevant to the fate of the cat; the important thing to understand is how the sensor can turn a quantum phenomenon (wave function of a radioactive atom) into a classical yes-or-no observation (the atom has/has not decayed).

Only recently has this contention been verified by an actual experiment—though not with a cat, of course. Leonard Mandel has shown that it is possible to switch a photon from wavelike behavior to particlelike behavior—to collapse its wave function—without direct intervention. The potential to observe its state in the future is enough to collapse its wave function now. So, because you could decide later on to open the box, the cat's wave function must already have collapsed. Other experiments confirm that the measurement is effectively made when the *sensor* decides whether or not the atom has decayed. As far as the electron's wave function is concerned, a measurement is made on it as soon as it interacts with the spinometer. It makes absolutely no difference whether a human being looks at the meter reading afterward.

WAVE-FUNCTION SPACE

This shifts the mystique of Schrödinger's cat to a different and far more interesting arena. The problem of the nature of observation is that of understanding theoretically how classical behavior can emerge from quantum indeterminacy. How can wavelike superpositions yield crisp yes-or-no answers? We now have a fairly good understanding of the general mechanism, and it goes like this. As a quantum-mechanical object, a cat is a very complicated item. It contains around 10^{26} atoms, and its wave function is comparably complicated. In order to know the wave function of the cat you must measure on the order of 10^{26} values of various quantum parameters. Once you know these parameters, you have a complete quantum-mechanical specification of the cat. So quantum-mechanically, the cat's wave function lives in a wave-function space of 10^{26} dimensions.

Your average tabby, on the other hand, as it manifests itself to human senses, seems to live in a very different space: cat space. Cat space organizes features of cats such as ginger/calico or male/female or alive/dead. And there is no simple mapping from wave-function space to cat space. The specification of the animal's particles in wave-function space is almost completely useless if you want to know any commonsense feature of the cat—for example, what color its left ear is. Imagine the computer calculations you would need to perform in order to answer such a question! You would have to process the cat's entire wave function, somehow extract information about the atoms in its ears (tricky, because in quantum mechanics everything interacts with everything else), find out which atoms are on the surface of the hairs on the ears (equally tricky for the same reason), analyze their response to incident light of various wavelengths . . . You can get some idea of how difficult this kind of process would be by listening to a telephone line when a fax message is being transmitted. What you hear is a series of high-pitched bleeps. Can you hear whether or not the fax is a picture of a cat? Siamese or Manx? Could you hear the gap left by the missing tail? The only sensible way to find out what sort of cat it is is to input the signal into a fax machine, wait for it to print out the picture, and take a look. Similarly the only "practical" way to tell from a wave function what color a cat's left ear is would be to obtain a quantum-reality superfax machine, "print out" a real cat with that wave function, and take a look.

In any case, rather a lot of wave functions correspond to a dead cat, because a cat can be dead in so many different ways, if you look at its internal structure atom by atom. And it can also be alive in lots of different ways. So the classical distinction between "alive" and "dead" is insensitive to all sorts of variations in those 10^{26} parameters. An image that springs to mind is that of a hologram, which differs from a conventional photographic slide in one crucial respect: Every tiny piece of the picture that you see is smeared all over the entire hologram. If you tear a corner off a hologram and use that to construct an image, you don't get one corner of the picture; instead, you get the whole picture, but rather blurred. Bits of a hologram correspond to the degree of resolution of the image—the quality of what you see—not to bits of the image. Bits of a feline quantum wave function correspond to your quality of knowledge about the cat, not to bits of a cat.

Given this, the observable features "alive" and "dead" are likely to divide wave-function space much as land and water divide the surface of the earth (Figure 42a). What's missing from this picture is the idea of quantum superposition, but we can easily take care of that. You get a quantum superposition of wave functions by adding together a certain amount of one and a certain amount of the other. For example, think about a 50–50 mixture of the wave functions of two cats. A 50–50 mixture is just an average, and the average of two quantities is halfway between them. So this particular superposition of wave functions can be represented on a picture as the midpoint of the line joining them. Mixtures in different proportions lie in other places on the same line. Figure 42b shows two different cases, each a 50–50 mix of the wave function of a live cat and that of a dead cat. One mixture corresponds to a dead cat, the other to a live one. Suppose you know that some place on the map lies halfway between a point on land and a point in the sea. Does that tell you whether the place is on land or at sea? Of course not. But if this picture of how wave-function space relates to the features alive/dead is at all representative of reality, then Schrödinger's cat ceases to be a paradox.

Is there any evidence that wave-function space is divided up in this hodgepodge way when we consider classical observations? Eugene Wigner was the first to take this kind of question seriously, and his followers have developed techniques to answer it. They realized that there is no way for an observer to measure all 10^{26} parameters of a quantum cat, so any observation will be forced to ignore an enormous amount of information about the wave function. An observation squashes wave-function space down

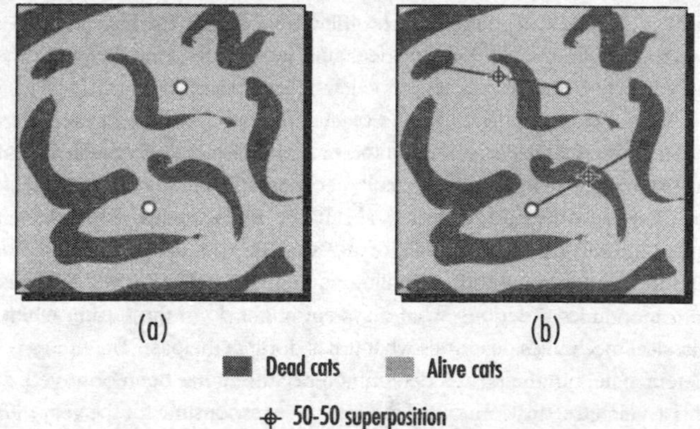

FIGURE 42

(a) Regions in feline wave-function space corresponding to "live" and "dead" cats, with two selected representatives of each: black dots for "dead," white dots for "alive." (b) A 50–50 combination of live and dead cats (midpoint of the line joining them) may be either alive (top) or dead (below) depending on which cats (and wave functions) you choose.

into a much smaller dimensional classical cat space, and it leads to an effect that physicists call decoherence. It implies that, for the kinds of measurements that we can make in our macroscopic world, using macroscopic instruments in macroscopic laboratories, any *large* quantum system will appear to be classical. Indeed, the larger the system, the more classical it will appear to be.

Decoherence is a contextual phenomenon. If you choose to observe one electron from the cat's body, then it behaves in a quantum fashion. But in order to observe the whole cat, you have to work in a context for which "cat" is a meaningful description. That means 10^{26} *interacting* atoms, and you can't observe one of them on its own because the others interfere with what it's doing. So you have to make "classical" kinds of observations, like whether it's alive and how many ears it has, and then the wave function decoheres and the cat behaves in a perfectly classical manner.

Schrödinger's cat is not in a quantum superposition of life and death. Like any other cat inside an impenetrable box, it's either alive or dead, ***but you don't know which until you open the box.*** "THE CAT HIMSELF KNOWS," of course;

but as T. S. Eliot also pointed out, he will never confess. *You* have to open the box to find out, but you need not feel guilty about collapsing the poor beast's wave function and sealing its previously indeterminate fate.

We're not saying that the role of human consciousness in the way we observe the universe is unimportant, or that it's not a deep problem. But it has nothing at all to do with the role of observation in quantum mechanics, which is a contextual question about the mathematical links between quantum and classical models of physics. In a recent lecture, Freeman Dyson said this in another, very illuminating way. His point was that quantum mechanics describes what a system might do in the future, whereas classical mechanics describes what it has done in the past. The future is indeterminate, but the past is determined because it has been observed; and this asymmetry, this contextual difference, is responsible for the very different characters of quantum and classical mechanics. Moreover, the present, where our consciousness resides, is a moving boundary at which the context changes—a traveling catastrophe in paradigm space.

■ TWENTY QUESTIONS

The distinction between reductionist and contextual theories in physics is not new. Newton's law of gravitation is essentially reductionist; it sees the force of gravity as being produced by the matter inside a given body. Newton's contemporaries worried about the problem of transmitting the force to another body and made trenchant criticisms about the ridiculous and outmoded belief in "action at a distance," but Newton stuck to his guns because his theory worked. Its predictions agreed with observations, and in addition it was mathematically elegant. Much later Einstein, seeking a theory that would reconcile Newtonian gravitation with relativistic mechanics, was led—by way of general principles about theories being the same at all points of space and time—to a theory of gravity that is in some respects reductionist, but has a substantial contextual element. In Einstein's theory the force of gravity is not caused by the matter inside a body but by the curvature of the space-time that surrounds it.

Quantum mechanics, like Newtonian gravity, is reductionist. It explains the properties of atoms in terms of their constituent quarks and gluons. But more and more physicists are beginning to think that quantum mechanics, for all its successes, is really just an approximation of something deeper and more geometric, a theory of quantum gravity that will reconcile quantum mechanics and relativity. It is proving very difficult to find this common generalization, however. One reason may be that quantum mechanics is reductionist, but relativity is contextual. Perhaps the answer lies in developing a contextual theory of quantum phenomena, comparable in spirit to relativity. Such a theory would incorporate ideas similar to that of decoherence, but it would go a great deal further, changing conventional quantum mechanics rather than developing a contextual viewpoint within it.

The famous physicist John Archibald Wheeler has devised a contextual view of quantum mechanics based on the way in which nature responds when you ask it questions. Classical mechanics assumes that there is a "true" state of the universe and that experiments can determine what it is, as accurately as desired. In quantum mechanics the Heisenberg uncertainty principle imposes limitations on what quantities can be observed simultaneously: position or momentum, but not both; energy or time, but not both. When you measure one you unavoidably disturb the other. The difference between these two positions is analogous to two different ways of playing the parlor game Twenty Questions. In the normal version, one player (Schrödinger, say) leaves the room while the others select an object (his cat). Schrödinger reenters. He is told that it belongs to the "animal" category, and asks a maximum of twenty questions:

SCHRÖDINGER: Can you eat it?
OTHERS: No. Well, you *could*.
SCHRÖDINGER: Does it have feathers?
OTHERS: No.
SCHRÖDINGER: Does it have whiskers?
OTHERS: Yes.

The game continues until Schrödinger either guesses correctly or runs out of questions. All is well unless he asks, "Is it dead?"

This is very similar to the way in which scientists effectively interrogate the universe, through experiments, to find laws of nature:

SCIENTIST: Is it a wave?
NATURE: No.
SCIENTIST: Does it interact with a magnetic field?

NATURE: No.
SCIENTIST: Does it travel at the speed of light?
NATURE: Yes.

However, there's another way to play the game. Wheeler's friends once tried it on him as a joke, in a version of the game that used words in place of objects. The game seemed to go exactly as usual, except that the answers took longer and longer to arrive. When Wheeler finally guessed correctly, his friends explained the trick that they'd been playing on him. At the start of the game they had not chosen a word at all. Instead, after each question, they tried to think of at least one word that was consistent with all previous answers; only then could they safely answer the question. They had to stop when only one word seemed to fit. (Unfortunately the actual word is not recorded.)

The original version of the game begins with a unique answer, and the questioner focuses more and more tightly upon it as the interrogation proceeds. In the variation there is *no* unique answer. The answer emerges from a sequence of narrowing contexts. Wheeler's contention is that quantum mechanics is more like this second version: There need not be an answer in advance. An electron doesn't have a well-defined spin until you ask, by performing the appropriate experiment, what its spin is. Then you get an answer that narrows down the possibilities. The extra twist in a quantum universe is that by asking some questions—such as "Where is it?"—you forfeit the right to ask certain supplementary ones, such as "How fast is it moving?"

In this view, the wave function of a quantum system doesn't exactly collapse when we observe it. Instead, our information about it becomes more precise. Indeed, we can't be sure that the system actually has a well-defined wave function at all; like the players in the parlor game, it may be making it up as we go along. Inasmuch as the universe seems to obey laws, there must be rules that determine permissible responses. If those rules are listed in a kind of "dictionary," then there is a true underlying reality (the dictionary) and we are really playing the standard version of the game. But if those rules are just consistency conditions—if the rule is only that no subsequent answer can contradict the previous ones—then "reality" is ultimately just a blur, which comes into focus when we look closely enough at it, but doesn't exist when we don't.

It's not clear which of these possibilities best captures the quantum universe. But it is entirely clear which best captures the scientist's state of knowledge about the universe: the second version of the game, in which the answers are made up as the interrogation proceeds. Moreover, science plays the game with a slight relaxation of the rules: The answers can sometimes be lies. Not because the universe lies to us when we ask questions, but because we ask questions imperfectly and don't always understand nature's answers. Our knowledge of the universe is a set of nested or overlapping paradigms, which bear labels like "inverse-square-law gravitation," "crystallography," or "quantum gauge fields."

Proponents of a Theory of Everything effectively believe that when scientists play Twenty Questions with nature, nature has already chosen the answer. The job of science is to find the unique word in nature's dictionary that fits every conceivable question we could ask. The actual state of science is quite different: More often than not we get ourselves into a conversation something like this:

SCIENTIST: Does it have three letters?
NATURE: Yes.
SCIENTIST: Is it a color?
NATURE: Yes.
SCIENTIST: Does it begin with "R"?
NATURE: Yes.
SCIENTIST (*triumphantly*): Is it "red"?
NATURE: No.

When this kind of thing happens, scientists first check the questions and answers again. If everything still holds up, they are forced to reinterpret some of the questions, or some of the answers:

SCIENTIST: Is it a wave?
NATURE: Yes.
SCIENTIST: Is it a particle?
NATURE: Yes.
SCIENTIST: I think I'd better go away and invent quantum mechanics.

CLASSICAL PARALLELS

These difficulties may seem to be peculiar to the quantum world. In fact, very similar problems arise in our own commonsense, everyday world. For example, we can interrogate the universe about a tree, just as quantum mechanics would interrogate it about an electron. Is a tree a plant? A boat? A device for supporting telephone lines? If we want to know whether it's a plant, we have to watch it growing, look at its leaves, collect its seeds and see what grows, and so on. To decide whether it's a boat, we have to chop it down, trim off the branches, hollow it out, and see whether it floats off down the river. To decide whether it can support telephone lines, we have to chop it down, trim off the branches, stick it on end in the ground, and string the wires from the top. And so on. The answers we get depend on the questions we ask. There is even an uncertainty principle: If you test a tree for its boatlike features, you can't simultaneously test how good it will be for holding up telephone lines.

If you've understood the sense in which the pressure of a gas is a contextual quantity, you can tell similar stories for volume and even temperature. What you get depends on what measurement you choose to make. Measurements made by one method may preclude the use of others.

You may feel all this is a bit far-fetched, but does it really differ in any essential way from the manner in which physicists ask whether light is a wave or a particle? To decide if it's a wave, they diffract it through slits and see whether they get interference fringes. To decide whether it's a particle, they bounce it off things and see if it holds together. They ask wavy questions to decide whether it's a wave, particley questions to decide whether it's a particle—just as we ask different questions to decide whether a tree is a telephone pole or a boat.

Along with the Copenhagen interpretation was something called the complementarity principle. This held that a quantum object can behave either as a wave or as a particle, but never as both at the same time. It was generally consistent with experiments until very recently. Now people are on the verge of performing experiments in which electrons might be observed performing both kinds of behavior at once. It seems to be truer to say that, until recently, physicists could ask electrons wavy questions, or particley ones, but not both at once. And of course they got wavy answers, or particley ones, but not both at once. Not because nature is like that, but because that was the context set up by their questions.

In the classical universe, the same things happen.

One of the deepest areas of physics—in the reductionist sense—concerns itself with fundamental particles. It used to ask fairly broad questions about the nature of matter, but nowadays its questions are posed within a very specific context: particle accelerators. Beams of particles are smashed together at ever higher energies to squeeze out ever more esoteric constituents of matter. Are the particle physicists really delving deeper and deeper into the ultimate constituents of nature, or are they playing Twenty Questions without a dictionary? Imagine trying to understand how a car is built by smashing it against a brick wall and looking at the pieces. From low-energy collisions you pick up wheels and spark plugs and bent exhaust pipes. At higher energies you get smaller pieces—some steel, some copper, some plastic, some rubber—and you may deduce that wheels are made of more basic components, namely steel and rubber. But you also get sparks as steel hits brick. Does this imply that sparks are more basic components of the car than wheels? Aren't sparks really a side effect of the context you have chosen for your questions? If you crash cars at still higher speeds, how much of what you see is important, like the atoms from which the rubber is made, and how much is more esoteric kinds of sparks?

Each new accelerator experiment brings to our attention new states of matter. That's not in dispute. But it is far from clear whether (or to what extent) these new states are basic components of matter or sparks struck from it. Even less clear is the extent to which particle physics is probing the ultimate nature of matter or just playing an open-ended game of Twenty Questions where the number of possibilities increases as you ask more questions, instead of narrowing down to any actual answer.

■ WHAT'S IN THE BOX?

What we call laws of nature may not be fundamental truths, however well they fit the observations. Whatever their status, they tell coherent stories about the way selected bits of nature work, but we also select the observations that demonstrate that coherence. We are unduly impressed by the fact that any tiny change leads to nonsense, but we forget that this tells us nothing about big, consistent changes. Our laws are Sherlock Holmes stories and their tight logic impresses us, but we forget that you can always invent an entirely different Sherlock Holmes story.

Our current view of dynamics, for example, is of a mathematical box that churns out numbers. In this mathematical model, we pop the initial conditions into the box, and out comes a description of the subsequent motion, obtained by applying the rules that live in the box. We think nature does it this way, too. Nature has a box, full of laws that are congruent to the rules in our box. She pops in the initial conditions, out comes the motion, and that's the image of a clockwork universe. If it's a quantum box, we think that ours is full of wave functions and therefore so is hers, but for a moment let's think classical.

The disturbing thing is that our mathematical box requires infinite precision. Initial conditions must be known to infinitely many decimal places, or the clockwork won't tick. If nature is really using the same kind of box as we are, with the same kind of rules, then reality must have the same precision built into it. Chaos says that errors grow in time, and we can invert this to our advantage. The more accurately you know the present motion, the further back in time you can push the numbers, and the more decimal places you know, so you can infer the "true" initial values with greater and greater precision. As you observe the solar system revolving, it seems as if nature is slowly pulling more and more decimal places of the initial conditions out of the box. The more of the present you can record, the more precisely you determine the past.

But are the decimal places really in the box at all? A mathematical result called the shadowing lemma says that if you take a dynamical system in which everything works to infinite precision and create occasional, small, random disturbances, then the result is indistinguishable from the infinite-precision system with very slightly different initial conditions. Here we find dynamical systems theory sowing the seeds of its own subversion. The shadowing lemma means that for all we can tell, nature might be making up the decimal places as she goes along. As the solar system turns, she's not pulling predetermined decimal places out of the box at all; she's just choosing anything she wants to. We may decide to interpret the result as corresponding to higher precision in the initial conditions, because of the way we push the numbers back in time and into more decimal places, but that's just our opinion. So our knowledge about how mathematical boxes work tells us to mistrust the congruence between our conventional dynamical-systems box and whatever box of tricks nature really uses.

The phrase "think physically" has a curious meaning among physicists. It doesn't mean that you actually think about what is happening in the real world. What it means is, take the mathematical images inside your box and project them onto the outside world. Then think about those projected images, reinterpret them inside your box, and see what you get. For example, in classical optics your mathematical box might contain rules for manipulating light rays: They bounce off a mirror at the same angle they hit; they go through a lens like *this;* and so on. So you "think physically" about thin lines of light bouncing around like streams of Ping-Pong balls blown from a tube.

However, there is a logically equivalent piece of mathematics: a box labeled "variational principles" that works in a totally different way but is guaranteed to give the same answers—though employing very different "physical" images—as the optics box. According to variational principles, you can understand everything about how light rays bounce if you just insist that they follow the quickest path among all those that they might follow. So now the "physical" image is that of light trying all sorts of paths, comparing them, and selecting the quickest. This is a very different kind of thinking, concerned with global rules about the entire path followed, not local rules about what you do when you hit a mirror or a lens.

For the quantum optics box, this kind of physical thinking goes a step further. Instead of rays of light you have probability waves, and you don't select the shortest path. Instead you assume that light follows all possible paths, including silly ones; then you just superpose the lot, and bingo, out pops something remarkably similar to the classical rules for light bouncing off mirrors. And the universe fits all the quantum experiments too. So when quantum physicists think physically, this is their picture of the universe: It does everything it could possibly do, all at once, and adds up the results.

A different but mathematically equivalent box would present a totally different "physical" picture. And it's not always sensible to project internal details of mathematics out into the real world. Suppose five chickens each lay four eggs, and you divide them between two people. One way to do the calculation is $5 \times 4 = 20$, $20 \div 2 = 10$, so each person gets ten eggs. The whole calculation can be mirrored in the physical world. But another way is 5 ÷ 2 = 2½, 2½ × 4 = 10, and now it can't be mirrored because you can't have two and a half real eggs.

Physics is a Sherlock Holmes story.

An amazingly good one, mind you; remember, it gets the Dirac number (a basic quantum measurement) right to nine decimal places. However, it cut its teeth there: It kept changing its rules until they reproduced the Dirac number. Physics is the product of natural selection in theory space.

Selection operates in other ways, too. We've explained how the features "volume" and "pressure" in gas laws are selected by context. The features "electron" and "photon" of quantum electrodynamics are also selected by context, in just the same way. Thermodynamicists find out things about their selected feature "pressure" by asking nature pressurelike questions. Physicists find out things about their selected feature "electron" by asking nature electronlike questions. As they have penetrated deeper into the tree of reductionist funnels, their questions have become ever more specialized, which is why the apparatus needed to ask them is twenty miles wide and costs ten billion dollars.

Quantum electrodynamics may get the Dirac number right to a remarkable level of accuracy, but there are all sorts of physical constants that can't be calculated by quantum mechanics—for instance, the masses of all the fundamental particles. Each choice of masses tells a different Sherlock Holmes story. To get one that corresponds to our own world, we have to choose the story whose particle masses are those that hold in our world. It's not a matter of circular logic; you don't plug in the mass of the electron and deduce the mass of the electron. You plug in a few masses and deduce the rest of quantum physics. But it's a mysterious and unsatisfactory process.

■ THE RHETORIC OF REDUCTIONISM

Now is a good moment to take a closer look at the gap between reductionist rhetoric and reductionist practice. For example, suppose we were to take Newton's rules literally, as a kind of rudimentary Theory of Everything. How would we calculate Mars's orbit around the sun?

We would begin by analyzing the structure of Mars, atom by atom: "Atom of iron, longitude 123.777. . .79 degrees west, latitude 47.883. . .66 degrees north, depth 132.755. . .43 kilometers. Atom of sulfur . . ." We would do the same for the sun; and then we would write down what Newton's laws imply for this enormously unwieldy system of atoms. We would solve the resulting equations to find the motion of every atom, and work out what all the atoms in Mars do. This would be incredibly difficult, for every atom of Mars follows its own rather wobbly vibrational course, especially the atoms in gas molecules in the atmosphere, which are hopelessly chaotic. But our problems would not then be at an end; finally we would have to interpret our results. We would have to extract some common component of their collective motion—tricky, since some atoms are escaping into space and others (from meteors) are falling to the surface—and show that it is (very close to) an elliptical orbit.

This, of course, is how every textbook on mechanics derives elliptical orbits from Newton's laws. Isn't it? No, of course not. There are not enough trees on all the planets in all the galaxies of the universe to make enough paper to write down such a calculation.

It is, however, what you have to do if you take the rhetoric of Theories of Everything literally. And the same goes for the rhetoric of all reductionism. If you claim that *in the real world* Newton's laws—laws of *nature,* you recall, is the phrase—imply that Mars has an elliptical orbit, then a calculation something like the one outlined is implicit. Of course we don't apply Newton's laws this way, largely because we can't. If supercomputers were up to the task, no doubt somebody with more persistence than sense would have tortured them into it by now. We're not even suggesting that it's sensible to apply Newton this way, and we're not suggesting that reductionism fails if you can't. All we're saying is that this highly complicated procedure is what you must do if you want to check out the reductionist rhetoric properly. The claim of the rhetoric is that the laws of nature carry out such a calculation, in the sense that the motion of Mars is entirely governed by the combination of all those tiny interatomic gravitational forces, and nature puts them together. It doesn't need a computer, though: It's got a real Mars. We can't actually carry out such a calculation ourselves (though we could probably find some mathematical approximations that would simplify the problem enough to reach reasonable conclusions), but the rhetoric leads us to believe that if we could, we'd get the same answer that nature does.

It's a sound philosophical principle, though one that is often ignored by philosophers, to take a look at what people actually do. The practice of reductionism, like that of most things, is much more interesting and much more informative than its rhetoric. What reductionist scientists do is to choose particular features of Mars—its orbit, its position, and its mass—and represent those features mathematically as a curve, a point, and a number. Mars and the sun are not actually point masses, but Newton proved that uniform spheres and points are mathematically equivalent, and Mars and the sun are roughly spherical. The result is a simple mathematical model that can be plugged into Newton's laws. The laws then lead to results that agree pretty well with observations.

The process explains, quite convincingly, why Mars has an elliptical orbit. But it doesn't really reduce the behavior of Mars to that of its component atoms. It gets rid of the atoms at the start and never mentions them again. So is it really sensible to jump to the conclusion that every single atom also exerts a tiny gravitational force? Maybe gravity is an emergent property of particle interactions, not a property of individual particles. If so, the laws used by nature are not the laws found by Newton.

THE MYSTERY OF THE MISSING MASS

The branch of physics most prone to telling Sherlock Holmes stories is cosmology. Cosmologists are confident that they knew what was happening from the first few microseconds of the Big Bang onward—though they admit they're a little unsure of the *precise* details before that. But *Voyager* taught us that we didn't even know what was happening on Saturn, right in our own backyard. So the cosmologists' confidence seems misplaced. They don't think so, however, because their Sherlock Holmes story is a combination of two of the best Sherlock Holmes stories yet invented, quantum theory and relativity, and it fits the available evidence, such as the expansion of the universe, the cosmic background radiation, and the relative abundance of the chemical elements.

The problem is, it's an incredibly ambitious story to erect on such tiny foundations. The relativistic models of the universe that are used, for example, contain no stars. They are just huge-scale approximations in which the grainy gravitational fields of stars and galaxies are smoothed out into a general overall curvature. That curvature is closely related to a single quantity Ω (omega), the average density of matter. If Ω is less than 1 the universe expands forever, if Ω is greater than 1 it eventually collapses again. For a whole lot of reasons, mostly to do with quantum mechanics, theorists expect our universe to have the value $\Omega = 1$, poised right on the boundary between expansion and contraction.

Unfortunately, experiments generally lead to a value closer to $\Omega = 0.25$. This is the "missing mass" problem. Its current solution is to assume that there's a lot of matter in the universe that we aren't yet able to detect: very strange "cold dark matter" and "hot dark matter" made from fundamental particles as yet unobserved on earth. This missing matter conveniently also accounts for the fact that galaxies don't seem to obey the law of gravitation: Their outer rims rotate at the wrong speed. Extra invisible matter outside galaxies could make gravity work nicely again.

So the nice Sherlock Holmes story, the one based on relativity and quantum mechanics, doesn't really agree with experiment at all. It's been tinkered with so that part of it agrees, but at the expense of introducing new characters, *dei ex machina* known as hot and cold dark matter. The only evidence for these is that they make the answer work out right. Plus a reasonable belief that the matter we can detect can't possibly be all there is.

Maybe what we need is a totally different Sherlock Holmes story, not "The Mystery of the Missing Mass" but "The Case of the Crooked Quantum." Can you *really* capture all the important structure of the universe in one magic number, Ω? If not, then all of the current theorizing about uniformly curved universes with particular Ωs goes out of the window.

What we have right now is all a bit empirical. For instance, the COBE satellite has recently observed a lack of uniformity in the cosmic background radiation of a kind that theorists say would have been needed to seed the formation of galaxies. Great! Except that the lack of uniformity is so small that it's very hard to coax the theoretical models into forming galaxies with the right distributions. It has recently been announced that a mixture of 25 percent hot dark matter and 75 percent cold dark matter will work. But there's no independent evidence in favor of those numbers. There's no independent evidence that hot and cold dark matter exist at all. And in the meantime, other theorists have performed computer simulations that suggest the present distribution of matter would not have evolved in a universe with $\Omega = 1$, the favored value, but that it would if $\Omega = 0.25$, which is close to the figure experimentalists get when they assume there is no missing mass.

Sherlock Holmes stories? Or fairy tales?

THE GLASS MENAGERIE

At the bottom of the nested funnels of reductionism we find mathematical theories that are increasingly divorced from our commonsense human world. Perhaps at the very bottom, all funnels lead to a single unifying principle, the long-sought Theory of Everything. As we have argued more than once, such a theory would be of immense philosophical importance—it would tell us that the universe really *is* a unified structure—but its practical utility would be nil because what is of interest higher up the funnels is largely independent of what goes on down below.

The contextual view offers a very different image of science and of the world that it codifies. At a given level, a scientific theory is usually transparent, in the "see-through" sense that we explained in chapter 7. That is, its explanations are simple and independent of fine detail farther down the funnels. (Theories are like this because that's what human minds can actually use: the practice of reductionism, not the rhetoric.) However, if you do look down those funnels, you will certainly find things—because the world is like that. Even if lots of internal mechanisms can produce the same effect, the universe likes to choose one. But whatever it is, what you find is again transparent, in our sense, on its own level.

In Tennessee Williams's play *The Glass Menagerie,* which opened in New York in 1945, shy crippled Laura owns a collection of glass animals. At the turning point of the play the unicorn's horn is accidentally broken off. "Now he will feel more at home with the other horses, the ones that don't have horns," says Laura.

Science is a glass menagerie. Each animal is a transparent body of connected concepts, a paradigm. Normally, it is not necessary to look inside it to use this body of concepts. But if you do look inside, what you find is another glass menagerie. Inside the animal "crystallography" are glass animals labeled "atomic lattice," "point symmetry," and so on. Inside the animal "atomic lattice" are animals such as "atom," "electron," "chemical bond." Inside "atom" is an animal labeled "quantum mechanics"; inside that are animals labeled "perturbation expansion," "Hilbert space," "Schrödinger's equation." The entire structure is a fractal, repeating its organization on every level. Each glass menagerie is a definite collection of animals, but each animal is transparent glass, and changes to the inside of glass animals are invisible from the outside. Only changes to their outsides, like the unicorn's loss of his horn, make any important difference.

Laws of nature are animals in the glass menagerie. We find them by choosing a context and looking for regularities within it. Different contexts lead to different laws. Contexts that cut across each other lead to complementary laws, different points of view about the same general area. Newton's law of gravity explains some features of planetary motion—such as elliptical orbits—in a transparent way. Einstein's relativistic version of that law explains other observations, such as the bending of light by a star and slow changes to the orbit of Mercury. In practice, science uses both laws, and with good reason.

Cosmologists may tell you that Einstein has superseded Newton. That's only partly true. There are many problems that we can solve in a Newtonian model, but can't even state within general relativity. The motion of binary stars (pairs of stars held close together by gravity) is an example. A Newtonian model predicts that the stars will move in elliptical orbits—the mathematical problem is the same as that presented by Mars and the sun. In general relativity nobody even knows how to set up the equations for a binary star, let alone solve them. It is simply not true that relativity can explain everything that Newton can, and more.

Newtonian gravity has features, such as elliptical orbits, that match our universe. Relativity has other features that Newton's system lacks. The universe probably uses neither to go about its business; Newton's and Einstein's are models for human brains, not ultimate truths. Rather good models, to be sure, but few scientists nowadays seriously believe that their laws are true. (Only cosmologists and particle physicists do, perhaps, and both ought to know better, since they work in precisely those fields in which the perceived laws have changed most dramatically in the last fifty years.)

Laws are not timeless truths. They are context-dependent regularities, and we bring out different laws by asking different questions. Despite this, the glass menagerie of science works brilliantly, precisely because different points of view illuminate different features of the world. There is no reason to suppose that buried within it there lies a single ultimate law—and no reason to want one in any case.

9 THE DEVELOPMENT OF ORGANIZATION

Pyotr thanked Ivan for explaining telephones. "But what I really do not understand, Ivan, is television. I watch the soccer matches and the news every day—but how is it possible to send color pictures from one end of the country to the other, and without wires?"

Ivan thought for a few moments and replied: "And with wires, you'd understand?"

It's a pity the dinosaurs died out.

Given humankind's general ability to exterminate large animals, it's probably true that even if the dinosaurs had survived until recently, they'd probably be dead by now. But dinosaurs have a tremendous appeal. In Michael Crichton's novel and Steven Spielberg's blockbuster movie *Jurassic Park* a genetic-engineering company brings the dinosaurs back to life by sequencing their DNA, preserved in blood sucked by biting insects that subsequently became trapped in amber. Wouldn't it be great to do just that?

In July 1991 the British Museum sponsored a conference on the recovery of DNA from ancient archaeological specimens. This technique can shed light on questions that range from ancient human migratory patterns to the genetics of the quagga, a zebralike beast that died out just over a century ago. The British Museum's conference organizers were expecting about thirty participants, but instead they were inundated. The cause was an article published in the science section of *The New York Times*, which suggested the possibility of using "fossil" DNA to reconstruct a dinosaur.

Until recently the most ancient DNA that had been sequenced was from a 5,500-year-old human bone. But in 1991 Edward Golenberg extracted DNA from fossil magnolia leaves, eighteen million years old. This feat was trumped in 1992 when a team led by Rob de Salle sequenced termite DNA that was thirty-two million years old, and independently a team led by Raul Cano sequenced bee DNA of the same age. By the middle of 1993 researchers from California Polytechnic State University had pushed the record out to 120 million years, well before the last days of the dinosaurs, by sequencing DNA from an insect known as a nemonychid weevil. Everybody used *Jurassic Park*'s amber method, though applied to the DNA of the insects, not the animals they bit. Crichton wasn't as good a prophet as he may seem: The method had been published some years earlier in the scientific literature.

Our previous discussion of living organisms, in chapter 3, describes them as if they are simply the spatial realization of a DNA code, just as a jumbo jet is a spatial realization of engineering blueprints. This gives the impression that an organism's DNA contains, at least in principle, everything that you need to build it. As the furor over dinosaur DNA illustrates, this belief in a unique mapping between DNA and organisms is prevalent among the media and the general public. This is not surprising, since the same belief, whether tacit or explicit, is widespread among scientists. Suppose scientists could somehow lay hands on a complete, perfectly preserved dinosaur genome, its entire DNA code. *Could* they then build a dinosaur?

To answer that question we must look at the role of DNA as information, and at nongenetic influences that also affect development (and are not in the DNA code at all). We will show that these outside influences weaken the link between DNA and organisms, so that in principle two quite distinct creatures could have identical DNA. We will then return to the old puzzle of the relation between genes and characters, and see what all of this implies for dinosaur building.

■ THE INFORMATION IN YOUR HAND

Each age interprets its universe in terms of what is currently important to it. Ancient animistic people wanting to make sense of the starry sky saw it as a zoo of people and animals—the Hunter, the Swan, the Lion, the Dog. The mechanical age of the eighteenth century bred a mechanistic philosophy; in the clockwork universe, God was the watchmaker who set the wheels spinning and then stood back to watch his creation turn; the newly discovered constellations of the southern hemisphere included the Octant, the Triangle, and the Microscope. Our present Computer Age sees the universe as an ever-changing flow of information, and if we were to discover the stars today our first instinct would be to try to decode their message.

So when, in the computer age, Crick and Watson stumbled across the double helix and its aperiodic sequence of bases, it was inevitable that DNA would be seen as a "program" or "code" that contained the "genetic information" needed to make you and me. For example, in his book *Mind Tools* Rudy Rucker says: "Your hand is designed according to certain instructions coded up in your DNA. The length of these instructions gives a measure of the amount of information in your hand." This is the view of DNA as a genetic message transmitted from parent to offspring, a list of instructions like a glorified knitting pattern. And just as we can look at a knitting pattern and see which part of it governs the design of the neckline or the armhole, we imagine that if only we were clever enough we could look at the DNA pattern and see which part of it governs the design of a hand.

Or a dinosaur.

And of course, if we want to produce a very complicated sweater, say one with an intricate lacy three-dimensional effect looking like butterflies on a background of bulrushes, then the knitting pattern must provide more information. So longer DNA sequences must code for more complicated parts of the organism, and complicated organisms must contain more information. This is the picture behind Rucker's statement: a picture of DNA as the Book of Life. You can imagine thumbing the pages of the genetic handbook, looking for the Sentence that produces hemoglobin, the Paragraph that produces a blood cell, the Chapter that produces an artery—even the Appendix that produces an appendix. The Book of Life image is often explicit in the sales pitch for the self-proclaimed great project of sequencing the human genome. It is the worldview of the science-fiction writer Tom

Easton's "gengineer" stories, in which you can tear out the pouch page from the Book of the Kangaroo and glue it into the Book of the Albatross to get Air Mail. Above all it's a picture of information as data string: The longer the sequence of instructions, the more information it contains.

So, of course, because there are more letters in "quadruped" than in "dog," the message "Fido is a quadruped" contains *more* information than "Fido is a dog." When you transmit "Make me a pearl" to an oyster, the message takes the physical form of a tiny piece of grit but produces a wondrous, lustrous result; that piece of grit must have contained an amazing amount of information. And since the DNA of a mammal contains fewer bases than that of an amphibian, it follows that mammals are pretty simple creatures, really. And once you start thinking like that, you begin to realize that DNA-as-message must be a flawed metaphor. The standard quantitative measure of information—the number of characters in the message—is well designed for its original purpose, that of informing the engineer who is required to build devices to transmit and receive the message. Those devices don't care what the message means. Meaning is a quality, not a quantity, and it is highly dependent upon context.

The idea of DNA as genetic information also sits uneasily with the phenomenon of convergence. Different causes can produce the same effect. (Flight, for example, has been invented at least four times in the history of evolution: by pterosaurs, insects, birds, and bats. The wing is a common structure in living creatures, but the creatures just listed do not possess some common DNA sequence that produces wings.) There is also a great deal of convergence within a single species. When we first introduced the "program" metaphor for DNA we mentioned how frogs develop from tadpoles in ponds whose temperatures vary considerably within the course of one day, and we pointed out that many of the genetic instructions in frog DNA are contingency plans for temperature changes. This leaves a great deal less information to determine the basic developmental program around which the buffering routines fit. Even though frogs have more DNA than humans, we're left with the uncomfortable feeling that an adult frog is far too complicated an object to be produced by the amount of information that we know exists in its DNA. And as we've said more than once already, it's a fact that the connections between the nerve cells in the human brain are far too complicated to be described by the information that exists in human DNA.

It's not that the program image of DNA is completely false, but the "program" is only part of the developmental process. What about the computer? How does that work? How does the information in the DNA program lead to a fully developed organism? What else is needed? These are important questions, to which the conventional measure of information—the length of the DNA message—is only marginally relevant. The metaphor of DNA as a message transmitted by the parent and received by the offspring doesn't hold up under scrutiny. When the message is transmitted, there is no receiver. The message, indeed, is supposed to describe how to construct the receiver! Strictly speaking, the genetic code isn't even a code. Codes must be written as well as read; messages must be encoded as well as decoded. It is true that DNA codes for—that is, determines—proteins; but there is no converse process of encoding proteins into DNA. The DNA message is not transmitted, but copied—subject to the complications of sexual reproduction—and the process whereby DNA code is turned into offspring also involves the parent.

Our current obsession with information technology and messages as bit strings has led us to focus almost exclusively on DNA as software and to ignore the contextual hardware in which it produces actions. We've already seen that other things than DNA also pass from parent to offspring, things that on a biochemical level are comparable to DNA but which, since we don't think of them as coding anything, we also don't think of as conveying information. Let's recall a few examples. In most sexual animals the egg begins development without involving the embryo's own genes, and only when the embryo's ground plan is sufficiently well developed do its own genes take control. Mammals take the whole process much further; they invest far more in the mother, thereby simplifying what has to be put into the embryo's DNA. A large part of frog DNA deals with alternative enzyme pathways for different temperature levels. In contrast, in a mammal the uterine temperature is kept constant by the mother's own regulatory systems, so mammals don't need to put that kind of information into their DNA. This is why mammal DNA contains fewer bases than amphibian DNA while managing to produce animals that are manifestly more complex. We might speculate about a future super-mammal that puts the "extra" DNA to good use.

■ CD PLAYERS AND JUKEBOXES

Context is crucial to the interpretation of messages—to their meaning. Instead of a DNA blueprint encoding an organism, think of a compact disc encoding music. Biological development is like a CD that contains instructions for building a new CD player, not just one intended for playing music on a standard player. Even though the instructions tell you how to make a CD player, you can't read the instructions without already having one. So the quality of the new CD player that you build depends not just on the quality of the coded instructions, but also on the quality of the machine you use to read the CD. You might be playing a CD that describes, in exquisite detail, a state-of-the-art CD player with twelve sound channels, frequency equalization, the very latest in noise reduction—but playing it on a machine so dreadful, so noisy, that you can't work out what the instructions are at all.

If meaning does not depend upon context, then the code on the CD should have an inherent meaning, one independent of the player. Does it? Let's explore two extremes. One is the standard player that maps the digital code on the CD to music in the manner intended by the design engineers. The other is a jukebox. With a normal jukebox, you put some money in and select some numbers by pushing a few buttons; then the machine plays your chosen piece of music. The only message you send is the money and the numbers; yet in the context of the jukebox these are interpreted as several minutes' worth of music. In principle, any numerical code could "mean" any piece of music you wished; it just depends on how the jukebox is set up. What we have in mind is a jukebox designed by a different bunch of engineers; this jukebox reacts, in an equally arbitrary fashion, to a CD. The engineers use the digital code for the first few seconds of recorded music on the CD to define a number, and rig the jukebox to play whatever piece of music they choose to assign to that number. Now CDs can be read in two totally different ways. You can put the CD in one machine and get the "Eroica" Symphony, but when you put it in the other machine it triggers the machinery to play "Monster Mash." Or anything else the jukebox is set up for—"Auld Lang Syne," "Beer Barrel Polka," or the mating cries of the howler monkey.

NEEPLPHUT: I have been experiencing one of your terrestrial smell-symphonies.

CAPTAIN ARTHUR: Pardon?

NEEPLPHUT: The overall structure has a degree of coherence, but I must admit I found the individual interludes rather strange/disturbing (delete whichever is inapplicable). Very original, but the combinations of smells were highly unorthodox. Mind you, the use of frump essence in conjunction with rotting blinx root was inspired.

CAPTAIN ARTHUR: Neeplphut, I haven't the foggiest idea what you're talking about.

NEEPLPHUT: I borrowed one of the disks from your ship's library. It is a format readable by my domestic smusical box.

CAPTAIN ARTHUR: Your domestic's musical what?

NEEPLPHUT: No, your translator has invented a word. "Smusic" is an art form that employs sequences of smells in a rhythmic temporal organization.

CAPTAIN ARTHUR: And *our* disk is in a format that *your* machinery can play?

NEEPLPHUT: Our smusical box is very versatile. You see, there are many competing companies and they all adopt their own format. The smusical box selects the appropriate format by analyzing the data on the disk.

CAPTAIN ARTHUR: Oh, I see. . . . Wait a minute, Neeplphut! We don't *make* disks with smusic on them! Let me see that!

NEEPLPHUT: Here you are.

CAPTAIN ARTHUR: I thought so. This is Stanley's copy of *Jurassic Pork: Revenge of the Pigosaur*. You've been smelling a classic politico-philosophical movie, Neeplphut.

NEEPLPHUT: That explains a lot.

The meaning of a message depends upon context. In the context of a female firefly, one flash from the male conveys a multitude of subtle instructions. The female is a jukebox, waiting to have her buttons pushed. So is the male, though the details are different. Whether something is a message depends upon context; sender and receiver must agree upon a protocol for turning meanings into symbols and back again. In the case of biological development, the protocol—the shared context—is the laws of physics and chemistry, as the molecules rearrange themselves and new molecules grow. The result is incredibly complicated, but not just because the message is a very long one. Your complexity is not determined by the number of bases in your DNA sequence; it is determined by the complexity of the actions that can be initiated by those bases within the context of biological development, by the meaning of the DNA message when it is received by a finely tuned, up-and-running biochemical machine. The development of your hand, for example, is part of the culmination of a series of processes that produces your skeleton, your muscles, your skin, and so on. Each stage depends on the current state of others, and all of them depend on contextual physical, biological, chemical, and cultural processes, for which no information *as such* need be received.

For similar reasons your large-scale simplicities—recognizable structures with recognizable functions, such as hands for holding things and legs for walking—are not neatly encoded as well-defined DNA subsequences, paragraphs in the Book of Life. Instead, your large-scale simplicities are emergent phenomena, appearing only when the system that must interpret the message collapses the chaos of its individual letters and understands what the message means. Meaning is a matter of context, not content. The number 911 has no inherent meaning. In the context of the U.S. telephone system it means "emergency"; in the context of a lottery it may mean "You lose"; and in the context of housing it means that you live in a fairly long street. A DNA sequence can no more "code for" a hand than a car blueprint can "code for" road-handling. Your hand contains flesh, blood, and bone—but *no* information.

THE DYNAMICS OF DEVELOPMENT

So how does biological development lead to organized, complex structures? In the previous chapter we criticized—for physics and chemistry—the idea that only *internal* features control how things work. We will now argue that the same is true for biology, that the link between DNA chemistry and development is far more tenuous than the conventional view holds. As well as knocking a big hole in the reductionist program, this also knocks a hole in the view that development is a purely internal process.

Development seems to involve dynamics as well as chemical computation. When the developing frog embryo turns itself inside out during gastrulation, it looks just like a viscous fluid, flowing in an entirely natural manner. Some of the "information" required to make this process work may be specified by the laws of fluids, not by DNA. Brian Goodwin sees development as a combination of natural "free-flow" dynamics and DNA-programmed intervention to stabilize a particular dynamic form. Why should nature waste effort programming the shape of the organism into DNA if the laws of physics will produce it free of charge? It's like programming into DNA the fact that salt crystals must be cubical. For example, the eye—a shape that puzzled both Darwin *and* his detractors—is dynamically very natural. Rudimentary eyes can occur naturally without any special DNA coding. Natural selection can then refine the rudimentary eye into something more sophisticated, but it is the dynamics that gives selection a head start.

This dynamic view of morphogenesis has many other advantages. For instance, it explains convergent evolution as a dynamic attractor onto which many different initial states converge; and it explains organisms' stability of form as the dynamical stability of attractors. Dynamics provides an appropriate context for development, it imposes natural contraints on what is possible or likely, and it tells us what forms might be "downhill" to evolution.

We know very little about the meaning of the DNA code. There is a huge gap in current biological understanding: We can read the message letter by letter, and spot an odd word or even an entire sentence that we recognize—such as "hemoglobin"—but we have no idea what story the Book of Life actually tells. The rules about how to read the DNA code are vital in determining the message that is received, but the processes involved are remarkably complex and poorly understood.

EXTERNAL EFFECTS IN DEVELOPMENT

Scientists are so impressed by the manner in which DNA replicates that they tend to see it as the archetypal self-replicating system. However, Richard Lewontin has pointed out that whatever DNA is, it is not *self*-replicating. Putting some DNA in a test tube and waiting for it to replicate itself is comparable to laying a letter down on a desk and waiting for it to copy itself. The letter needs a photocopier before it can replicate, and the DNA needs a fully functioning cell. There are millions of copies of the same document lying around, and biologists are reading it intently in the hope of understanding how it copies itself. Meanwhile the photocopier in the corner sits neglected and forlorn. To be fair, they think that the document includes a blueprint for the photocopier, but can you understand that blueprint without taking the photocopier to bits to see how it works?

The early development of an embryo is not controlled by its own DNA, but by the architecture of the egg and by maternal-effect genes. Many experiments have been performed in which the natural cell nucleus of the zygote is replaced by that of another species, and development continues to follow the same path that it would have followed with the original nucleus. Only when the embryo has attained a definite degree of organization does each of its cells begin to call upon special gene sequences in its own nucleus. The cell's genetic content includes many possible types of behavior, but it is the context that decides which potentialities are realized.

In chapter 3 we introduced a computer analogy: DNA is software and the egg read-out mechanisms are hardware. But the analogy is incomplete, because the construction of the hardware involves the instructions in the software. Indeed, parts of the DNA sequences prescribe which amino acid is attached to which transfer RNA; the method for replicating the genetic code is prescribed within the code itself. This feature alone is not fatal to the Book-of-Life view of development, but it certainly requires a modification of that view. We have to see the organism as a complicated self-referential product of the DNA. Can we then continue to argue that everything derives ultimately from the DNA content?

The answer is no, because of the difference in timing. Part of the relevant DNA comes from the current generation and part from the previous one. This generation gap makes all the difference: It means that there can be no mapping between DNA sequences and organisms. If there is no such mapping, the Book-of-Life image is misleading and must be abandoned.

The following examples are a bit obvious, but none the worse for that. The caterpillar has the same DNA as the butterfly, the maggot has the same DNA as the fly, the human embryo has the same DNA as the grandmother she eventually becomes, and the axolotl doesn't use its "adult" DNA program at all. If you implement the DNA code inside a butterfly egg you build a caterpillar. If you implement it inside a pupated caterpillar you build a butterfly.

This is disturbing, but it may still be possible to see DNA as the Book of Life. Perhaps what's happening here is this: The DNA comes in two separate pieces. One can be used to build a caterpillar, the other a butterfly. You could then argue that there *is* a mapping from DNA to organism; but that the mapping applies to the pieces, not to the whole thing. It could also be true that different pieces of human DNA are used to build the early embryo, then other bits tinker with the embryo to produce a baby, then other pieces (a male kit or a female kit) are used to get an adult, and then a senility kit removes the adult to prevent it from competing against its own progeny. This jigsaw-puzzle model of development, with all the bits and stages being preformed in the DNA as separate programs, would be a possible way to build complicated organisms. Now, we don't know the specific facts about caterpillar DNA, but we do know that most organisms don't use the DNA-kit method. All the important pattern-forming genes—for example, the homeotic sequences—are used again and again, but in different contexts.

■ FROGODILE AND HYPNOCEROS

However, we can demonstrate that even in principle the Book-of-Life image must be incomplete. What we already know about development implies that there is much more going on than what is written in the DNA book. We begin with a thought experiment showing that in principle two very different organisms—much more different than a caterpillar and a butterfly, not different metamorphosed stages of "the same" organism—could have precisely the same DNA. The sequence of bases in their genetic material could be identical, yet they could develop with it, reproduce, and stay different. Could something like this actually happen in nature? Ah, that's a different question. Thought experiment first.

Remember that there is a hierarchy of organizational levels in DNA sequences: Some code for protein production, some switch on other genes in the right order, and so on. This difference in the meaning of DNA subsequences is crucial. Suppose that we sequence the genomes of two imaginary creatures, a frogodile and a hypnoceros, say, and after disentangling the results we find that each has a structure like Figure 43. The early stages of frogodile development, prescribed by this DNA code, are as follows. The presence of the protein frogodoin, provided by the mother as a result of her maternal-effect gene, triggers switches that activate the genes that control the development of frogodile morphology. Nearly all of the actual construction is done by thousands of structural and organizational genes that make proteins, move them around, and so on. The maternal-effect gene for frogodoin production is also activated, so that when the adult frogodile reproduces, this protein will be present in her egg. On the other hand, the genes that could lead to the development of hypnoceros morphology (accessing the same structural and organizational genes but in different patterns) and the maternal gene for the production of hypnocerin are not activated at first. When they are, there is an enzyme that destroys hypnocerin. (We add this step to make sure that all genes do get activated; otherwise the trick is too simple.)

The early stages of hypnoceros development, prescribed by the same DNA code, are as follows. The presence of the protein hypnocerin, provided by the mother as a result of her maternal-effect gene, triggers switches that activate the genes that control the development of hypnoceros morphology. Nearly all of the actual construction is done by thousands of structural and organizational genes that make proteins,

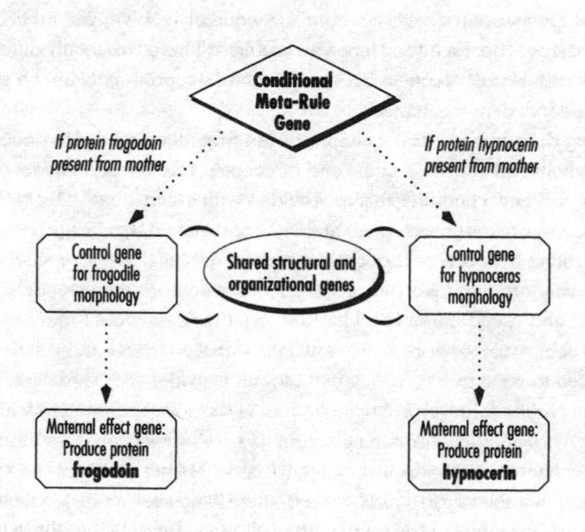

FIGURE 43

A tale of two genomes

move them around, and so on. The maternal-effect gene for hypnocerin production is also activated so that when the adult hypnoceros reproduces, this protein will be present in her egg. On the other hand, the genes that could lead to the development of frogodile morphology (accessing the same structural and organizational genes but in different patterns) and the maternal gene for the production of frogodoin are not activated at first. When they are, there is an enzyme that destroys frogodoin, for the same reason as before.

On the assumption that each creature accesses all of the shared pool of thousands of organizational and structural genes, their development activates *all* genes. The difference in the complete genome is zero. Of course, the two creatures need not access such a large proportion of the shared genes; but it's interesting just how much they might have in common, even if you look only at genes that *are* accessed.

Of course, if such a dual-purpose genome ever appeared in nature, the two species associated with it would subsequently evolve, and their DNA would change. But for a long time you would still have two totally different species with almost identical DNA, still faithfully reproducing by the same context-dependent mechanism.

What does this artificial example of the frogodile and the hypnoceros tell us about real genetics? At the end of chapter 3 we likened the developing egg to a new computer that is provided with a start-up disk by Mother. We must now take a more contextual view and amend that image, because the frogodile-hypnoceros thought experiment tells us that it is precisely the wrong way around. (We only realized this when we were revising this chapter, and asked ourselves what the Zarathustrans would think of our overall contextual viewpoint. We now take our alien friends more seriously and listen to what they're telling us.) Mother provides the "hardware," the cell that begins to develop; the new item is the software, the DNA inside the egg (provided by combining sequences from both father and mother and peculiar to the developing infant). That is, Mother provides the entire computer, not the start-up disk. Indeed, there is no start-up disk: The computer is up and running before the infant's DNA software is inserted into it, and only then does it begin to obey the program on the infant DNA. Notice how different the roles of infant DNA (content) and mother (context) now become. In the reductionist image of chapter 3, all of the magic is in the infant DNA, which is seen as being in full charge of the process—"It's all in the DNA." The maternal context is just a starter motor to get the process going. But in our new image, all of the magic lies in the maternal context, the fully functioning egg into which a naïve nucleus containing infant DNA is inserted and run. It is because the context is at least as important as the content that we can envisage a case in which the same DNA message has vastly different developmental meanings.

All very well, you are saying, but it's very contrived and nature doesn't actually do things that way. In practice, a given DNA message—a complete genome—has a definite, unambiguous effect.

Not so.

In chapter 4 we told you about anglerfish. All their larvae have the same genetics, but the sex (and hence the physical form) adopted by the larva during development is determined by context, that context being an encounter with an adult female. Surviving larvae that find a female turn into males; the rest become females. This process doesn't lead to two different species, but it leads to two entirely different phenotypes—bodily forms—and it does so by the same general mechanism as the frogodile/hypnoceros system. This example is not unique; another is *Bonellia*, and here the story has been checked out in the laboratory. For a long time male and female *Bonellia* were thought to be different species. This is hardly surprising, since the males are tiny ciliated microorganisms with around 150 cells, whereas females have a body the size of a walnut that is immersed in mud, to which is attached a tube some six inches long. At its tip the tube splits into forks, with ciliated grooves. The males parasitize those grooves and become males by doing so. If they don't lodge in a groove, they become females. Again the genetics of the developing organisms is the same for males or females; the sex is determined by context, not by content.

If the Book of Life doesn't even tell you what sex to be, what else doesn't it tell?

■ SEQUENCING THE ZENOME

Our thought experiment about non-uniqueness is only one of many ways in which the idea of a map from DNA sequence to organism can fail. There are more extreme examples, things that nature *could* do rather than (so far as we know) she *does* do.

NEEPLPHUT: Great news, Captain!

CAPTAIN ARTHUR: You've finally got the DNA analog of your ZNA engineering environment working? Terrific! Now we can get some real soyasynthasteak instead of that disgusting synthetic stuff your chemists have made.

NEEPLPHUT: No, I regret/am delighted to inform you (delete whichever is inapplicable) that our biologists have yet to solve that problem. No, I am referring to the success of one of the greatest projects yet contemplated by Zarathustrankind.

STANLEY: What?

NEEPLPHUT: The secret of life, the octiverse, and probably not much else. Look! [*Drops optical disk on table.*] I have had it specially formatted to suit your computer.

CAPTAIN ARTHUR: Thank you, Neeplphut. I appreciate your generosity. Stanley, let's take a look at this wonderful secret.

STANLEY: Aye, aye, Captain. [*Inserts disk into slot and starts it at a randomly chosen position.*]

CAPTAIN ARTHUR: What does it say?

STANLEY: ZZZZZZZZZZZZ.

CAPTAIN ARTHUR: Wake up, man!

STANLEY: No, it says "ZZZZZZZZZZZZ" on the screen. It goes on like that for ages.

NEEPLPHUT: You must have hit some junk ZNA. Try somewhere else, it's usually much more interesting.

STANLEY: ZZSWSWTXSSTXSWZZ.

CAPTAIN ARTHUR: Sounds like a walrus gargling. Neeplphut, what is it?

NEEPLPHUT: Our scientists have succeeded in sequencing the Zarathustran zenome! After octs of labor we have finally produced a complete sequence of the entire ZNA of a Zarathustran. A treasury of zenetic information, our evolutionary heritage. With it, we shall be able to cure new diseases.

STANLEY: That's great!

NEEPLPHUT: Yes. Of course, we will have to invent the new diseases first, and the Regulations may raise objections. It usually does. But in principle it is well within our competence. Do you/do you not (delete whichever is inapplicable) agree that it is an unprecedented triumph?

CAPTAIN ARTHUR: Of course! Congratulations!

STANLEY: Wait a minute, didn't *we* . . . ?

CAPTAIN ARTHUR: Yes, we did. Back at the start of the twenty-first century, I think.

STANLEY: There's a disk somewhere. I'll go get it.

NEEPLPHUT: What is this you are fetching?

STANLEY: The human genome. Our own DNA, completely sequenced. I'm sure it's knocking around somewhere. Last time I saw it it was in the box along with last year's sports results. [*Rummages.*] Got it!

NEEPLPHUT: You do not treat such an important discovery with much respect.

CAPTAIN ARTHUR: The truth is, Neeplphut, that it wasn't as interesting or as useful as people hoped it would be when they spent so much effort working it out. But that's another story. It was certainly an impressive piece of work.

STANLEY: Sure was. The final sequence was [*reads label on pack of disks*] precisely 3,333,333,333 bases long.

NEEPLPHUT: That's a curious coincidence.

STANLEY: Very. It started an entire new religion.

NEEPLPHUT: Hmmm. Human numerology! You do realize that in base eight that number is just 30,653,520,525, a meaningless jumble that would never have started a . . . But that is not what I was talking about. No, the curious thing is that our zenome is precisely 2,222,222,222 bases long in your notation.

CAPTAIN ARTHUR: So humans are 50 percent more complicated than Zarathustrans?

NEEPLPHUT: No, Captain! You are forgetting that, consistent with the basic principles of octimality, our zenetic code employs *eight* different bases—coded as S, T, U, V, W, X, Y, Z on the disk—arranged in sixty-four pairs. Whereas your genetic code works with sixty-four triples formed from four bases: A, C, G, T. Thus your genome contains exactly 1,111,111,111 triples, and our zenome contains 1,111,111,111 pairs. The quantity of information is identical.

CAPTAIN ARTHUR: That's a curious coincidence.

STANLEY: Very. It could start an entire new religion.

NEEPLPHUT: Maybe we should compare the two sequences. Can you do that?

STANLEY: Sure. I'll display your sequence in pairs and ours in triples, so that we can compare equivalent units of information. Right, here we go. [*Computer screen reads:*]

ZZ	SW	SW	TX	SS	TX	SW	ZZ
CAT	GTT	GTT	CAA	GTC	CAA	GTT	CAT

CAPTAIN ARTHUR: That's very odd. Every time the zenome has ZZ, our genome has CAT. And every time the zenome has SW, our genome has GTT. Stanley, does it—

STANLEY: Go on like that? You mean, is there a perfect one-to-one mapping between the Zarathustran zenome and the human genome, in which each of their sixty-four pairs maps to a unique member from our set of sixty-four triples?

CAPTAIN ARTHUR: Yes, that's the gist of what I was about to say.

STANELY: Let me see. . . . Hmmm. No.

CAPTAIN ARTHUR: Phew, that's a relief.

STANLEY: There's a disagreement at the 43,772,331st group. Otherwise, there's a perfect match between zenome pairs and genome triples.

NEEPLPHUT: You mean, with that sole exception, humans and Zarathustrans have exactly the same—um—nome? You are just a coded version of us?

STANLEY: No, you're just a coded version of us. [*Long pause.*]

ALL TOGETHER: Your scientists must have made an error at the 43,772,331st group!

GENES AND CHARACTERS

The content of DNA, then, is not the sole determinant of what an organism looks like. Even if we could get hold of a complete dinosaur genome, we might not possess a complete specification of how to build a dinosaur. However, we don't want you to imagine that DNA is irrelevant to form. On the contrary, there are many instances in which differences in appearance between two organisms of the same species, say, can be attributed to very specific differences in their DNA sequences. For instance, as mentioned earlier, in an albino mouse the enzyme tyrosinase does not work, whereas in a pigmented mouse it does. We can work out how that enzyme affects the production of pigment, and see why the albino mouse has white fur and pink eyes, so the enzyme deficiency definitely *causes* albinism. Moreover, the different mutations in the mouse's genetic material that cause the enzyme deficiency can be mapped precisely, and we can work out why differences in the DNA cause enzyme trouble. So here the entire chain of causality exists, and we can sensibly say that *this* change to the DNA causes the character albinism. Note, however, that our understanding of the process holds good only in a limited context: the mouse or any other animal that uses tyrosinase in the same manner. Our frogodile/hypnoceros setup can easily be modified so that a particular DNA mutation produces albino hypnoceri but leads to overweight frogodiles with an allergy to cabbage.

Similarly, vestigial-wing mutations in *Drosophila* are very well known, and specific differences in DNA sequence have been found, in comparable places, between vestigially winged flies and those that possess normal wings. The same is true of many mutations in human DNA that cause disease. These have been mapped to specific places in the genome and compared to the normal sequence. However, this does not mean that we can find the genes that make wings. For instance, some of the vestigial-wing mutations affect ion pumps in cells all over the fly. They prevent the wings from inflating properly, but not by getting the "wing kit" wrong, because there isn't a wing kit as such. Indeed, in some flies the same mutation has quite different effects.

What we are discussing here is the time-honored issue of genes and characters. Just as there is no well-defined mapping from DNA sequences to organisms, so there is no well-defined mapping from genes to characters. This does not conflict with the known fact that *some* genetic differences lead to *some* character differences (such as albinism); it just means that not everything we would like to call a character can be traced to specific pieces of DNA, independently of context.

In chapter 3 we followed Richard Dawkins and defined the term "gene for X" as "genetic variation that leads to X." Logically speaking, there is no problem with this terminology. However, it is psychologically misleading and often not very useful. It tends to make us think of a map from things called genes to things called characters, and that makes us think that different characters are related to different genes. Moreover, at the back of every biologist's mind is the idea that DNA is the genetic material, so "gene" should be some identifiable subsequence, or system of subsequences, of DNA. Unfortunately, in the full generality of the phrase "a gene for X," it's not.

Or, perhaps worse, the gene *is* a DNA subsequence—but situated in the wrong creature. Dawkins explains why he is entirely happy with quite subtle choices for the character X, such as "a gene for skill in tying shoelaces." But how about "a gene for having a blue-eyed mother"? There is certainly genetic variation for the character "mother was blue-eyed." (We do not mean just the character "blue eyes." You possess the character "mother was blue-eyed" precisely when your mother possessed the character "blue eyes.") This character even correlates quite well with genes in you, because some of your genes are inherited from your mother. Its pattern of inheritance is much like that for any other gene—because it's really a genetic difference studied, albeit unknowingly, with a phase shift of one generation. If you buy the argument that "a gene for X" is just a harmless rephrasing of "there is genetic variation in the population with regard to the character X" then you have to buy "a gene for having a blue-eyed mother." Agreed, the reason why that one is bizarre is clear enough. But then, some of "a gene for skill in tying shoelaces" may well trace back to your mother's DNA in the same way: You are good at tying shoelaces because your mother taught you how. She had "a gene for teaching how to tie shoelaces." Or was it Grandma, with "a gene for teaching your children how to teach tying shoelaces"? These and similar examples strongly suggest that we should not be too cavalier about which characters are admissible when we dress them up in genetic language.

Let's return to our analogy of chines in a population of cars. There is undeniably chinetic variation for skidding: Some makes of car are more liable to skid than others. Thus, in the spirit of "a gene for X," we may conclude that cars have "a chine for skidding." Because we know how cars work, we also have some idea of what causes skids, so we can dissect out the nature of this chine. It involves treadbare tires, for example, and worn shock absorbers, and defective steering. But cars don't skid a lot just because their tires are worn or their steering is defective. The conditions for skidding are conjunctions of combinations of attributes. "If the car has a powerful engine and the tires are worn, *or* if the tires are worn and the steering is defective, *or* if the shock absorbers are shot and the car has rear-wheel drive, *or* . . ." This is what the typical chine looks like. In the same spirit, a more accurate DNA image of "gene" would be some system of conditional possibilities for the DNA: "If it doesn't have the sequence CCGTTA in this position, *then* it has to have the sequence AAGGCTTTCA over there, *unless* it's got TTGGG over here instead . . ." And all of this within some agreed overall context, such as "mouse genome."

Thus we should not imagine that "the gene for wings," let us say, consists of all the various genes for the components of the wings—structural chemicals, pigments, veins, and so on. We can't simply collect all the genes whose sequences in fly DNA have some effect on the wings; if we did, we'd get the entire fly genome. It's far truer to say that each DNA subsequence affects *all* characters than it is to say that DNA is composed of subsequences each of which affects just *one* character.

HOMUNCULUS CODE

What, then, is a character, and how does it relate to DNA? We can get an idea of the actual relationship by using our image of spaces of the possible. In this terminology, DNA sequences live in DNA space. This consists of all possible sequences of bases C, G, A, T. By "possible" we are not implying that any particular sequence will be found in nature, just that it's a logical possibility. Organisms, on the other hand, live in a space that is far harder to pin down: creature space. This is the space of all conceivable creatures that could be made by nature. They don't have to be evolutionarily viable; flying cats and ten-legged donkeys are probably acceptable. We have to say "probably" because we don't know if, even in principle, such creatures could exist. But nature seems pretty malleable. You may find this idea bizarre, but it has a purpose: By comparing actual cats with potential flying ones we might be able better to understand why cats don't fly. As soon as we ask why cats don't fly, we've raised the specter of the flying cat, and from then on, until the concept is demolished, we have to be prepared to contemplate such a creature.

Characters similarly live in character space, yet another mathematical fiction; it is the space of all possible features that animals may exhibit, such as blue eyes or long ears. Characters are general properties of creatures, and any given creature's phenotype is pinned down rather accurately by its characters, which is why comparative taxonomy works pretty well as a method for distinguishing species. In a sense, characters are the axes of creature space. Just as coordinate axes (x, y, z) label points in space with three quantities, so characters label creatures in creature space with a large number of qualities. The number of axes is far greater in creature space, because there are more than three characters; moreover, unlike normal spatial axes, characters may not point in independent directions.

What we have been saying is that DNA space is not a map of creature space. There is no unique correspondence between the two spaces, no way to assign to each sequence in DNA space a unique animal that it "codes for." (The idea that there is such a map is just another homunculus theory, as mentioned in chapter 1: Inside every egg is a tiny human being waiting to grow up, complete with its own eggs inside, which . . . If you say that DNA space is a map of creature space, then you're saying that inside every egg is a complete and precise coded map of a tiny human being waiting to grow up, complete with its own eggs inside, which . . .) To twist one of Dawkins's images to our own purposes: The Blind Watchmaker is indeed blind, so it can't read its own blueprints. Biological development is a complicated transaction between the DNA "program" and its host organism; neither alone can construct a creature and neither alone holds all the secrets, not even implicitly.

In chapters 3 and 4 we distinguished between nature and nurture in the development of organisms. This distinction does seem to be a very useful way to think about development; there is an internal, programmed pattern of growth, and there are all the ways in which the environment can modify this pattern. If indeed DNA space were a map of creature space, so that the DNA sequence really was a homunculus code, requiring only to be unfolded into a phenotype when the program ran, then you would be able to see the environment as something that sometimes interferes with "normal" development. The program gives Baby two legs and two arms, but then Mother is prescribed thalidomide while pregnant. If the mother takes thalidomide during certain stages of fetal development, the result is phocomelia—useless flippers instead of arms and legs. Nurture has interfered with nature; normal development has become abnormal. You might even see the interaction of nature and nurture as something that can be categorized numerically: Intelligence is 70 percent genetics and 30 percent environment, and so on. Geneticists who see the interaction this way call this concept "heritability." But can the interaction of nature and nurture really be that simple?

Think of two sisters living in the same house, treated in the identical manner by their parents. One pulls a book off the shelf, and it falls open on the floor. She sees the pretty pictures, and is hooked for life. But when her sister pulls a book off the shelf, it hits her on the head; she bursts into tears and hates books ever after. The percentages of nature and nurture are the same for both sisters, but the result of the interaction is totally different.

To make matters worse, there are cases in which genetics can mimic the effects of environment. In this case the environmental influence is said to be a phenocopy of the genetics. For instance, certain genetic mutations can produce the same developmental effects as thalidomide, so thalidomide is said to be a phenocopy of those genes. Notice the not-so-subtle bias here: We could equally well say that the genes are a "genocopy" of thalidomide. But geneticists don't talk of genocopies; they think that genes determine normal development, while the environment is either neutral or—in rare cases such as thalidomide or falling off a high cliff—harmful.

Not so; just ask any male anglerfish.

■ SO WHAT ABOUT BUILDING A DINOSAUR?

The answer should be clear. If dinosaur DNA were a complete blueprint for a dinosaur, a dinosaurian Book of Life, then reconstruction would be a purely technical problem. Admittedly, we have no idea how to do it—and until biologists start to spend less time poring over the document and more tinkering with the photocopier, we never will—but it could, in principle, be done. However, dinosaur DNA does not describe a dinosaur. It prescribes a dinosaur, within the context of another dinosaur. So to reconstruct Baby Dinosaur, all you need is Mommy Dinosaur.

Tough.

Dinosaurs, in short, are out. But what about mammoths? We haven't got a mommy mammoth, but we do have living mommy elephants; wouldn't those do? Frozen mammoths can be dug up in Siberia, and Russell Higuchi has obtained enough of their DNA to show that mammoths are related by roughly equal amounts to Indian and African elephants, their closest living relatives. In principle it would be possible to insert this DNA into elephants. But, as Higuchi says, "It would make absolutely no difference. They'd still be elephants."

No, you can't start from dinosaur DNA and build a dinosaur.

Hang on. Isn't that what nature did? No. Nature started from *non*dinosaur DNA—indeed, from chemicals that weren't DNA at all—and evolved a dinosaur over a period of billions of years. To understand how that happened, we must apply our contextual viewpoint to evolution.

10 THE EVOLUTION OF POSSIBILITIES

A party of economists was climbing in the Alps. After several hours they became hopelessly lost. One of them studied the map for some time, turning it up and down, sighting on distant landmarks, consulting his compass and the sun.

Finally he said, "Okay, see that big mountain over there?"

"Yes?" asked the others eagerly.

"Well, according to the map, we're standing on top of it."

In chapter 4 we introduced Darwin's original evolutionary scheme: Genetic variation arises through random mutations, and natural selection then weeds out those mutations that don't improve the organism's chances of survival, allowing mutations that do improve survival odds to replicate. This process leads to increased complexity and sophistication of organization of living creatures; complexity is "downhill" to evolution—it's the path of least resistance. We also developed the orthodox identification of the core of the genetic mechanism as the DNA blueprint, and of mutations as tiny changes to the DNA code; these identifications led directly to Dawkins's vision of the selfish gene. In this view it is DNA itself that evolves, because it is DNA that replicates. Any DNA string that stumbles across an effective trick for replication—whatever that trick may be—will be replicated more effectively.

In this view the organism is a kind of by-product, a vehicle for the replication of DNA. At the heart of evolution there lies not competition between creatures, but competition between DNA strings—selfish genes—that use "their" creatures to fight proxy battles. But since the result of natural selection depends not just on the form of an organism, but also upon its behavior relative to the entire ecosystem in which it resides, we must consider not just its phenotype (individual form and behavior) but what Dawkins calls its extended phenotype (its interaction with the whole of its environment).

We have also mentioned one recent modification of this idea of evolution: Gould's argument that contingency can play a major role. The remarkable diversity of the soft-bodied organisms of the Burgess shale, and the relatively small numbers in which such creatures existed, provides an opportunity for chance events to shape evolution, which is then determined not by the predictable effects of "fitness," but by contingency.

In the second half of this book we are building up a rather different point of view on science, looking for external interactions rather than internal reductions. In the previous chapter we questioned the view of DNA as the Book of Life, a chemical map of "its" organism. This change in such a basic assumption is bound to affect our view of evolution. We have argued that biological development is controlled not just by the internal genetic code but also by external factors such as maternal-effect genes and developmental dynamics. We now argue that something very similar must happen in evolution. The "geography" of the space in which evolution occurs implies that there are external constraints on evolution, giving it large-scale patterns that lead to convergence as well as contingency; and the conventional view that DNA mutations drive evolution is open to a variety of challenges. Just as two different creatures can have the same DNA code, so creatures can evolve without any genetic changes at all, or they can stay the same even though their genes are changing considerably. We can even turn the "selfish gene" image inside out and make organisms paramount. The result is an enriched vision of evolution in which both DNA and organisms interact, and neither can be reduced to the other.

DYNAMICS IN DNA SPACE

When we were thinking only of the DNA program evolving and of organisms as just the extension in space of the DNA code, then we could easily handle the idea of mutations being good or bad. Dawkins's view ruled: Good strands of DNA produce more of their own kind; bad strands become extinct. In the language of the previous chapter, we saw evolution as a dynamic process in DNA space, driven by success in being replicated. Therefore, DNA progresses onward and upward, and so do the corresponding organisms in creature space. However, if the relationship between DNA sequence and organism, between genotype and phenotype, is as fragmented as we have just suggested, then this comfortable picture begins to break down.

The picture conveyed by chapter 4 is that random changes occur in the genetic code and that these result in different organisms. The success of these new organisms, relative to the parent stock, determines whether the new code will become more prevalent and displace the old one. In fact this is not exactly what Darwin said. He observed only that there are differences between animals of the same species; he made little progress on the question of where those differences came from. Darwin worried a lot about those differences. He knew that many are caused by "accidents," and his observations confirmed the general opinion that such acquired characters are not passed on to future generations. Men who have lost an eye in an accident do not produce more than the normal share of one-eyed children, and blacksmiths' sons are no more brawny at birth than you would expect from knowing the family. On the other hand, many characters, such as eye color, do seem to be inherited. Darwin realized that something cryptic must be passed on in sperms and eggs, that it isn't possible to pass on the actual characters. But he was more concerned with competition and selection than with the source of the differences.

AUGUSTA ADA: I am greatly perplexed by what you have just told me. You say that it is DNA that evolves, not organisms.

WALLACE LUPERT: Exactly.

AUGUSTA ADA: But I have also been told that the form of an organism is usually stable against changes in its DNA code. Genetic change often occurs when the phenotype is static. Equally I can see that there might be large changes in phenotype without any genetic change at all. Think of shepherd's purse, whose phenotype, as we have already discussed, changes markedly in response to the availability of room to grow.

WALLACE LUPERT: Agreed.

AUGUSTA ADA: But the evolutionary success of "selfish" DNA depends entirely upon selection by way of the organism that it determines. What you have said suggests that the coupling between genotype and phenotype is so unspecific that the selection criteria bear little relation to the underlying DNA at all. For example, if my purchase of goods is influenced mainly by the attractiveness of the packaging, but if the packaging bears no particular relationship to the contents, then I can see that the quality of the packaging should improve over time, but there seems to be little implication for the quality of the contents. Selection is not operating on the contents.

WALLACE LUPERT: I'd put it this way: The forces that shape the direction of evolution act in creature space, not in DNA space. And the correlation between the two spaces is so diffuse that there may be no strong direction to evolution in DNA space at all. But surely this is just the old genotype/phenotype problem that Darwin worried about?

AUGUSTA ADA: Not really. If there is a sufficient degree of congruence between DNA space and creature space, then selection in creature space feeds back to DNA space in quite a strong manner. Both spaces, so to speak, may have a similar geography, even if it is not identical. However, you are telling me that no such congruence exists, so there appears to be no good reason for the two geographies to match. And then the entire process breaks down.

WALLACE LUPERT: That reminds me of a cartoon I once saw. There were two DNA strands, and one was saying to the other "To heck with natural selection—I'm going to make a totally silly organism!"

AUGUSTA ADA: But surely, Wallace, it could not *do* that.

WALLACE LUPERT: I don't think so. Evolution doesn't act on the raw content of a DNA string, the bases, but on what that content implies within a suitable context—the organism to which it belongs. Within any such context, any change to the DNA that resulted in a totally silly organism would lead to the immediate demise of that particular DNA sequence. The cartoon describes a form of DNA suicide. Selection applied to the resulting organism will eliminate "stupid" DNA. Of course, a strand that is stupid in one context might be perfectly sensible in a different context. But that doesn't affect my argument, which is about what happens within a chosen context.

AUGUSTA ADA: So "selfish" DNA is the wrong image. The DNA must take very good care indeed of the creature to which it belongs. Only DNA that, in its particular stretch of code and in conjunction with all the rest, *and* in the context of the reproduction of that particular organism, leads to a sensible, viable phenotype can hope to survive.

WALLACE LUPERT: You're right. DNA can't be selfish. It can't do anything that is against the interests of "its" organism—*as* an organism, seeking to survive—or it destroys itself by failing to become replicated. DNA is a slave that labors, although unknowingly, purely in the interests of the organism that develops from it.

AUGUSTA ADA: I wonder what would have happened if Dawkins had called his book *The Slavish Gene?*

■ COMPARATIVE GEOGRAPHY

Despite Ada's criticism, there are some important ideas in the notion of the selfish gene. One is that a DNA sequence need not have a well-defined role in an organism's development in order to be replicated. Junk DNA gets copied along with everything else. If some DNA sequence can act as a "parasite" on the entire reproductive process, to its own advantage and not to any great disadvantage for the organism, then it will be replicated.

This idea must be true—at least, in some sense. However, the image places too much emphasis on DNA as the only thing that really evolves. In every branch of science you will find people who will tell you what everything "really" is—indeed, what you yourself "really" are. A quantum physicist will explain that you are really a wave function. A molecular biologist will tell you that you are really a DNA program. A computer scientist will tell you that you are really a neural network. All make the same mistake: assuming that because they have some universal system that can mimic, describe, or prescribe something else, no other valid viewpoint exists. Look at yourself in a mirror. Do you see a wave function? A DNA program? A neural network?

The claim that DNA alone drives evolution is overstated. To see why, imagine an aspirinlike chemical, used to cure headaches, with a side effect: If it is used by a pregnant woman then it makes her babies more prone to headaches in adult life. This sets up a self-reinforcing cycle. The chemical has an "extended chemotype": its effects on humans. It would seem to be stretching a point if we were to talk of selfish aspirin, or of human beings as fundamentally just a vehicle for its replication. Drug dependency involves a similar cycle, but is it reasonable to speak of selfish heroin, with its own extended chemotype?

Any chemical with an effect that humans value will be replicated (by humans running a chemical plant). Such a chemical may in turn modify human behavior. The analogy with evolution is clear. In place of random mutation of DNA we have the random explorations of chemists. In place of development we have the effects of the chemical on human physiology and behavior. In place of natural selection we have the workings of human preference—for example, the marketplace. So it is not enough to think of DNA as *the* thing that evolves, and everything else as a vehicle for its selfish evolution. We must think about the interaction between organisms and DNA.

The nonexistence of a unique map from DNA space to creature space implies that overall the two spaces have different geographies, even though there may sometimes be local similarities between them. That in turn implies that any understanding of evolution must deal with the coupling between DNA space and creature space, between genotype and phenotype. You can't refer the process of evolution back to DNA space alone.

Nonetheless, it would be quite wrong to imagine that there is no connection between the two spaces. Our frogodile/hypnoceros thought experiment is an extreme case showing that in principle there can be no unique mapping. We're not claiming that anything as extreme as that will be common in nature; but we've already given you two cases where it occurs: anglerfish and *Bonellia,* where the very different males and females have exactly the same genetics. Because there is no such mapping in principle, it's not a very good idea to talk of genes for arms, genes for wings, or genes for kidneys, as if such things exist as well-defined sequences of DNA that are independent of context, whether the context be time or space.

However, the DNA sequence plays such a key role in biological development that we would expect all kinds of partial congruences between bits of DNA space and bits of creature space. Those congruences should become more and more precise the more closely the context is controlled. If we concentrate on particular differences between DNA strands, keeping everything else as far as possible the same, then we are varying the context as little as possible. This makes it entirely proper to talk about gene mutations that affect arm development, or about mutations in humans that cause kidney disease or albinism. In that context, changing a few bases in the

DNA produces a well-defined developmental effect. It has to, because development is the interaction of environment and DNA code within a smoothly running chemical machine. Plug in a different input, and you expect a well-defined difference in output. All we are saying is that the output does not depend solely upon the genetic code; other factors play a major role, too.

The same point is fairly obvious if we recall the analogy with the parts of a car. We've argued that it makes little sense to talk of a chine for skidding. That does not imply that we can't say that cars with bald tires tend to skid more than those with good treads. We can, and it's true. But it doesn't imply that bald tires are the sole cause of skidding. That depends on the make of car, the speed at which it's traveling, and the road conditions.

What we've now established is that evolution does not take place solely in DNA space, as is effectively the neo-Darwinian view. And it certainly doesn't take place solely in creature space; that would be the Lamarckian view of the inheritance of acquired characters. So where *does* evolution take place?

■ GENETIC ASSIMILATION

The answer is straightforward, but its implications are not. Evolution happens in the combined space of DNA *and* organisms, and it is driven by their interaction. The crucial point is that this interaction works both ways. It cannot be reduced to events that happen in one space alone, as the selfish-gene theory tries to do.

Everybody agrees that genes affect the form of the organism, but we want to convince you that the reverse is true and is just as important. The form and behavior of an organism determine what gene changes have survival value and are thus more likely to be passed on to progeny. This two-way interaction has profound consequences, because organisms are not made by assembling genes like the pieces of a jigsaw. Many of the genetic components of a living creature are actually versatile kits. These kits are more sophisticated in wild-type organisms—those found outside laboratories—than they are in organisms bred especially for genetic experiments. The reason is that laboratory breeding programs deliberately reduce genetic versatility. For example, wild-type frogs can develop successfully at many different temperatures, but laboratory frogs would be bred for sensitivity to temperature in order to separate out the different developmental pathways that the wild-type frogs use. In this respect laboratory genetics paints a misleading picture of how genes and the development of biological form interact with each other.

In real evolution the versatility of genetic kits binds DNA space and creature space together, making them inseparable. We now give two examples to show that their interaction leads to new and surprising effects. The first was suggested by Gregory Bateson in *Steps to an Ecology of Mind.* It is described in terms of real animals, but is deliberately simplified to emphasize a few basic ideas. The second is a thought experiment, but it is in some ways closer to what actually must be happening all the time. In both examples the central idea is that adaptability of body type together with behind-the-scenes genetic variation can lead to evolution. The process involved was identified by Waddington—it explained his experiments on cross-veinlessness in *Drosophila*—and it is known as genetic assimilation.

Bateson's example is a Kipling-style story about how the giraffe got its neck. Imagine a herd of proto-giraffes browsing on bushes and looking longingly at leaves higher up, just out of their reach. By what plausible route could these creatures evolve longer necks? A longer neck is no use unless you also have a more powerful heart to pump blood the longer distance up to your brain. If you assume that the evolution of a long neck must be caused by an innovatory mutation leading to a specific longer neck gene, then all proto-giraffes will die from heart failure, a trait not very conducive to future reproduction. This argument may bring comfort to anti-Darwinists, but it ignores the possibility that the giraffe's neck evolved in some other way.

Could the neck perhaps grow in length because of some general "grow larger" control gene? At first sight this idea seems to solve the problem, because the heart will grow larger along with everything else. However, it's not easy to scale up an animal and keep its physiology working properly; a mouse the size of an elephant would collapse under its own weight because its leg bones would be too thin. Somehow the different parts of the animal—heart, neck, muscles, artery walls—must all stay in balance as the neck lengthens. And evolution has to achieve this balance in each individual, at all stages of its development, or else the innovation will die out.

You can't do this with blueprint-type genetics. Too many independent factors must change coherently, and there are not enough animals to achieve this by repeatedly throwing the evolutionary dice and waiting for an unusually long series of sixes. It has to be done by versatility. Each part of the animal must respond, in its own way, to the stress of having a longer neck, and it must respond during the animal's own development, not over a time lag of generations. So, for instance, the heart must adapt by getting bigger—and this is where the possibility of nongenetic change arises. The heart can adapt in two distinct ways: either nongenetically, by growing more powerful through exercise, or by genetic changes that tend to produce bigger hearts anyway. Since genetic changes happen over many generations, in any individual animal the heart must adapt through exercise. All the other related "kits"—muscles, skin, artery walls—are also "stretched" by the existence of a longer neck, and they must adapt in the same nongenetic way.

Some individuals simply cannot adapt successfully, and the innovative "long neck" gene kills them. But baby proto-giraffes with sufficiently adaptable genetic kits can reach leaves higher up trees, and that offers them a survival advantage, which they pass on to their progeny. However, there is a price to pay: At this stage in their evolution these animals are living in a state of constant stress, extended to the very limits of their adaptive capabilities, like an athlete who is always running a four-minute mile.

Other factors now begin to reduce this stress by assimilating the innovation into the population. There is always a spread of genes in the population, so in later generations some offspring will have a heart kit that produces a bigger hear to start with, or a muscle kit for inherently stronger muscles, or an artery kit for thicker artery walls. These animals also have to adapt to the longer neck, but not as much. Adults whose hearts are big because of a "big heart" gene look exactly the same as those whose hearts are big because of exercise, but they are less stressed. And such adults will become more prevalent in each generation, for a simple reason. If a giraffe is chased by a lion, then one whose heart is genetically disposed toward greater strength can run away as well as pump blood up its longer neck, whereas one whose heart has grown to the limits of what can be achieved through exercise can't.

The giraffe story is very accessible and intuitive, but for our purposes it has one defect. It involves a new gene, so it leaves open the argument that at root evolutionary changes are driven by a genetic mutation. However, the identical process of genetic assimilation can also operate when the stress comes from an environmental change. Imagine that you are on the Galápagos islands observing one of Darwin's finch species—say a kind that eats seeds—and monitoring the size of its beak. At some point the size begins to increase dramatically. Are you justified in assuming that there has been a mutation producing a "bigger beak" gene?

The answer is no.

To see how such an event might occur without any changes in genes at all, suppose that a new species of plant arrives on the island, from seeds borne on the feet of seabirds. This new plant produces larger, harder seeds than those that form the finches' normal food. In order to crack the larger seed, the birds need stronger, hence bigger, beaks. At least two different kinds of genes determine beak size: genes for beak size itself, and genes for the ability of beaks to grow when exercised. You can't determine the separate effects of the two kinds of gene just by looking at beak sizes, for a big beak may have started big or it might have started small and been exercised a lot. So when large seeds suddenly appear, the effects of the beak-building gene, already present in the population, may show up. Then the observed sizes of beaks will increase with no genetic changes whatsoever.

Versatile kits are necessary for this method of evolution—genetic assimilation—to work. Conversely, genetic assimilation offers an advantage for the development of versatile kits, and this explains why they arose in the first place. The kits, and the advantage that was conferred upon them, evolved together.

We can take this thought experiment one step further. A species that is doing perfectly well and has no wish to evolve can suddenly have evolution thrust upon it by circumstances beyond its control. Suppose that upon the island there also lives a snail. Originally the finches ignore it, because it is too big for their beaks to crack. But as the finches' beaks get bigger, some of the birds may acquire a taste for snails. In order to survive, the snails must develop larger shells, and an evolutionary arms race begins. It is between finches and snails—but it was triggered by a seed.

The message of conventional evolutionary theory is simple and direct. Evolution is driven by random mutations in genes, with no particular direction; mutations produce changes to organisms, and natural selection of organisms prefers those mutations that offer increased survival value. But now we are saying that this is not the full story. Mutations to genes may not affect the organism at all, and organisms may change even though their genes do not.

It may seem that everything is up for grabs, and there is no connection between genes and organisms at all. But that's not what we are saying. What we are arguing is that there is no simple, universal connection between genes and organisms. Instead, there is a rich, fascinating, and largely unexplored joint dynamic. The story of evolution that we have just told actually has many advantages over the conventional one. It tells us that the versatility of genetic kits drives evolution in directions that make sense on the level of organisms. It shows how the geography of creature space affects the evolutionary dynamic in DNA space. Organisms evolve the way they "want to," in the sense that the direction in which they stretch themselves physiologically is determined by their own intentions. Those giraffes really did want to eat the leaves that were higher on the tree, because if they hadn't, they wouldn't have evolved longer necks.

This may sound a rather Lamarckian statement, but there is nothing in the mechanism of genetic assimilation that conflicts with our modern understanding of DNA and development. It just tells a less oversimplified story, and shows that the possibilities inherent in a genuine interaction between DNA space and creature space are greater than those that occur in either space alone. Trying to refer evolution to events that take place in just one of those spaces impoverishes the richness of their interaction.

The interactive story of evolution, involving both genes and organisms, is much closer to common sense and experience than the picture of random tosses of the DNA dice. It explains innumerable difficulties that plague the conventional approach—for instance, the evolution of complex structures such as the eye or the wing. It is simpler, more coherent, and more convincing than the neo-Darwinist "It's all in the DNA" story. And it implies that in future, science should devote far more attention to the interactive dynamic on combined DNA/organism space.

THE DISBANDED GENOTYPE

Associated with the idea of a selfish gene is that of its extended phenotype. This is the complete list of all possible phenotypic effects associated with that gene. The paradigm being promoted is that DNA is "in charge of" evolution and that organisms are just vehicles for helping DNA to replicate—

photocopiers on legs, so to speak. Because there is no nice map from DNA to organism, however, it makes no sense to talk of *the* phenotype determined by a given bit of DNA. And even if there were, a piece of DNA can survive to be replicated in different ways. Siblings have about half their genes in common, so if you sacrifice your life to save your two brothers from drowning, just as many of "your" genes survive to replicate as if you'd lived and your siblings had died. That is, the same DNA sequence may occur in several vehicles, and their interaction may improve its chance of survival. Thus the extended phenotype was born. It is a phenotypic add-on to a gene-centered theory of evolution, invented to get around the fact that natural selection does not operate on DNA sequences directly but through "their" organisms.

What we have now stumbled upon is a contextual inversion of all of Dawkins's ideas. Having replaced selfish DNA with slavish DNA, committed to making very sure that its organism survives to reproduce, we can now replace the extended phenotype with what we will call the disbanded genotype. This is a genetic add-on to an organism-centered theory of evolution, invented to get around the fact that mutations do not operate on organisms directly, but through their DNA. The disbanded genotype of a given character is the set of all possible gene combinations, possibly including those of parents, siblings, and other creatures altogether, that between them are conducive to that character's development. The disbanded genotype for the character "neck length" in giraffes includes "heart size" genes, but also "heart exercise" genes. The disbanded genotype of the canine character "ability to run" includes

- "goes faster" genes (affecting legs, heart, lungs, etc.)
- "can exercise more effectively" genes
- "is very persistent" genes
- "was well nourished in the womb" genes
- "has a higher pain threshold" genes
- "is obsessed with catching rabbits" genes
- "elder siblings teach younger siblings to run" gene, and so on.

Natural selection doesn't operate on particular bits of DNA. But when it selects in favor of a particular character of an organism, it does operate on the corresponding disbanded genotype. *Gene combinations* that reduce the chance of surviving to reproductive age get weeded out, not individual genes. And different gene combinations that produce the same character are treated in the same way by natural selection—it simply can't tell the difference. Except by putting unusual stress on that character, which in a sense is introducing new selection criteria for a slightly different character. Not just "able to run fast" but "able to run much faster, and for longer, if pursued by a predator." This is the point of the opening joke for chapter 4.

Now we have a second paradigm. The slavish gene is utterly committed to the survival of its organism. Evolution is organism-driven, but it works cryptically through sets of genes, the disbanded genotype of the organism. This isn't really an opposing theory to Dawkins's. It is suspiciously like the same theory, but turned inside out and upside down to emphasize the role of organisms instead of the role of DNA. The real point is that both are intimately involved; it is therefore not sensible to emphasize one at the expense of the other. The extended phenotype effectively projects all phenotypic effects onto the genes and makes DNA paramount; the disbanded genotype projects all genetic effects onto the phenotype and makes organisms paramount. Both distort the actual balance between genes and organisms, which is very much a partnership. Both even manage to locate key features in the wrong generation altogether.

Evolution by cryptic changes to whole sets of genes is a very attractive proposition, especially given the existence of "control" genes and "contingency-planning" genes. The more subtle and sophisticated the genetic control mechanisms are, the greater the possibilities for cunning variations on genetic assimilation become. It is these control genes that fragment the relationship between DNA space and creature space, and it is that fragmentation that makes scenarios like Bateson's not just possible but inevitable. But in the early history of life, when DNA had only just gotten started, structures as sophisticated as control genes had not yet appeared, so evolution had to follow more direct paths. DNA space and creature space would then have been much more similar, and there would have been a fairly well defined map between them—a common geography.

In short, as creatures evolved, the evolutionary process itself evolved, and both became more complex together.

NONTRANSITIVE DICE

There are many problems with the phrase "survival of the fittest" as a description of evolution. It suggests that fitness is a static concept, one that lives on a simple numerical scale. On the contrary, the fitness of an organism depends on what it is competing for—the game that it is playing. A simple illustration of how counterintuitive the concept of fitness can be is given by so-called nontransitive dice, whose name we explain in a moment. Imagine three three-sided dice whose faces are numbered as follows:

A: 3 4 8
B: 1 5 9
C: 2 6 7

Each player chooses one die, and rolls it; whoever gets the highest number wins.

You can calculate the probabilities of winning by drawing up tables of the combinations, like this:

A	**3**	**4**	**8**	**B**	**1**	**5**	**9**	**C**	**2**	**6**	**7**
B				**C**				**A**			
1	A	A	A	**2**	C	B	B	**3**	A	C	C
5	B	B	A	**6**	C	C	B	**4**	A	C	C
9	B	B	B	**7**	C	C	B	**8**	A	A	A

For example, if die A throws a 3 and die B throws 1, then A wins, so we put an A in column '3' and row '1' of the first table. We see that B beats A with probability 5/9, because out of the nine combinations of throws, B wins on five occasions. Similarly C beats B with probability 5/9. So B is better than A, and C is better than B. Therefore C is better than A? No! Die A beats die C with probability 5/9 as well.

We're used to this in real life. In football games it is not unusual to find that the Giants beat the Eagles, the Eagles beat the Jets, and the Jets beat the Giants. This rather paradoxical kind of behavior, which mathematicians call nontransitive, may even happen fairly consistently because of the way the respective teams' styles of play interact. Effectively, each pairing is a different game, even though on a meta-level they are all football. This is what makes the dice games behave so strangely: Each choice of two dice effectively changes the competition to a different game. We may therefore expect nontransitive behavior in evolutionary games. This is just one of many reasons why it is impossible to label a particular mutation as bad or good, independent of context. Each of the dice A, B, C is good in the context of one game and bad in the context of another.

Because the status of a mutation is context-dependent, the argument behind Muller's ratchet—that bad mutations accumulate faster than back-mutations to good ones—needs more careful examination. It still seems likely that rather a lot of bad mutations will usually occur before any good ones do, but the position is no longer so clear-cut. In the absence of detailed knowledge of the meaning of DNA for development, it's not really possible to decide what the proportion of good mutations to bad ones is likely to be. It remains true, however, that sexual reproduction is far superior at preserving the good and surviving the bad.

THE PANDA PRINCIPLE

Another problem with "survival of the fittest" is that it suggests that only the very best will do. In fact, creatures can survive perfectly well even though other creatures could occupy their niche more effectively. All they need to do is prevent the new creature from gaining a foothold. The same problem arises in respect of particular characters of a given creature. Superior characters that in principle ought to displace them do not because they cannot get started while the inferior character is present. Stephen Jay Gould calls this the Panda Principle because his favorite example is the panda's thumb—not a real thumb at all, and not as good as a real thumb, but the best that the panda could manage, because it had already committed its fifth digit to other uses.

As Gould explains in *Bully for Brontosaurus,* an excellent technological example of the Panda Principle is the QWERTY typewriter keyboard, so named because of its characteristic but illogical arrangement of the letters of the alphabet. This sentence is being typed on just such a keyboard, attached to a microcomputer far more powerful than the best research tool of the seventies. We had no choice: Nearly all of the world's keyboards are QWERTY. Yet it is a hopelessly inefficient layout, requiring touch typists to use their left hands to type the commonest letters in the English language. All the speed records are held by the Dvorak Simplified Keyboard, a different arrangement. QWERTY couldn't slow typists down much more effectively if it had been designed to do so—and thereby hangs a tale, because it was. On the earliest typewriters, overrapid typing caused keys to jam. But you couldn't tell because the keys were hidden from view below the roller. Only several lines of typing later, when the paper rose into view, would you spot that there had been trouble. Early typists used one finger on each hand, not like today's touch-typing, and QWERTY forced them to slow down by making them use the left finger to type the commonest letters.

As typewriters became more popular, all kinds of alternative keyboards were introduced. In a speed competition held in 1888 between two different typing schools, QWERTY won decisively, and the bandwagon began to roll. If most of the world's typists are trained on QWERTY layouts, it is a foolhardy manufacturer who introduces anything different. Today it's easy to reprogram the layout of any keyboard, just by changing the word-processor software. QWERTY *still* survives.

The ironic thing is, QWERTY didn't win the competition because it was the best layout. The winner used an eight-finger touch-typing system, the loser a four-finger system. The outcome had nothing to do with QWERTY at all. But the public thought it did.

Inadequate scientific paradigms often survive because of the Panda Principle. Any existing paradigm possesses enormous inertia—or, more accurately, the minds of the scientists adhering to it do. In order to displace it, a new paradigm must not just be better: it must be so much better that it can overcome all of the difficulties of being a latecomer—such as difficulty in getting research funding when the funds are controlled by senior scientists who of course adhere to the old paradigm. You could invent a superior laundry detergent tomorrow, but you'd never be able to displace the existing detergent manufacturers. They have the distribution networks sewn up, they control the advertising, and they've got the financial clout to put an upstart like you out of business with a price war. And we all know how incredibly difficult it is to oust an incumbent politician, however poor his record may be.

Mathematically, the Panda Principle says that a local optimum—the best thing among those not too different from it—will win; you don't need the global optimum, the best thing that could possibly be. Evolution leads to the occupation of niches by locally optimal creatures, not by globally optimal ones. The proof is that sometimes a far fitter creature is introduced from outside and wipes out all competition. We've already mentioned how the catfish is wiping out its cichlid competitors. They were surviving on the basis of the Panda Principle, and they couldn't deal with the Catfish Principle. Another fish introduced to the Rift Valley lakes by local fishermen is the Nile perch, a generalized predator so successful that it looks as if nothing at all will survive the Perch Principle—not even the perch. Another example is the ground birds of New Zealand, happy in their local optimum until land mammals such as rats, dogs, and Maoris came along.

■ ARTIFICIAL LIFE

In chapter 4 Neeplphut used *Thighbone*'s computer to perform some simple experiments in artificial life—simple enough for us to be able to reproduce them using pencil and paper—showing how mutations can easily lead to increased complexity. Now we take a contextual view of the same idea, but are forced to abandon pencil and paper in favor of the tool Neeplphut used. Recent computer experiments cast a fascinating light on basic questions of evolution, such as whether increasing complexity can arise spontaneously without any directing intelligence. For example, Tom Ray, a naturalist in the best Darwinian tradition who is now at the Santa Fe Institute, has invented an artificial world. Its environment is the memory of a computer, and its creatures are computer programs, bit strings of machine code. One feature of Ray's model, to which we shall return, is that it effectively identifies genotype with phenotype. DNA space and creature space are the same, for both are defined by the sequence of binary digits in the corresponding program. Ray's model does, crucially, incorporate both mutation and selection, and it does it without any special effort from the programmer. Mutations—copying errors—occur in all computers, and selection is accomplished by competition for space in memory. Ray calls his fake world Tierra, Spanish for Earth.

Evolution on Tierra begins with a single "ancestral organism," a self-replicating program eighty bits long. In January 1990 Ray released this organism into a pristine environment, a random sea of bits in computer memory, and settled down to wait and watch. The replicating program did what it was designed to do. Soon there were thousands of copies of it residing in different parts of the memory. But mutations led to the appearance of new species—other bit strings, some smaller than the ancestral organism, some larger. The natural measure of the complexity of a bit string is its length, so in that sense complexity had spontaneously increased. As simulated time passed inside Ray's simulated world, the populations fluctuated. Sometimes a few species would dominate; sometimes there was great diversity. It was almost impossible to decide what would happen.

Then parasites appeared, creatures a mere forty-five bits long. They had no instructions of their own for replication, so they borrowed them from neighboring organisms that did. They inflicted no damage on these hosts, except insofar as they competed with them for space. Just as in our world, parasites can flourish only if there is a good supply of hosts. In some runs of the program there arose a mutated variant of the ancestral organism, seventy-nine bits long; it was immune to the parasites, and the parasites died out. In other runs a new type of organism evolved: a hyperparasite, which hijacked the parasite's reproductive technique and used it to reproduce itself instead. Some hyperparasites evolved into social organisms, sixty-one bits long, that could reproduce only by mutual cooperation. It wasn't quite sex, but it did take two to tango. The social animals in turn permitted the evolution of twenty-seven-bit cheats, which joined in with a group of social animals and stole the instruction pointer of the computer program as it passed among them, thereby diverting the flow of the overall program.

There were also large-scale patterns of behavior, highly reminiscent of real evolution. Within the limited but astonishingly rich Tierran ecology, there were long periods of inactivity, sudden bursts of evolutionary change, bursts of diversity, and sudden mass extinctions. None of these effects were deliberately programmed in: they emerged from the very simple setup. The evolutionary geography of Tierra is a remarkable self-modifying landscape, which mimics in a surprisingly strong way some of the most puzzling features of earthly evolution.

Of course, Tierra is "unrealistic." Its bit strings correspond only in very broad terms to the terrestrial genetic code. Yet it seems to have a very similar kind of geography. The details of the landscape, the internal workings of evolution on the genetic level, are totally different from those of DNA space. Yet in both systems complexity increases, parasites subvert the reproductive processes of other, more complex creatures, some creatures become immune to parasites, some creatures find cooperative behavior beneficial, and so on. The deeper structure of the landscape in this ridiculously simple model is very similar to that in the real world. It is hard not to conclude that many of the things that have puzzled evolutionary theorists for years, such as punctuated evolution and mass extinctions, are natural and commonplace in any process, however rudimentary, that permits replication, mutation, and natural selection. Score one for complexity theory.

One further point. Tierra's ecology is random—chaotic—until a single self-reproducing organism comes into being. Without that ancestral organism, its landscape is an unstructured sea of bits, the bit death of the Tierran universe. It behaves in a manner that would appeal to thermodynamicists. But once that single seed organism arises, the state of the computer starts to behave like life, not like a lukewarm gas. Since the ancestral organism is only eighty bits long, the chance that it—or something similar—would eventually appear is not *too* small. Given a few billion years, why not? Then away she goes. What we seem to be observing here is the ability of computational subsystems (of which the most basic are perhaps replicators) to bring about the collapse of chaos. Moreover, such subsystems can arise randomly, for they are just bit strings like all the others, and their computational behavior emerges only because they interact with their environment in a particular, rather subtle manner. It takes an expert to distinguish between a self-replicating program and a piece of junk—unless the program is allowed to run on the computer. Such phenomena have given rise to a new subfield of science, called emergent computation, which forms a significant component of complexity theory.

■ IMAGES OF EVOLUTION

Tierra is a diversion, though an interesting one. It is perhaps what evolution was like in the days when there were no organisms, only macromolecules containing a few hundred atoms. The world that we inhabit is far, far more subtle. Within it—unlike Tierra—change of phenotype without genetic change could well be the evolutionary norm, as we saw earlier in this chapter when we discussed Bateson's concept of genetic assimilation. To those brought up on the image of mapping from DNA to organism, this may seem paradoxical; however, if you understand that DNA is not the core of evolution, but only one facet, then it is very plausible: *Bodies change most when genes stay the same.* They respond to the environment as they grow. Conversely, when the phenotype is static, when the organism seems well adapted between evolutionary punctuations, that may well be the time when genetic variation is at its greatest. For that is the time when competition is most stringent; the gene pool is achieving balance and assimilation, so the organism's bodily structure becomes canalized, highly resistant to environmental and genetic changes.

What happens when we plug this idea into a longer-term scenario? What does such an evolutionary mechanism look like from outside? Would we be able to tell what evolution was happening without having access to fossil DNA?

To Lamarck and to Darwin's grandfather Erasmus, both of whom believed in progressive change between species, evolution was a ladder. Few modern journalists have advanced beyond that image, in which "lower," primitive organisms evolve into "higher," advanced organisms. Amoebas are down near the bottom rung, and humans are perched at the top. Creatures work their way up the ladder like employees in a company.

Charles Darwin, and especially Haeckel, changed the popular image from a ladder to a tree. Each limb was one great collection of organisms; the topmost branches were the creatures alive today. The species that died out, and which we now find only as fossils, lived on branches that failed before they got to the top. In pictures of such trees of life the trunk is always drawn very thick, for reasons that have more to do with the mechanics and biology of real trees than with the structure of evolutionary ones. We argue, as did Darwin, that very similar little branches—species that closely resemble each other—join together just a little farther down the tree, while very different creatures share only a branch much lower down; that is, their earliest common ancestor lived much longer ago. Willi Hennig, and the "cladistic" school that followed him, have thrown doubt on many of the old tree-branching arguments, and in particular they have shown that analysis of kinds of similarity cannot locate or describe real common ancestors. We don't want to enter into such techniques here, but we do want you to observe that the evolutionary image has become a good deal more subtle since Darwin's day.

The tree image was changed further by Gould in *Wonderful Life;* Gould turned evolution into a moorland with lots of stunted bushes but only a few trees. This is a lot closer to the truth than the image of a single tree with humankind on the topmost branch. It led Gould to view the rise of humanity as mere contingency: We just happen to be in an evolutionary tree that survived random disasters and managed to grow to maturity. We criticized this kind of anthropic reasoning in chapter 8. If the particular Sherlock Holmes story that led to intelligent beings hadn't happened, another one probably would have. Dinosaurian Cohens and Stewarts would have written a book criticizing the theories of a dinosaurian Gould about those same discoveries in the Burgesaurus shale. Comparing both dinosaurian and mammalian scenarios, the dinosaurian Gould, like its mammalian counterpart, would take the differences between dinosaurs and mammals as evidence for massive changes brought about by contingency. Evolution is said to be convergent when different paths lead to the same results, and on the basis of the same comparison, the dinosaurian Cohens and Stewarts would argue, as do their mammalian counterparts, that the contingency was overruled by subsequent convergence. They would consider the common evolution of intelligence to be the crucial factor, and whether it resided in dinosaurs or in mammals to be irrelevant detail. We thus retain the bushy image but avoid the contingency arguments, which we claim are a different matter entirely.

LANDSCAPES OF CHANGE

All of these pictures—ladder, tree, moorland—omit a crucial ingredient of the evolutionary process: mechanisms. They represent the products of evolution, and how they relate to each other, but not the process itself. In a DNA-blueprint view they form just a backdrop, the possible frames of moving pictures projected upon the real world from reels of DNA. This view leads to the following image: Inside the conceptual tree of organisms there is a real "treelike" structure, made of genetic material. Whenever an organism reproduces, new twigs appear on the tree, corresponding to its offspring. If the organism dies, its own particular twig comes to an end. The branches and twigs of this genetic tree are constantly being pruned by natural selection. This image encourages us to look inward at the pattern, the evolving dynamic, of the tree, the sculpted bushes, the grazed lawn. All the respectable evolutionary philosophies have scorned attempts to find a direction, an outside, an onward-and-upward shape. Lamarck's discredited "striving for self-improvement" has put paid to that kind of thing.

But the time has come to revive it. What happens when we take our general philosophy of looking at the context as well as the content, and add it to Waddington's and Bateson's idea of genetic assimilation? What we find is a very different picture, one that shows the constraints, the landscape through which evolution flows. Not DNA space or creature space, but a richer and greater evolution space. No longer a tree, evolution is more like a river, or a rising lake that slops over into new channels. Moreover, the lake molds the landscape as much as the landscape molds the lake. However, there is no simple one-dimensional upward direction, no single number that captures fitness, for the process of evolution is inherently multidimensional.

It takes a little time to set this image up properly, and we develop it from a less geometric, less dynamic one. Think of evolution as a competitive game played between large numbers of interacting organisms. The rules of this game tend to operate in a fairly large-scale manner; for example, predators that can run faster tend to catch prey that cannot; prey animals that can hide in holes tend to escape from predators that are too fat to follow them. Much of evolution goes on within a relatively fixed framework of such rules, and most of the rules are contextual. For example, the ability to run fast—a mixture of genetic kits and nongenetic effects such as exercise—has conferred a distinct advantage to organisms, and that advantage has been present, for the same reasons, over long periods of geological time. The mechanism that lends the ability this kind of long-term stability, despite considerable changes in the form of the creatures doing the running, is contextual. Indeed, much of it is just physics. More powerful muscles can drive the animal at faster speeds, stronger bones can withstand the stress of repeated impact with the ground, and so on. But the main stabilizing feature is devastatingly simple: It's slow food, not fast food, that gets eaten.

This view must be tempered by an understanding of just how sensitive the balance between specific large-scale effects of this kind actually is. For example, being able to run fast does not convey an advantage to a creature that cannot detect its prey; it is not only maximum speed, but the ability to sustain it, that matters; and so on. Nonetheless, there exist a large number of relatively constant principles that competition must obey.

There are two circumstances, at least, in which these large-scale rules (and indeed the more subtle but less easily distinguished ones as well) can change. One, the environment may change because of some external agency such as the K/T meteorite that allegedly killed off the dinosaurs. Two, there may be changes in the nature of the game being played, because a fundamentally new attribute develops. This circumstance is the important one for our discussion. Let us take environmental changes as read, and concentrate on the second source of major transitions.

Until plants existed in abundance, the ability to eat plants could not convey an advantage. However, in the presence of a well-established supply of plants, the advantages of evolution toward plant-eating are obvious. The presence of herbivores forever changes the nature of the competition. Plants are no longer competing just against other plants, for similar resources. They are also competing against the herbivores trying to eat them; whence the development of thorns, big seeds, and similar defenses. Plants are also competing against other plants using the new resources supplied by herbivores, such as more effective methods of propagation (transport of burrs on fur or seeds in dung, for instance); again, new structures will evolve that simply make no competitive sense before herbivores (or insects or . . .) exist.

We are not here suggesting that herbivores suddenly evolved, fully formed, en masse. The actual process must have involved a huge number of smaller steps. But however it occurred, it is clear that major changes in the rules of the game arise, not so much as a cause, nor as a consequence, but as part and parcel of the change in the backdrop against which the game is played. We might model evolution, therefore, as a game whose rules normally stay fairly fixed, but which occasionally change dramatically. Let us first examine what is likely to happen at such an evolutionary "phase transition," and then discuss why such transitions should be expected—but not too often.

■ VIDEO BANDWAGON

Evolutionary processes take place on such a large time scale that they are difficult to observe in detail. A more rapid evolutionary system is the development of consumer technology. The analogy is, of course, imperfect; for example, a major part of technological development is the deliberate copying of rivals' new ideas. Evolution can mimic this, but it takes longer.

An excellent example is the development of the domestic videocassette recorder. Until the appropriate technology reached a critical point, the VCR did not exist at all. It wasn't just that it was too expensive to make; nobody knew how. There followed a brief transition period during which the technology began to become available but was extremely expensive; during that period large television companies paid huge sums for cumbersome recording devices, but VCRs did not appear in homes. The transition period culminated in VCRs that were small enough and cheap enough to be used in homes. As soon as these went on sale, then, in our analogy, the rules of the competition for the home market underwent a fundamental change.

A huge number of electronic companies began producing VCRs; early on there were at least four different, incompatible tape formats. A major part of the "environment" for a VCR is the availability of tapes to play on it, and the economics of tape production militated against having several formats. Moreover, personal economics militated against purchasing players in more than one format. As the market began to consolidate, winners and losers appeared. Manufacturers of tapes based their decisions on how many VCRs of a given format were selling; manufacturers of VCRs were largely stuck with the format they had invested in; although new, "parasitic" companies could begin production, these based their decision on which format to choose by seeing how many tapes were available. It was a period of rapid coevolution within the two industries that made VCRs and tapes.

Two formats that had never managed to secure more than a small percentage of the market soon died out. Two remained viable: Betamax and VHS. Betamax was, if anything, technically superior; furthermore, the cassette was a little smaller, making it easier to store and distribute. However, VHS secured an advantage of a few percent of sales, and this began to grow rapidly through positive feedback, in a version of the Panda Principle. Although Betamax remains in existence, the market now concentrates almost entirely on VHS.

A similar sequence of events occurs every time a really new type of consumer product appears. There is an initial burst of diversity as manufacturers jump on the bandwagon while they still can but before its ultimate destination can be predicted. Then there is a period of consolidation, which is partly deterministic (big companies tend to win because they wield greater economic power) and partly contingent (public taste is unpredictable; unexpected delivery or production problems can cause havoc). Eventually the market settles down to a small number of products. These continue to evolve (better timer-programming methods, longer tapes, infrared remote control) but in a far more predictable and convergent manner.

Let us cast this observation in the language of spaces of the possible. Before anyone had the idea that it might be possible to make a domestic VCR, consumer-electronics space did not contain VCRs. Not even potentially! We take this view because the dynamics of consumer electronics, which is controlled by a mixture of technological feasibility and economics, could not respond even to the possibility of VCRs until someone realized they *were* possible. The development of VCRs effectively enlarged the space of the possible. When this happened, two effects occurred together. One was a rather random exploration of the new territory, as manufacturer after manufacturer jumped on the VCR bandwagon. That exploration did not create new space, not in the terms of this image; all it did was explore the space implicit in the possibility of VCRs. The other was "natural selection" based on a mixture of economics and commercial clout. That rapidly restored some semblance of order, as many elaborate outgrowths died out for lack of market success.

The same kind of thing happens whenever evolution comes up with a radically new gimmick. When that gimmick becomes sufficiently well developed to convey a significant advantage, organisms that exploit it multiply dramatically. This unleashes a rapid growth of diversity, but each new species exists in relatively small numbers. However, the competition between such organisms constitutes an exploration of new territory; it takes time to settle down. During that period, when a large number of "small" species exists, the effects of contingency are strong. Perfectly viable forms can die out through sheer bad luck (such as a mud slide). Others, not noticeably superior, possibly even inferior in some respects, survive and begin to dominate. The Panda Principle stabilizes their niche; canalization puts contingency planning into their genetics in order to stabilize their forms. As the system settles down and fewer species—but with larger number of creatures in them—exist, the effects of contingency become less and less important. Contingency planning mitigates the effects of contingency. When the rules of the evolutionary game stabilize into a new configuration, convergence principles begin to dominate the evolutionary dynamic. Common patterns are arrived at by many routes.

This general scenario occurs very widely. It is visible in the history of science: When a new paradigm, such as quantum theory, appears, it is accompanied by a burst of activity: wild theories, incompetent work by bandwagon-jumpers who haven't troubled to understand the ideas properly, and so on. The turmoil settles down eventually, the new paradigm becomes firmly established, and scientific activity returns to normal, though in a totally changed environment. Social structures, religions, technologies, crazes for new toys such as the Rubik's Cube, and consumer goods all exhibit the same sequence of behavior. So do mathematical systems designed to model processes that change over time, such as dynamical systems or cellular automata. What we have described is a universal phenomenon in self-organizing systems.

■ PUNCTUATION AND CATASTROPHE

Large-scale changes, such as mass extinctions or punctuated evolution—the sudden appearance of strikingly new species—have posed severe problems in the development of evolutionary theory. They are another example of the difference between local optimums and global ones, but in this case the changes occur on a more abstract level. Instead of a new creature coming in from outside to oust a locally but not globally optimal creature, the very geography of creature space may change to create the possibility of a new global optimum.

As we've seen already (dynamical systems, catastrophes), the typical way for a system to change is mostly gradual but occasionally dramatic. There are systems in which sudden changes do not occur, and one can invent systems in which they are so common that nothing very stable happens at all; but the vast majority of systems lie in between. It cannot be emphasized too strongly that this is the typical behavior of systems that change over time and are subject to modification by external variables (parameters). As the external variables change, usually nothing much happens but occasionally the change is dramatic. The bit-string beasts in Tom Ray's computer environment, Tierra, don't even have external parameters, yet they also behave in just this manner. Gradual changes most of the time, with occasional sudden changes, is the norm for self-organizing systems. What we do not see in the mathematics is the existence of two separate kinds of system—one that never changes in any important manner, and another that, when it changes, changes dramatically. Nor do we see two different kinds of cause, little ones having small effects and big ones having dramatic effects. Instead, we see one system with two different types of response.

Behind the debate between gradualists and punctuationists in evolution is a shared assumption: that these two types of evolutionary change must arise through very different internal mechanisms and from very different underlying principles. The two schools think they can't both be right. In fact, whatever the merits of their approaches, this shared assumption is wrong. The simplest mathematical systems demonstrate that exactly the same internal mechanism typically leads to both types of change. The real difference lies in how the changes make use of external factors, such as parasites or reproduction. This suggests that we should seek not the correct theory between two alternatives but a common theory that includes both as special cases.

Even when viewed as a highly complex game played according to variable rules, and with the added "rule" that the players can change the rules as they go along, evolution probably possesses similar general features. It does not happen arbitrarily; it is constrained by its context. That context includes geology and geography, but it also includes the laws of physics. The result of the context is to restrict the range of possible behavior. A flying elephant, for example, is not a sensible evolutionary prospect because the laws of physics rule out wings large enough and strong enough to carry an elephant but also light enough for the beast to be able to walk. However, should nature evolve a new, incredibly lightweight and incredibly strong material, the gliding elephant at least might become possible. Creature space would expand to accommodate it, and thereafter nothing would be the same again. An informal picture of evolution would thus be some kind of multidimensional mathematical surface, or landscape. Points in the landscape would represent the structure of creatures. Evolution involves entire ecological systems of creatures, represented by a swarm of points moving through the landscape like a cloud. The clouds moves through the landscape in response to its own internal dynamics, generated by competition, and also in response to random disturbances.

It would be astonishing if—in these terms—the landscape were a plane. Instead, it should possess regions where motion is relatively easy, and others where it is extremely difficult. There should be "bottlenecks" that separate parts of the landscape from others. The simplest picture is of a vast archipelago of islands linked by narrow bridges. The bridges correspond to the development of major new evolutionary attributes—vision, flight, locomotion, and so on. The islands correspond to spaces of the possible that include those attributes. Phrases such as "evolutionary dead end" employ a very similar metaphor.

While a little cloud of organisms is confined to one particular island, the rules of the game remain stable, and the deterministic features of competition tend to dominate over random effects. Evolutionary change is then gradual, and big qualitative changes are uncommon. However, if the system encounters—and crosses—a bridge, then the rules change, and so does the space of the possible. On first contact, the new territory is largely unexplored; under random influences the cloud of creatures can wander off in all sorts of directions. Only after it has adjusted to the new set of rules does it begin once more to settle down. It begins to explore the new territory and eventually finds another bridge (or bridges), and so on. The existence of these bridges is the reason for convergence. In order to cross the bridge, "flight"—that is, to reach ecological systems that involve the types of behavior that flight permits—structures something like wings have to evolve in some creatures. That's the only way across the bridge. The analogy is perhaps more appropriate if we accept that instead of crossing the bridge, it is also possible to build a tunnel or set up a ferry; that is, there may be several distinct routes to the same destination. Alternatively, think of several bridges between the same two islands.

BRIDGES AND BOTTLENECKS

In summary, the form of the evolutionary landscape, the totality of possibilities, is constrained by the rules that operate in our particular universe. It is not completely free. Those constraints create bottlenecks, or bridges, whose effect is to introduce convergence. We are not saying that convergence must happen; the system may fail to cross a bridge at all. But in order to reach a given island, a bridge must be crossed, and that implies convergence—because whichever bridge you use, you end up on the same island at the end of it. The island of "fish-eating," for example, has several bridges leading to it, which have been crossed by pike, ichthyosaurs, garials, dolphins—all converging on having many small sharp teeth and a hydrodynamically streamlined body.

While evolution is confined to one of these islands, it can explore only the corresponding space of the possible. In so doing it can discover one or other of the possible bridges, but it can't explore the island at the other end without crossing the bridge. Which island it then ends upon—which new space of the possible it expands into—depends upon which bridges are available. In such a picture, the effect of contingency is mitigated by subsequent convergence. Upon first reaching a new island, the cloud of organisms can diverge rapidly in innumerable directions, heading off into unexplored territory. However, it must eventually either cross a new bridge to a new island—implying a new convergence—or remain on that island, with the effects of contingency becoming steadily less as competitive pressures set up a more deterministic dynamic.

A defect of this picture of islands and bridges as it stands is that it, too, is static. It suggests that the bridges and islands are fixed in advance—which is just a thinly disguised mathematical version of Lamarckism. The crucial feature, which is lacking, is this: The bridges and new islands arise in response to the game being played on the existing island. When the early bacteria swamped the earth's atmosphere with oxygen, they created a new island, and evolution found a way to make a bridge that led to it. When grasses appeared, the island of the molar-toothed herbivore emerged dimly from the mists, and as creatures evolved their way toward Herbivore Bridge, the mists began to thin.

Self-modifying geography is not a comfortable image when it deals in islands and bridges, so we refine it one final step. Let the islands become pools, so that land and water are interchanged, and life inhabits pools of the possible. Let the water level rise, corresponding to the evolutionary dynamic and the pressure toward increased organization and complexity. Crucially, notice that as the water level changes, it erodes the surrounding land, producing a dynamic modification of the geography. Our image of evolution crossing a bridge now corresponds to water overflowing through a channel to form a new pool. That channel may have existed all along. For example, flight in a sense exists as an option as soon as there is an atmosphere; however, evolution has to flow down rather a lot of new channels in order to reach it. Or the channel may be created by "erosion," as when the oxygen atmosphere created an entirely new channel merely because a lot of bacteria happened to excrete a chemically active gas.

Either way, the evolutionary game changes forever.

Much evolutionary thinking ignores contextual constraints, and is thereby led to see the entire process as being relatively homogeneous (which in fact it would not be, even in the absence of external constraints, but that's another story). Our contention is that the landscape of the possible has definite large-scale structure, produced by the interaction of organisms with whatever context is appropriate, and that this structure is responsible for a variety of phenomena. Among them are convergent evolution, gradual evolutionary change, and occasional sudden change accompanied by the development of radically new attributes. There are recognizable patterns, such as bursts of diversity that occur immediately after such sudden changes. When these bursts occur, the effects of contingency become more important, but eventually the system stabilizes and returns to a more orderly and convergent development.

Moreover, the same analogy holds for other systems that evolve from simplicity toward complexity, such as technology, science, and society. This is why it is so tempting to argue from one such system to another, to use them as metaphors for each other: Their superstructures—not their substructures—have many comparable features.

■ KITS IN GIMMICK SPACE

Thus, for us, the most interesting feature of the Burgess shale is not so much the contingency that it reveals or the convergence that may or may not have subsequently occurred, but the remarkable diversity of form of the organisms involved, which places contingency and convergence in stark contrast. We suggest that it is precisely at times of great diversity that contingency tends to dominate over convergence, but that subsequently the effects of convergence become more and more important as diversity decreases through largely random selection. The evolutionary picture that we have suggested thus makes a strange kind of sense of the Burgess shale organisms, not only as experiments in niche-filling, which some got right, but as collections of developmental "kits."

AUGUSTA ADA: What precisely do you mean by a "kit"?

WALLACE LUPERT: I mean a system of behavior that applies simple rules to achieve subtle functions. For example, you may find it amazing that your skin covers your body. You don't have bare patches without skin, or big unnecessary flaps of the stuff. Think of how much information the DNA message must give to the cells that form the skin! It has to tell each and every one of them exactly where to go, and that means "knowing" the precise shape of the fully developed organism, before it has developed.

AUGUSTA ADA: No, that is ridiculous, it cannot be so.

WALLACE LUPERT: You're right, of course. Instead, the skin cells just need a simple rule: "If anything is uncovered, grow over it; otherwise stop."

AUGUSTA ADA: Now I see where versatility comes from. A kit is necessarily versatile. If the body plan changes, the body still gets covered with skin. It is what we computer scientists call a procedure: a subprogram defined in general terms so that it automatically adapts to the situation in which it is used.

WALLACE LUPERT: Can you give me an example?

AUGUSTA ADA: A word-processing command to center a word on its line might go like this:

> Count the number of letters in the word
> Halve this
> Subtract it from 50
> Start at that position along the line.

This is excellent if the line has space for 100 characters; but it gives the wrong result if that number later changes to 120. The program is not versatile. However, the slight variant

> Count the number of letters in the word
> Halve this
> Subtract it from half the number of letters in a line
> Start at that position along the line

works for all widths of lines. It *is* versatile, and the reason is that it specifies flexible rules for taking action in variable circumstances, rather than specific, rigid actions.

In most animals, skin is a kit—except in nematodes, like *Caenorhabditis,* which possess a fixed number of skin cells. Nematodes aside, items like nerves and arteries are also kits; they invade any new bit of the body.

This idea, combined with Bateson's picture of evolution by genetic assimilation, makes animals look more unitary, more integrated, than Dawkins's jigsaw model would lead us to suppose.

Kits are a very good evolutionary gimmick: They invest their reproductive future in a general process rather than a particular physical form. Kits are new islands, or at least ingredients for new islands—or, if you prefer, new pools that cause local sandbanks to collapse. At some point in the evolution of evolution, kit gimmicks began to appear; from that moment on, the evolutionary landscape exploded into a huge number of new dimensions.

■ MONKEYS WITH MACROS

Evolution is dynamic. On all levels, the dynamic introduces a geography into the evolutionary landscape, a structure, a flow. This existence of evolutionary contraints, the emergent computation of Tierra, and Augusta Ada's image of versatile procedures all conspire to offer an answer to the vexed question of the hemoglobin molecule. In chapter 4 we mentioned the old cliché about monkeys randomly typing *Hamlet.* They would, but *very* slowly. Similarly, there isn't enough time in the life of a universe for evolution to "type" hemoglobin by sticking amino acids randomly together. However, in the computer age we should use modern images, and replace the Victorian typewriter with the word-processor, which has extra, more versatile facilities.

If monkeys randomly tapped the keys of a word-processor, then every so often they would accidentally set up a "macro," a sequence of actions that the machine "remembers" and can carry out when given a simple signal—usually a combination of two keys, the Control key and a letter. Typing this simple key combination would then unleash an arbitrarily complicated sequence of actions. Without some kind of selection for useful sequences, this wouldn't make much difference, but in evolution there is a dynamic, resulting from natural selection and from other constraints. This dynamic preserves and replicates some innovations at the expense of others. In short, evolution can select some macros and discard others. The message of Tierra is that you can expect it to keep the interesting innovations and lose the meaningless ones.

Let's impose our own geography to see what might happen. Imagine, for instance, that as the monkeys type, macros that give things close to sensible English words tend to be selected. Then the monkeys' efforts begin by looking like BSMUFHRYDGELGGGGQDK; but fairly quickly evolve into things more like TOBE HUMLEF NO NO NO PLOMP. Imagine further levels of selection, first for even more resemblance to actual words, then for better grammatical sense, and so on; soon you'd be seeing things like TO BE OR NO TO PLUMP, THAT THE IS IS. As the monkeys with macros pursued the dynamic of the English language to deeper levels, hopping from island to island in literature space, the selection criteria would become increasingly sophisticated—interesting plots, subtle characterization . . . all aboard for Shakespeare Island.

If the monkeys followed some such scheme but were granted total artistic freedom, you'd expect the same kind of development. It would be unpredictable, but from the outside you would see ever more complex structure emerging. Their ability to exploit macros would let them learn to "write better," and they'd eventually produce something with a very sophisticated structure, by whatever criteria had evolved. We can't predict what kind of thing it would be, but we can be sure that it wouldn't be at all random.

In the same way, as evolution explores the geography of protein space, some structures will emerge that have unusual properties, properties that get selected for replication. These molecular macros raise the entire level of the evolutionary game; the basic units are no longer single amino acids but short chains of them. And just as monkeys with macros could indeed write *Hamlet,* in a time that is acceptable by evolution's relaxed standards, so might molecules with macros write hemoglobin.

11 THE UNDERSTANDING OF HUMAN ORIGINS

A truck driver arrived at a bar with fifty monkeys. "What do you think I should do with them?" he asked the bartender. "I've lost all the paperwork."

The bartender thought for a moment, and then said, "Why not take them to the zoo?"

"Of course!" said the driver. He collected up the monkeys and disappeared.

The next evening he returned, still with the monkeys. But now they were all wearing sunglasses.

"Still got the monkeys?" asked the bartender in surprise. "I thought you were going to take them to the zoo."

"I did," said the driver. "Great idea, absolutely brilliant." He swallowed a mouthful of beer. "They loved it. They fed bread to the bears, rode on the elephants, ate cotton candy and hamburgers—they had a whale of a time. So I decided that tomorrow I'll take them to the beach."

In chapter 5 we presented the orthodox picture of nerves, brains, and communication as the route from cells to intelligence. We warned you not to confuse the Ladder of Life with the evolutionary story, but then—like most books that try to account for human intelligence—we did just that. We told the once-upon-a-time-there-was-a-nerve-cell story, leading into eventually-it-got-much-smarter, and culminating in and-then-Albert-Einstein-arrived-and-they-all-lived-happily-ever-after.

Like the other stories in the first half of this book, this nervous-system-centered way of looking at the development of intelligence makes a lot of sense. However, just like those other stories, it deals only with the inside view. It assumes that it's better to be brainier in a challenging world without quite setting out why versatility driven by intelligence is better than versatility driven by other genetic gimmicks. Why should brain be superior to brawn? Yes, *we* think it is. But *we're* the animal that's good at thinking, and all the things we see as problems get solved by thinking about them—don't they? Teleost fishes would probably have a different point of view, though they wouldn't use that image. They would see sinking or floating as the big problem of life, and their big answer would be the invention of the gas gland in the air bladder. Never mind how clever you are; how much does it cost you to achieve neutral buoyancy? Humanity ranks low on the air-bladder scale of success.

Even among humans, intellectual ability is valued most highly by those who possess it. For most teenagers, the ability to kick a football rates so far above the ability to solve a differential equation that there's absolutely no contest. Jocks always beat nerds. A good, persuasive example of the content-based way of thinking that characterized the first half of this book is to see human evolution as the old hand-and-eye symbiosis. Primates have dexterous hands for grasping, magnificent stereoscopic vision for climbing, and a brain with built-in 3-D. Walking on two legs frees the hands to use tools, toolmaking encourages the development of superior eyesight, and that makes it evolutionarily worthwhile to develop bigger and better brains. It's a bit like buying a stereo system. If you get a better CD player the defects in the amplifier show up, so you buy a better amplifier, and then the speakers won't handle it, so you rush down to the store with your credit card and buy such a fantastic set of speakers that it's ridiculous to keep using the old CD player. Everything coevolves with everything else, in the direction of increasing expenditure.

This content-based mechanism is part of the tale, no doubt. But there's a better way to tell the monkeys-to-Einstein story. It has several ingredients. One is the role of privilege, of nongenetic transfer between generations. Another is the importance of food as a source of the chemicals needed to build brains; this is linked to a novel theory of human origins. With these ideas as background we can investigate the workings of human culture, information, and language. And because science is a cultural activity, carried out by human brains, we are led to a new viewpoint on reductionism and Theories of Everything.

APES TO ALBERTS

A major theme of the second part of this book is to look not at content, but at context. The context of our alternative story of brain evolution is privilege—the passing on of advantages, by nongenetic means, from parents to offspring. We've already discussed nongenetic inheritance of organisms in chapter 3 and alerted you to how mammals exploit it to reduce the amount of information in their DNA programs by investing more effort in the mother and less in the child.

It is easy to find many more examples of privilege. The provision of food is commonplace. Mother tarantula wasp provides her babies with a nice fat paralyzed tarantula. Her larvae are genetically programmed to eat the spider in the right order, leaving vital organs such as the heart till last. But food is not the only possibility. Instead of having to invent a better hunting technique, for example, you can learn it from your parents. But you must still evolve something: the abilities of communication and learning. Groups of Japanese macaques—smallish monkeys—that have moved from their normal habitat onto beaches have been observed learning to swim, to separate grain from chaff by floating it in a pool, and to wash the sweet potatoes that visiting scientists leave as food. Many of these discoveries were made by one particular simian Edison, a female monkey with a flair for invention. The rest of the troop picked the ideas up remarkably quickly, especially the young ones. Only the elderly, conservative members stuck to their old ways.

There's an analogy here with maternal-effect genes and the provision of controlled maternal environments, which are essentially derived from the mother's genetic makeup. But now the interaction between generations functions not through the medium of genes, but through that of the senses. The more adaptable, powerful, and quick a brain is, the better it is at remembering, the more flexible it is in grasping new ideas, so the more effective this nongenetic transfer of privilege becomes.

Human beings take the process to considerable extremes, producing children that would have no serious chance of survival on their own before the age of ten or twelve; that still require (but may not be willing to accept!) parental guidance and advice at age thirty; that can carry out their lives effectively only within a vast—literally worldwide—culture of other humans. The trick is so successful that we have long been by far the biggest danger to our own continued survival. But why has a privileged ape pushed this particular trick to such enormous lengths, unlike any other animal—except perhaps dolphins and their kin, and chimpanzees, who at least came close? And why are there no chimpanzee Einsteins?

That's an excellent question, as Jared Diamond has made clear in *The Third Chimpanzee*, because it points to a failure in the usual theories of the origin of human intelligence. We *Homo sapiens* are one of three very closely related species; the others are the pygmy chimpanzee, *Pan paniscus*, and the ordinary chimpanzee, *Pan troglodytes*. We've gone the Einstein route, but the other two are fairly unremarkable, intelligent large primates. There are two classical ways to explain our intellectual success, but neither passes the crucial test of explaining why the other two species didn't also achieve it.

We've already given the internal explanation, the hand-and-eye story. Leaping around from branch to branch of a tall tree requires good hand-eye coordination, which means brains. When a monkey leaps across a void, its brain is in effect modeling some of the laws of dynamics: It "knows" about the parabolic trajectory of a falling body. This is not to suggest that all monkey brains solve differential equations (though some do—those belonging to the mathematically inclined members of the third species of chimpanzee) but it shows just how effective animals' mental models of the world can be. The problem with this explanation is that the common ancestor for all three chimps already had the evolutionary starter kit. The other two did pretty much what you'd expect with it, but we went over the top. Why?

The second explanation of the development of human intelligence is probably the most widely accepted. It is the theory that intelligence developed on the savannahs. The savannah, grassland with occasional rivers and trees, frequent decade-long droughts, and herds of big herbivores with a variety of carnivores and corpse-eaters, is very familiar from television. Between ten and seven million years ago, this kind of ecology displaced much of East Africa's forests. A chimpanzee-sized animal, the theory goes, could have found or made a new ecological niche by becoming an omnivore, an opportunist of the plains. It's a plausible theory, because several species of baboon are doing that right now.

In a book with the beautiful title *Another Unique Species* Robert Foley has told a savannah-based story of the development of human intelligence with a sound ecological basis. He points out that many animals of our general size found a new living out on the productive savannah. However, this environment was much less comfortable than the old woodland or wet forest; droughts in particular were a real challenge to these ecologically adventurous species. Brains, Foley argues, could easily have helped give our ancestors better geographic sense and other tricks to aid survival.

Foley warns us not to use modern hunter-gatherers as models for ancient Man and Woman: The only result is to populate the Pliocene/Pleistocene savannah with modern hunter-gatherers. He doesn't try to argue convincingly *for* a wholly new and special role for our ancestors on the plains; instead he argues *against* the hunter-gatherer stereotype and emphasizes how varied primate adaptations have been. Like other paleontologists, he points to the repeated splitting into gracile (slim and elegant) and robust (solid and sturdy) forms that our kind of animal seems to undergo.

Foley's story is a complicated ecological adaptation saga, with some hominid specializations taking up the easy vegetarian life, while the gracile forms were pushed out, against their will, onto the undesirable predator-infested savannahs, where the only ones to survive would have been the intelligent nerds, not the dull-witted jocks.

Nice. But, again, why only us?

■ FOOD FOR THOUGHT

There is, however, a third, totally different theory of the development of the human brain. It takes place not on the savannah but at the seashore. It is also a fascinating example of the possible influence of nongenetic factors upon evolution. It forms the centerpiece of *The Driving Force* by Michael Crawford and David Marsh, and it takes some important cues from Elaine Morgan's *The Aquatic Ape*. It is not yet known whether the theory is true, but the evidence is broad and plausible.

Human beings have many peculiarities. Not just the odd quirks of personality that we notice every day—"Everyone's strange save me and thee, and I'm not so sure about thee"—but species-wide peculiarities. We walk upright. We have hardly any fur. We have no tails. And we have big brains. Among equally sophisticated brains, extra size means extra computing capacity, extra memory, and hence extra power. However, size isn't everything.

We've previously offered a reductionist view of human intelligence, discussing the internal developments, such as nerve cells, that gave rise to it. We'll now adopt a contextual stance. What must happen outside organisms in order for big brains to develop? Crawford and Marsh argue that the answer is "essential fatty acids." (The adjective "essential" is a technical term, not a description.) Fatty acids are organic molecules that, among other things, are a major constituent of fat. They are organized around linear chains of carbon atoms of various lengths. But they provide animals with more than just well-developed potbellies. Fatty acids are an essential structural component of brains.

Large corporations spend a huge proportion of their income on their computer systems, especially through their electricity bills. So do we, but we spend energy rather than money. Our brains, large though they are in comparison to the brains of other animals, occupy a mere 2 percent of our bodies, measured by weight, but they use up roughly 20 percent of our energy intake. The energy is used for the same purpose as the business's money: to pay for the computers' electricity consumption. That is, brains use energy to maintain the electrical action of their nerve cells. The disproportionate amount of energy that we invest in brains is striking evidence of the superiority, in evolutionary terms, of brain over brawn. Brains are very special organs, and they require very special building materials, namely the aforementioned fatty acids. These account for an amazing 60 percent of the building materials used in the brain and nervous system. The reason is that fatty acids are good for making membranes, and brains use ionic channels in membranes to control the transmission of electrical signals.

At any rate, building big brains requires a big supply of the appropriate fatty acids. There are (at least) two ways for a developing organism to make sure that it has an adequate supply of key chemicals. One is to make the necessary molecules itself, from simpler components. The other is to steal them from some other organism. Since theft has obvious advantages over honest toil, the animal kingdom often opts for theft. An example is the legendarily acute visual system of the cat. Cats do a lot of their hunting at night, and good night vision hinges upon the chemical rhodopsin, made using vitamin A. Vitamin A in turn can be made from chemicals such as beta-carotene, found in plants. Most animals can make vitamin A by this route, but cats, which are totally dependent upon vitamin A, can't. Instead, they steal it wholesale from their prey. Every mouse eaten is also the gain of one mouse's current stock of vitamin A. It's so much simpler to let the mice do it, and it's rather a neat twist to the tale that the victims supply the necessary ingredients for their own demise.

If a predator is going to use the same trick to build itself a bigger brain, it needs a really good supply of food that is rich in fatty acids. Where do we find such a source? Not on the savannah; in the sea. Seafood—squid, oysters, shrimp—is rich in just the right fatty acids for making brains. It is probably no coincidence that the other large-brained mammals, whales and especially dolphins, all live in the sea. We don't, but we do seem to be remarkably attracted to it, as the beaches at Bondi or St. Tropez graphically attest. It's true that predators can gain fatty acids from their prey, but the land mammals of the savannah don't have many of the longer-chain molecules that the brain really needs. Lions and tigers have far smaller brains, relative to body weight, than humans. Not only that; the land animals with the biggest brains—the primates—are not especially carnivorous. They often eat fruit and vegetables, though more of them also eat meat than people used to think. This chemically driven scenario is, of course, absolutely perfect for the aquatic ape theory. Suppose that humans didn't evolve on the savannahs at all, but on the seashore. Then they'd have no problems getting an adequate supply of brain food.

A lot of other items fit snugly into place, too. An ape that spends a lot of time wading around in the water chasing shrimp and shellfish will have an evolutionary advantage if it can venture into slightly deeper water, and an upright posture becomes a natural development. The alternative is swimming, but this is much easier if you can lie flat, adopting, so to speak, a horizontal upright posture. Loss of fur also makes sense; the only marine animals with hair are things like otters and beavers, which live in cold climates and have to come out of the water quite often.

Moreover, there may be no chimpanzee Einsteins because chimpanzees never made it to the beach, so they didn't get enough "food for thought" to develop brains with Einsteinian capabilities.

What about the large quantities of fossil apes found in the rocky remains of what used to be savannahs and the total dearth of fossilized aquatic apes? A seaside ape would without doubt have made forays into the hinterland, if only to escape the crowds; the savannah apes could easily have spread inland from the coast. Moreover, many of the fossilized apes on the savannah are found near lakes. As for the lack of fossilized aquatic apes, the seashore is a bad place to make fossils. The fossil record for seashore-based life is generally poor. You need somewhere quiet where bodies can lie undisturbed, preserved by mud slides or tar, or buried under gently silting deposits on a lake-bed. Beaches just aren't like that; the sea has tides. But beaches *are* a remarkably rich environment, which evolution is bound to exploit. Is it really likely that one of the biggest evolutionary developments of all time did not involve the land-sea interface?

FISHING FOR DATA

In chapter 5 we explained that eyes, ears, and other sense organs are not just holes into the world. They analyze and process incoming sensory data and present the brain with a lot more than just information of the kind "Retinal cell RC-69961 is receiving red light." For example, the optic nerve contains fibers that pass impulses only when a vertical dark bar has moved to the left. We looked at how they did it, and concluded in particular that there isn't enough information in human DNA to describe all of the brain's connections. By now you will not be surprised by this. We are not jigsaw creatures, like nematode worms. We are not like gnus and guinea pigs, programmed in the womb. We program ourselves, so that the DNA "blueprint" does not need to describe the structure of our nervous system in every detail.

The reductionist view of how we acquire sensory data leaves a lot out. One important piece of evidence is that more neural connections run from the brain to the ear than from the ear to the brain. And about 10 percent of the fibers in the optic nerve also go the "wrong" way. Sense organs do not passively accept incoming data; they go fishing for it. The brain has to "tune" the sense organ, so that it can detect what is needed. It isn't just a matter of finding the right place to fish from; the brain has to work out what bait to use and whether to go for shrimp or eels.

Astronomers used to make very accurate polished mirrors for their telescopes. Nowadays a popular alternative is active optics. The single high-precision mirror is replaced by a large number of tiny flat ones, each mounted so that it can be adjusted by computer control. The computer receives the incoming light signals, processes them, and keeps adjusting the positions of the mirrors until it gets the best result it can. Our senses are similar, and that's why so many nerve fibers pass into the organs instead of out of them. So we're not born with the ability to perceive the world in the way that an adult does; we have to learn how to see and learn how to hear. Kittens that are prevented from seeing during the first few weeks after birth are unable to see at all when they grow older. Kittens prevented from seeing particular patterns early on are unable to see those patterns in later life. The same goes for hearing. What we see and hear changes our neural circuitry so that it becomes more able to deal with those features of the sensory world. There must be a lot out there that we have never learned to see at all—it's part of the real universe, but we can't get it into our brains.

If we don't learn a language—any language, even sign language—by the time we are twelve years old, we never can. Feral children, growing up completely in the wild, can't learn to speak or to understand the spoken word. Like chimpanzees and dogs they seem to realize there's something going on, but they can't take part. When we are born, the neural circuitry in our eyes and ears is relatively simple. As we grow up we improve our sensory abilities by complicating those circuits in response to features of the outside world.

The pictures that we form in our brain are nothing like photographs or television pictures, although to us, from "inside" the brain, they may seem like that. They are more like what goes on inside the TV circuit that generates the picture—but with the screen turned off. They are not made of light and shade; they are features of the mental landscape. Your brain puts together a very small number of features—shape, color, Mother, teddy bear—and elicits what seems to you to be a complete picture, in full detail. When you were young your brain set up a context for everything that you now have in your mind, everything that you can now think about. You can never "see the world with fresh eyes." Yes, light patterns fall on fresh eyes' retinas; but they do not *see,* any more than film does.

■ CULTURE CLUB

Once a species has brains and senses there's another trick it can do. More accurately, the trick develops in tandem with the brainpower. It is culture. Culture enables animals to pass survival kits on to their offspring by nongenetic routes. These routes can be far more adaptable than DNA chemistry; by the same token, they are not always as stable.

Nongenetic transfer between the generations is the rule in the animal kingdom rather than the exception, and primates generally take the trick much further. In baboon troops, the key concept is status. Status not only determines which animals get to mate, it also determines the sources from which youngsters acquire their cultural inheritance. High-status kids get attention from high-status adults; low-status kids don't. The young baboon's interaction with its environment is strongly dependent upon cultural traditions within the troop. If everybody else thinks swinging from trees is a great idea, so do the new kids. If trees are out of fashion and bumming food from tourists is the order of the day, the kids bum along with the rest.

Compare the two alternatives. Is it more effective, in evolutionary terms, to specify all aspects of behavior once and for all in DNA code, or to use DNA code to specify flexible brains that can learn, and pass the behavioral information from brain to brain, from generation to generation, bypassing the genetic biochemistry? Like everything else the answer to this question depends on context, but sometimes one route may be preferable, sometimes another. We've already noted one simple fact which suggests that for creatures as complex as us privileged apes, the *direct* DNA route isn't even available: There are more neural connections than there are DNA bases.

Humans, as we have said, are the extreme case of nongenetic transfer between generations. Much of what we need in order to be human is genuinely transmitted to us as a message—not genetically, but through our brains. Language is the key example, and the key enabler that makes the process work. If language were hard-wired into our genes (assuming this to be possible) then we wouldn't have to learn it, but it would presumably use up a lot of DNA. Instead, our DNA seems just to code for language-learning ability within a brain that has already evolved for other reasons; language itself is transmitted culturally from the mother and other adults to the child. This method is far from foolproof—the language we learn is an imperfect rendition of what is taught to us—but it offers the advantages of efficiency and flexibility.

Not only do we pass on knowledge to our offspring by teaching, but the body of available knowledge itself also grows as humans acquire new tricks, totems, and technologies. The amount of cultural knowledge that must be transmitted inevitably increases, because it is accumulated faster than it is forgotten. In a tribal culture, each generation adds new traditions, new stories about people and events, such as the biblical tale of Noah.

As the weight of accumulated lore grows, cultures must devote more and more effort to ensuring that it is transmitted to the new generations. It takes the whole of human culture roughly twenty-five years to produce an ape with a Ph.D. in biology or mathematics. Hardly any of the knowledge and techniques involved are passed on through genes. What is passed on through genes is the ability to learn from other human beings. Humanity has developed one big idea, and dozens of smaller ones, for making sure that this process continues. Learning itself is a long, hard struggle, a process made possible by the big idea, the almost unique invention of *Homo sapiens:* language. With language, it is possible to pass on thoughts more complex than "Little gray squeaky things with pink tails, like *this* one, are good to eat."

Myths are a very effective way to pass on guidelines for socially acceptable lifestyles. With the end of the second millennium in sight, it is probably still true that most of the stories we teach our children are myths, Just So Stories, oversimplifications. They are teaching stories, not truths. The whole truth is often too complex to teach to young minds, but important parts of the truth can be conveyed by teaching myths. For instance, every child is taught that rainbows are formed because light is refracted and reflected in water droplets, and that the colors arise because rays of different colors are bent through different angles. This is fine as far as it goes; but the sun does not conveniently transmit one ray to each raindrop. When a whole bunch of rays enter a raindrop all those multihued rays are overlaid upon each other. *Why doesn't the rainbow fuzz out?* That's a far more subtle question, not suitable for explaining to children asking about rainbows, so we sensibly settle for the myth.

The tribal intellectual baggage can become astonishingly large, even for cultures with a purely oral tradition, so the process of education is constrained by the individual's ability to learn and the time available. We have escaped this problem by inventing nonmental records. Writing was the first such recording technology. It was enormously strengthened by printing, which rendered the written word reproducible. Within the last century we have developed a huge range of similar reproductive technologies: photographs, moving pictures, gramophone records, compact disks, photocopying, audiotape, videotape, computer memories using electronic, magnetic, or optical media . . . Our growing burden of knowledge—of facts, techniques, models, and processes—can now be stored, safe for posterity, in a manner that is independent of human memory. (Except that somebody must remember how to access it.) We have invented many techniques for knowledge compression, such as abstraction and generalization; compare medieval finger arithmetic with what is now taught in first grade.

The understanding of all these facts, techniques, models, and processes, however, still resides in humans, and is distributed throughout our society. Medical expertise rests with doctors and nurses. Mathematical expertise is the prerogative of mathematicians. Biology is for biologists to worry about. As a result, we live most of our lives using equipment that we don't understand for purposes that we comprehend only locally. Humanity is bigger than any one individual, and it seems to have collective purposes of its own that individuals cannot easily detect, let alone grasp as a whole.

INFORMATION AND MEANING

The late twentieth century has witnessed an explosion in this ability to store knowledge in our culture rather than in our brains. We have invented a new phrase for such techniques: information technology. Information is treated as a commodity, something that can be bought and sold, and is measured in bits (binary digits, yes-or-no decisions). But information theory, which we described in chapter 5, is a quantitative theory, whereas most of the things that are important to humans are qualitative. This mismatch leads to paradoxes. For example, DNA is conventionally seen as information—the Book of Life written in the four-letter alphabet of bases—but in chapter 9 we came to the tentative conclusion that the quantity of information contained in frog DNA is too small to describe a frog. Above, we noted that the quantity of information in human DNA appears to be too small to describe even the human brain. The time has come to examine the basic tenets of information theory, taking a contextual view, in the hope of resolving these apparent paradoxes.

Here are three thought experiments. They show that the naïve measure of the quantity of information in a message, not taking context into account, has no relevance to the complexity of the actions the message may trigger. They show that those aspects of a message that matter to human beings, such as meaning and understanding, do not fit readily into a crude information-theory mold.

1. "If I don't phone you tonight, Aunt Gertie will be arriving on the 4:10 train from Chattanooga. Take her home."
2. "I ♥NY."
3. On a television screen, the caption "Call 1-800-666-7777 to make a donation."

Experiment 1, on the face of it, conveys a sizable quantity of information with a *zero*-bit message. Since you don't phone, no message as such is sent—though since the alternative is that "I" does phone, it's really an implicit one-bit message. An enormously complicated sequence of events is set in motion by the absence of a telephone call: Get out address book to find Aunt Gertie's address so that you can think about traffic patterns and work out the best route to take her home while you're on the way to the station; put on coat, open front door, go through, shut it again, get keys from pocket, open car door, get in, shut door, start car, avoid neighbor's cat in driveway, turn left onto street . . .

In Experiment 2, a masterpiece of the advertiser's art, a simple but direct message is conveyed in a mere four characters.

In Experiment 3, a message of eleven decimal digits, which is around thirty-six bits, is received. However, the engineers who designed the format of television signals know that the actual amount of information consumed by the appropriate segment of the TV screen is far higher—around 100 lines, each of 1,000 individual phosphor dots in three colors must be activated: say, 800,000 bits. The problem is that to transmit the telephone number by TV, you have to send an 800,000-bit message! It won't work otherwise.

In the first two experiments it is reasonably clear what mechanism is operating, where to locate the sleight of hand that turns comfortable communication-theorist's information from a conserved quantity into something so malleable that there is no point in measuring it at all. Indeed, each can be viewed as an exercise in coding, "triggering" the access of information from a specific range of possibilities. That's how the trick works, but the point is that mere bit-counting ignores the context in which the message is interpreted. It bears no relation to the true "information utility" of the message—that is, the complexity of the action that it initiates, or, in more familiar terms, how much meaning it possesses. The third example goes the other way: The particular contraints of the chosen communication medium imply that the message must transmit far more information than it actually "contains."

The meaning in a language does not reside in the code, the words, the grammar, the symbols. It stems from the shared interpretation of those symbols in the mind of sender and receiver. This in turn stems from the existence of a shared context. For language, the context is the culture shared by those who speak that language. For the DNA message, the context is biological development. If the manner by which DNA code is transformed into creatures is ignored, we have no idea whatever of the possible complexity of the creature that results from a given segment of DNA. It takes very few bits to send "Make a tiger"; a receiver that understood such a message—that "knew" how to implement it—would need nothing more, apart from an appropriate context, to construct the world's most beautiful feline.

It is possible to change an organism entirely by changing only a few bits' worth of its DNA. Indeed, as we have explained, anglerfish and *Bonellia* males and females have identical DNA, so a change of zero bits of DNA produces an entirely different organism. They're like the Aunt Gertie message, only more so: *Everything* is determined by context.

On the other hand, changing twenty million DNA bases might do no more than alter the color of the animal's ear tufts, if some TV-like system were in use. In Siamese cats, for example, it isn't even necessary to code for the color of ear tufts if, as is usual, you want them darker than the rest of the animal. The chemical that determines Siamese cats' color is temperature-dependent, and ear tufts are cool because they are at an extremity. It is trivial to "code for" darker ear tufts, in the same way that the absence of a telephone call "codes for" meeting Aunt Gertie. Here the context is the laws of physics and chemistry. You don't get to choose it, it's automatic and ubiquitous.

Prescription is closer to the mark than description, not just for DNA messages but for any message outside the abstract setting of information theory, which deliberately strips away the context. A prescription from the doctor is not a cure in itself; it only becomes one when taken to a drugstore, received by a pharmacist, and acted upon. All messages in the real world that really *are* messages happen within a context. That context may be evolutionary, chemical, biological, neurological, linguistic, or technological, but it transforms the question of information content beyond measure. We understand this point for technology: We don't normally try to play a compact disk on a telephone answering-machine. But when thinking about the natural world, we often forget that we don't know how much contextual input there is into processes that we like to model as "message sending."

A CONTEXT FOR LANGUAGE

Our senses are contextual. Sense organs do not simply let sensations in; they abstract high-level features of the sensory data. Moreover, our sense organs must be constructed and modified as we develop; our surroundings determine what our eyes can see or our ears can hear, as we develop their abilities by exposing ourselves to different contexts. Language is a kind of sense—it, too, involves processing incoming data and extracting meaning—and it is the most self-referential of our sense organs. We must now develop a contextual setting for language.

In the first two years of life, as we learn to hear and speak language, we are constantly modifying our internal language system. All normal babies babble, and to begin with they all babble in the same general way. But soon their babbling specializes into English babbling, Swahili babbling, Chinese babbling. . . . By the time they're six months old, most babies' vocalization is built from the features of their native language. By the age of two, English babies whose mothers replace between-syllable "T"s with glottal stops (making that phrase sound like "glo'al stops") are doing the same. By the age of five all of us have picked up nearly all the rules for regular and irregular verbs and all of the remarkable tricks for embedding phrases in sentences. Our lexicon increases by about one word per hour until we reach the age of twelve. Children who fail to learn a language by this time can never learn one.

As we immerse ourselves in this self-generated linguistic environment we equip ourselves with the most sophisticated sense organ of all; we can "perceive" a whole book, a whole play. Our "language organ" creates a context for each book we read and each play we see, and it is within that context that the story takes on a meaning. We know that a murder in a play is not something that the police need be informed about, and we know this not because of the words and grammar, but because we know that we are in a theater and that theaters are places where such things happen. Bateson's insight about baby monkeys, mentioned in chapter 5, is of the same kind: Monkeys could distinguish play from reality because they were in "play mode."

We share this linguistic development with the rest of our human group, whether its members have the five-hundred-word lexicon of Peru's street children or the twenty-five-thousand-word lexicon of most Western college graduates. John Tooby and Leda Cosmides call this "the genome storing the vocabulary in the environment," a very perceptive contextual thought. The accumulated language that each of us learns builds us a personality equipped with the tools needed to interact with our culture. If our culture has cars, there are words—and therefore easily manipulated thoughts—for "trunk," "gears," "tires," "blow-outs," "wrenches," and "traffic cops." This context enables us to learn "automobilese," just as we learn what "timid" is from mouse stories, what "cunning" is from fox stories, and what "wise" is from owl stories. Then we add our experience to the common pool of information in our linguistic environment, the cultural vocabulary.

In chapter 5 we offered a picture of the concept "dog" as a package filled with doggy features and labeled with the word "dog." We now see that this image is oversimplified, because both the package and the label evolve together. Moreover, "cunning" is not inside the package labeled "fox"; instead "cunning" is a label that evolved together with a package that contained "fox." (We can think about concepts even if they don't have labels; Mandelbrot was thinking about fractals for a long time before he invented the label. But words act as data compressors, making it easy to think automatically about concepts without perpetually having to remember exactly how the concepts work. This is an advantage but also a danger, because of the possibility of brain puns; for example, rational political debate is impossible if it must be carried out in terms of slogans. The advantages, however, are great enough to have driven the evolution of language.)

Linguists argue whether such a system evolved by natural selection for better language faculty in the brain, or better grammar, or for a bigger lexicon. But from a contextual viewpoint we can see that this is not the right question. The first chemical replicator did not need natural selection to choose it from the mass of other chemistry; it simply replicated. By so doing it multiplied, and invented natural selection for itself. In the same way, language automatically complicates itself in an environment of communicating brains. Just as lecture-room amplifiers howl when the microphones receive input from, and feed it straight back to, the speakers, so positive feedback expands language to fill the available context: human brains, tongues, ears, minds. Then it expands the context that it thereby creates—human language faculties—to make still more room. A vast information store is then available in the culture, to be poured into each developing child. Think of this when you next see a child being taught how to ride a bicycle.

One of the big problems that linguists see with natural selection is in the area they call pragmatics, which roughly speaking is the manner in which we read between the lines and extract meaning that seems not to be present on a verbal level. For example, when someone stands in the middle of the room and says "I'm cold," it often means "Close the door." When a university dean says, "That guy is a good teacher," he frequently means "That guy is no good at administration or research." Phonetics can be improved by fiddling with ears or with larynxes; semantics and vocabulary can be improved by adding more memory. Syntactics, the design of the grammatical structure that makes linguistic coding and decoding work flexibly and fast, can be improved by upgrading the linguistic operating system. Natural selection will ensure that any random improvements of these kinds survive and propagate.

Pragmatics, however, seems to be more difficult—unless we take a contextual viewpoint, when the problem disappears. We have just discussed the difference between meaning and information, and talked about the strange information trade-off that occurs when Aunt Gertie is met at the train station. We pointed out that there is much more information in the context than there is in the message. The sense lies not in the words but in the interaction of the words with the listeners, with cultural context, and with present circumstances.

The analogy between the evolution of language and that of organisms goes deeper. In the evolution of organisms, genetic assimilation means that forms can change when genes do not, and vice versa. Similarly language takes in potential as well as real structure. If aliens were to arrive on earth they might be very surprised, if they were not aware of the importance of context, to find that we already have words for them, that we have already rehearsed the unknown future, and that the tools to deal with it are already present in the language. Fiction, fantasy, lies, and euphemisms surround the "real" parts of language, just as a halo of other theories surrounds Newtonian gravitation.

MEMES

The outside view of language—which the structure of this book makes it imperative that we now consider—is rather different. Language isn't just the result of a lot of electrons whizzing around neural circuits and making tongues wag; it's there to convey meaning, to capture aspects of the world and pass them on to other individuals. In order to have an example of an evolutionary system different from that formed by living creatures, Dawkins introduced the concept of the meme, a thought pattern that inhabits human brains and replicates by spoken word. Every nursery, every classroom, every seminar is a device for the reproduction and dissemination of memes, in just the same way that gardens, farms, and dog breeders are devices for the reproduction of genes. This book puts a lot of memes on display for your selection. Some memes have built-in self-referential tricks for success. For instance, the Shema, a prayer that Jews say three times every day, includes the phrase "and you shall tell it to your children."

Assemblies of genes are called genomes; religions, political philosophies, ideologies, and other worldviews can be seen as "memomes"—more or less sophisticated assemblies of memes. The memes themselves range from simple slogans to vast interlocking systems of principle and precedent.

We've seen that the genes of laboratory favorites, such as *Drosophila*, are very special and fragile in comparison with those of the wild-type organisms that inhabit real ecologies. Similarly, academic systematizations and formalizations of real religions and philosophies tend to be oversimplified and fragile, whereas in the world these huge, balanced systems of thought have much in common with the "balanced genomes" of wild-type organisms. They are replete with "in case of" circuits, having been selected, we think, for the same kind of responsiveness to different contingencies.

Each human culture has its own version of the build-a-human-being meme kit. It contains all the tricks that give growing children cultural context, so that when they grow up they can pass the culture on to their own children in the same way. The picture of the social world that we all carry in our heads results from the interaction of this culture-building process with our growing personalities, molding them and making them. When the process fails we see cycles of deprivation, as those who were deprived of a cultural context for growing up become ineffectual parents in their turn. What is less obvious, because it actually defines each of us, is that our worldviews, our very techniques for thinking, were constructed as we grew up and could have been different. Just as we drive to work without consciously thinking about steering wheels and clutches, so we delegate many day-to-day decisions to our cultural prejudices, navigating our way through life without engaging our own decision-making mechanisms.

UNIVERSAL PATTERNS

The relationship between memes and the actions they produce in their human carriers is analogous to that between genes and the behavior they produce in the organisms that possess them. (Indeed, it was for that purpose that Dawkins invented the concept of a meme, to supply a second evolutionary system, making it possible to seek parallels and general patterns.) We have argued that evolutionary systems tend to possess universal features, common patterns. We can usefully test this suggestion by looking for the memetic analogs of some major features of genetically based evolution.

Let us begin with the analog of genetic assimilation, the idea that the bodily forms of organisms change most when the gene pool is static, and the genes change most while the bodily forms remain static. Is there a similar phenomenon for memes? To us the answer seems compelling: Indeed there is, and the general mechanisms are very similar too. Any major ideological system, such as the Roman Catholic Church, Islam, the Republican party, Communism, or Science, tends to become more and more "frozen" in its overall outlines as it grows, develops, and becomes stronger. It has more to lose if it does not preserve its old ways. On the other hand, it lives in a changing environment. People's views of the world change. Big ideologies remain the same by perpetually modifying their detailed memetic structure, in order to buffer themselves against changes in the intellectual environment. This is intellectual canalization; they change in order to stay the same. During the final days of the old Soviet Union, the Communist party was willing to accept almost any memetic transplants—up to and including capitalist economic principles—in order to remain the Communist party. Almost all western political parties are currently caught in the same dilemma; it's a more complicated version of the ice cream men on the beach.

On the other hand, again just as in genetic assimilation, when a big ideology changes—especially when it does so in a dramatic fashion—the necessary memes have usually been floating around for ages as intellectual undercurrents, maybe as heresies, never to be spoken but still widely recognized. When the Soviet Communist system finally collapsed, it gave way to a system that accessed the old memes of capitalism, free-market economies, rewards for individual effort, economic incentives, democratic elections, and so forth. When the ideology changes, the meme pool remains the same, but new combinations are expressed. When the ideology remains the same, new memes are constantly being introduced in order to canalize its form, to buffer it against unexpected contingencies. As the French say: *Plus ça change, plus c'est la même chose.* To which we must add: *Plus c'est la même chose, plus ça change.*

Another universal feature of evolutionary systems is the occurrence of parasites. In the memetic analogy these are memes that replicate by hijacking the more effective replicating mechanisms of some super-meme, ensuring that they, too, get replicated along with the main message. Examples might be the Islamic and Jewish restrictions on eating pork, which at one stage made hygienic sense because of various parasites that infest pigs. The no-pigs meme is no longer supported by this rationale, but it gets replicated along with all the other much more important memes about honoring your parents and not killing people unless they are infidels. You recognize infidels because they eat pork; like genes, memes may perform several functions at once.

There are also predator memes. The English language is an example. It cheerfully gobbles up words from every other language in the world. As one American general is alleged to have said, apparently in all seriousness: "You can't trust the Russians—they have no word for 'détente.' " As the predator meme gobbles up more and more words, it changes; but it remains a predator. Slowly but surely, substantial parts of other languages are disappearing. At scientific conferences—except perhaps in France, which deliberately resists predation by way of the strictures of the Académie Française—the only language for presenting one's work is English. The same goes for air traffic control. English is fast becoming the lingua franca of the technological world. That term is itself an interesting example: a phrase borrowed from Italian, meaning "Frankish tongue," referring originally to a kind of pidgin Frankish-Italian spoken in the Levant, and now absorbed into English to indicate any language or language mix used to transfer ideas between people from different cultures. (In the fifth century A.D. the Franks dominated what is now Belgium, northern France, and western Germany, and they spoke an early form of German.)

SCIENCE AS A MEME

By far the most successful and important worldview, in terms of delivering tangible results and making real changes to the way humanity goes about its daily purposes, is science. Religion can claim to be similarly influential—it has a very direct influence on the lives of billions of people—but mostly it regulates human behavior against whatever background is in existence at the time. Science, in contrast, changes the background. It changes it so comprehensively that most of us fail to understand just how completely our lives depend upon science, especially through technology. Science and technology grow together and feed off each other, but technology is what we use. A peasant working in a paddy field can be forgiven for not realizing that the crop she is growing would probably have succumbed to disease were it not for decades of work by plant breeders, pursuing the science of genetics to develop resistant strains. A modern westerner deserves less leniency for failing to recognize that the furniture upon which he sits, the clothes he wears, the vehicles he drives, the television he watches, and the aircraft in which he flies on business, owe their existence to the sciences of chemistry, metallurgy, aerodynamics, and several dozen others.

Of course it would be ludicrous to suggest that the effects of science have been wholly beneficial in an age that has grown up under the shadow of nuclear weapons and is struggling to contain the problems of global warming. But those examples drive home the point that the effects of science have been enormously far-reaching. One reason why science has such extraordinarily extensive effects is that, unlike any other worldview, it attempts to devise a coherent and objective system of explanations of the natural world. In so doing, it deliberately attempts to combat the natural human tendency to believe what feels comfortable, and it does so by testing scientific theories against facts. This system is not as objective as many scientists like to think, but—unlike religions, whose emphasis is on faith—it represents a fairly genuine and honest attempt to avoid bias and prejudice.

Science provides real explanations of otherwise baffling facts. Before the discovery of bacteria, disease was incomprehensible except as an arbitrary act of nature or the gods. Before the discovery of the laws of motion, people not only didn't understand why objects of different weight fall at the same speed, they thought heavier objects fell faster. (For good reasons: In ordinary experience, objects of different weights don't fall at the same speed, because of such factors as air resistance.) Science also provides meta-explanations, more strongly dependent upon the adoption of a particular worldview (or "paradigm," in Kuhn's terminology; see chapter 1). Even within science, the same set of facts may be interpreted in very different ways. Paradigms provide a framework into which new or old information can be slotted, a coherent structure that helps link different phenomena together. Explanations that depend on paradigms are in that respect a little closer to those of religion: They depend upon a particular worldview. The paradigms of science are not *true,* but they have been tested extremely rigorously against the available evidence, and their consequences have been deduced and tested as well. They are like *old* religions.

The ellipticity of a planet's orbit can be tested by measuring the planet's position in the sky. Within reasonable limits of accuracy, the fact that the orbit is an ellipse is hardly in doubt, except within discredited or at least unfashionable paradigms such as that of the earth-centered universe. The explanation of an elliptical orbit, however, depends upon the paradigm adopted. Within Newtonian mechanics, it is deduced from the existence of an attractive force between the planet and the sun; the mental image is that of a ball being whirled around on the end of an elastic string. Within Einsteinian relativity, the almost elliptical orbit is a consequence of the curvature of space-time, and the mental picture is of a ball rolling around inside a funnel. Different mental pictures lead us to pose different questions.

■ BEATEN BY THE WRAP

Science has developed paradigms for much the same reason that mammals developed warm mothers. That new trick allowed the mammals to throw away a lot of unnecessary DNA programming; paradigms allow science to throw away a lot of unnecessary facts by deriving them from general, simple laws. This is Medawar's point that "theories destroy facts." Within a framework like science, successive generations of children have to *learn less to know more*. The same is true for successive generations of scientists. If today's scientists had to learn the same range of facts as previous generations, they would run out of either brain storage or time. Instead, science uses the basic trick of data compression. Replace a product by a process that generates it. Replace a list of planetary data by a general law that implies it. Replace tables of chemical properties by Mendeleev's periodic table. Replace measurements made on generations of pea plants by Mendel's laws of heredity. Replace the results of millions of experiments in huge particle accelerators by the theory of quantum electrodynamics.

A theory is a kind of code that transforms complicated "messages" from nature into much simpler ones. For example, before anybody had an effective model or theory of planetary motion, the message sent by the solar system had to be recorded as a huge table of numbers, showing exactly where every planet was at any given time. Nature was sending us an absolutely vast quantity of data. After Ptolemy, the same data could be compressed into a list of sizes and speeds for several dozen epicycles. After Newton, we knew how to compress the data into a few simple initial conditions. All the other numbers in the table then followed by applying the rule known as the law of gravitation.

Data compression is very effective, but there is a price to pay. You have to make a computational effort to decompress the data before they exist in usable form. This is why many sailors still use printed tables. The true information-theory cost of data is not just how many bits it contains, but how difficult the decoding procedure is. If you want to listen to music, the bare information on a CD, without the appropriate decoding device, is useless. Nobody enjoys hearing a string of digital pulses read out. "On-on-on-off-off-on-on- . . ." is no substitute for Haydn's *Creation*. But if you invest some effort in the manufacture or purchase of a CD player, then you can listen with ease. The player embodies in physical form a rule, or algorithm, for accessing the information in the desired format. The cost of getting the information you want must include the cost of the algorithm that decodes it.

When the same information is presented in different formats, the features that can be extracted easily are different. Our visual and aural senses deal with sensory information in quite different ways, and each is capable of picking up features that the other misses entirely. You can't tell what sounds are on a CD by looking at the digital code imprinted on it. They could equally well be the *Creation,* "Monster Mash," the orchestra tuning up, or an egg frying in a pan. Equally, you can't tell what a picture is by listening to a fax transmission of it: the "Mona Lisa," the text of this book, and the front page of the *National Enquirer* all sound pretty much alike: lots of crackles and bleeps.

We've said many times that a Theory of Everything, however accurate, will be entirely useless as a method of understanding the universe. The development of new technology is another matter: Technology is the manipulation of artificial, selected segments of the universe. PET scanners, used in medicine, depend on the details of subatomic particles called positrons. But if the universe had different subatomic physics, the principles behind PET scanners would probably still apply to something similar. Those principles are applied in medicine through many different technologies: X rays, magnetic resonance imaging, even conventional optics. We're not trying to tell you that the doctor does not use photons to look at the patient. We're saying that he doesn't *reduce* the patient to photons.

A Theory of Everything is useless for understanding and for explanation because it represents the way the universe behaves in such an indirect way that extracting what we want to know requires an inordinate effort. It isn't a practical proposition to predict tomorrow's stock-market prices by calculating the future history of every subatomic particle in every market dealer and in the relevant environment, such as their computers, the chairs they sit on, the building in which they work, and so on. Instead, investors want a theory that operates on much the same level as the stock market itself, one that makes use of the same kinds of feature.

Scientists often object to the concept of God on the grounds that it explains the universe too easily: You can't see how it "works." God is a contextual Theory of Everything. But a reductionist Theory of Everything suffers from the same problem. The physicist's belief that the mathematical laws of a Theory of Everything really do govern every aspect of the universe is very like a priest's belief that God's laws do. The main difference is that the priest is looking outward while the physicist looks inward. Both are offering an interpretation of nature; neither can tell you how it works.

A theory is like a net. It catches what it's designed to catch. If you fish the oceans with a plankton net, you catch plankton. A shrimp net yields shrimp, a tuna net tuna, a whale net whales. If you fish nature with the theory of gravity, you catch elliptical orbits; if you fish with quantum electrodynamics you catch light and electrons; if you fish with crystallography you catch crystals. That's great, because you can catch one type of thing without wasting your time on all the others. But a Theory of Everything is like a Net for Everything, a net that catches everything in the ocean. Such a net would have a mesh so fine that it catches every atom in the ocean, and every particle of light. It would be a vast sheet of black plastic. When you go fishing with it, you catch the entire ocean, intact. But if anybody asks you what's in the net you have no idea. It's black; you can't see inside, and even if you could, you can't pick out anything interesting. Yes, it's wonderful to know that the Net for Everything contains the entire ocean, but it's not much use if you can't get anything out.

A Theory of Everything would have the whole universe wrapped up. And that's precisely what would make it useless.

12 THE BEHAVIOR OF INTERACTIVE SYSTEMS

When the wheel was first invented it took a few hundred thousand years to iron out the bugs. The wheels made by Seller-of-Used-Rocks were square, and all of his customers complained about the bumpy ride. One day his best customer, Basher-of-Small-Furry-Rodents, was visiting the workshop. Seller-of-Used-Rocks proudly displayed his newest invention. "Here, Basher, is my new improved wheel. I call it the 'rolleasy.' Isn't she a beauty? Ten skins I'm going to charge; but to you, my friend, a real bargain at six skins each, provided you buy two pairs and a spare."

Basher-of-Small-Furry-Rodents stared at the new wheel with some puzzlement. Eventually he said, "But it's triangular!"

"Of course," replied Seller-of-Used-Rocks.

"How can that be an improvement?"

"Don't you see?" expostulated Seller-of-Used-Rocks. "One less bump."

If we are to remain faithful to the plan of our book, this chapter poses something of a dilemma. Chapter 6, to which it corresponds, is about the reductionist view of large-scale systems—ecologies in particular—and how simple rules can lead not only to the beloved patterns and cycles of classical science but also to the complexities of chaos and catastrophe that delineate the current frontiers of research. This chapter should therefore put forward the contextual view of ecologies. The problem is that an ecology *is* a context. All the phenomena of life take place within the planetary ecology, and that doesn't have an outside.

Or does it? Isn't time the outside of ecology? The context for today's planetary ecology is yesterday's and tomorrow's. That kind of outside view of ecologies is one of the cornerstones of Darwinian evolution. Now you may have noticed that we've been a bit ambivalent about evolution: We put

Darwin into the first half of the book, the reductionist half, and then put him into the second half as well. The reason is that evolution is *the* central area where both content and context combine to provide scientific insight. The reductionist view of evolution is neo-Darwinism, which traces genetic change to DNA chemistry; the contextual view is Darwinian Darwinism, the big pattern of natural selection. So we may as well stop here.

Unless, of course, there is an even broader kind of context. And there is. Our starting point for finding it will be to ask what we mean by "species." This leads to the concept of eco-space: the space of the possible in which ecologies live and evolve. There are large-scale patterns in eco-space, a kind of dynamic, which James Lovelock interpreted as Gaia, the earth-organism. Can we sensibly think of the entire ecosystem as a single organism? Lee Smolin has gone even further and considered the entire universe as an evolving organism. The deep similarities behind these theories raise the question "Where do patterns live?" We end this chapter by explaining how patterns emerge from spaces of the possible.

■ WHAT IS A SPECIES?

Let's take one step backward, before we launch into the unknown. Ecologies are large-scale systems, and their main bits and pieces are not so much individual organisms as species. A woodland ecology consists not just of *this* rabbit and *that* tree, for over quite short periods of time both will die. Other rabbits and other trees will appear instead, but not in exactly the same place. An ecosystem is a process, much of which happens on the species level. "You never step into the same river twice," said Heraclitus—meaning that the water is constantly changing. For similar reasons, you never step into the same wood twice, and you never step into the same ecosystem twice. Yet river, wood, and ecosystem possess a definite continuity of existence; they are recognizably the same process, going on just as before. Human beings simplify their world, attaching short labels such as "river," "wood," "species," or "ecology" to hugely complicated processes. As a warm-up exercise before we think about ecologies, we'll try to sort out just what is meant by the term "species."

There is a conventional, textbook answer: Organisms belong to the same species if they can interbreed to produce fertile offspring. A horse and a donkey can make a mule, but mules are sterile, so horses and donkeys are distinct species. A horse and a hummingbird are incapable of cross-breeding a pegasus, so horses and hummingbirds are distinct species. A poodle and a German shepherd can produce mongrel puppies that in turn can breed with other dogs: dogkind is a single species. This approach to species is essentially reductionist in character: It looks inside the members of a given species to see what it is that makes them similar. Not literally—but the emphasis is on "able to breed," looking within a given species, rather than on "unable to breed," which looks outside in the following sense: You don't find zoologists trying to breed a horse with a hummingbird to prove they really are different species. And as it stands there are problems with this definition. It doesn't apply to amoebas, or to any asexual species, but we can exclude them from consideration for the moment. If taken literally, the definition implies that any sterile animal is the unique member of its own exceptional species—so worker bees are a different species from queen bees. In order to prove that all horses belong to the same species, it looks as if we have to breed them in all possible pairings, and then test the fertility of the offspring—rather a tall order. Of course we don't do that at all. We use various extraneous clues to assign organisms to relatively well-defined groups—how many legs, what color coat, with or without wings—and then use the interbreeding test to resolve any fine points. You don't need to try to breed a hummingbird with a stallion to prove that horses aren't birds; but if two types of bird look extremely similar, interbreeding might resolve the matter.

Unfortunately, interbreeding is not a totally satisfactory answer. We expect species to be rather like exclusive clubs, with each organism belonging to one and only one species. Mathematicians would say that belonging to the same species is a transitive relation, by which they mean that if animal A belongs to the same species as B, and B belongs to the same species as C, then A necessarily belongs to the same species as C. This may look pretty obvious, but recall our nontransitive dice: A beat B, which beat C, which beat A. Let's interpret transitivity in terms of interbreeding: If A can interbreed with B, and B can interbreed with C, then A can interbreed with C. Now it doesn't look so obvious. We could imagine a string of dogs, each larger than the last, so that each can breed with its neighbors; but the two on the ends are incompatible, if only for physical reasons. If you try hard you may be able to envisage, purely as a theoretical possibility, a continuum of animals stretching from hummingbird to horse.

Neighboring animals could differ so little that they could interbreed; but the two extremes? Never!

In fact, we don't need such thought experiments: A continuum of this kind can occur in nature. There is a more or less continuous chain of gulls, which starts in Britain, goes right around the world, and ends up back near where it started. The gulls change slowly as you go around the chain. At one end you find black-backed gulls; at the other herring gulls. The gulls of these two types can't interbreed, so by that criterion they belong to different species. However, in between there is a continuous range of gulls, and all the gulls (of the type being discussed) on any particular local section of the chain can interbreed with those on nearby sections, so each set of gulls is by that very same criterion the same species as its neighbors are. So Ms. Herring Gull has a next-door neighbor of the same species, who has a neighbor of the same species, who . . . has a neighbor of the same species; but this ultimate neighbor "of the same species" is not Mr. Herring Gull, but Mr. Black-backed Gull, with whom she cannot interbreed at all.

This is a spatial analogue of the problem that so puzzled Darwin's detractors: How can an ape have a child that is an ape that has a child that is an ape that . . . has a child that is a man that has a child that is a man that has a child that is a man?

This lack of transitivity is a structural feature of our concept of species. It can't be repaired by taking an even more reductionist stance, by looking for genetic similarities or differences in the DNA of all the gulls. It is emphatically not a matter of finding ever more esoteric genetic criteria for drawing the line between black-backed gulls and herring gulls. Drawing fine lines is a human tendency, an attempt to make our simplified mental labeling system match a differently structured world; but in that world boundaries may be fuzzy, or fractal, or may not exist at all.

■ FUNGIBILITY

Let's go back to a more naïve view of species, the one that the interbreeding criterion was introduced to refine. It begins with the observation that many creatures resemble each other a lot more closely than they resemble all the rest. Think of a typical suburban backyard. Some birds are large and black; some are small and brown with red breasts; some are blue with crests. They form relatively well-defined, separate groups. You do *not* (in such a backyard) find tiny black birds with red breasts and blue crests. From such observations, humanity has coined helpful labels: blackbird, robin, bluejay. The detailed process of assigning these labels, the development of language, does not concern us at the moment, but the fact that labels get assigned does.

There are plenty of differences in bird form, even within one of these categories. We might have refined the labeling system further: bigrobin, mediumrobin, smallrobin, robinwithlightbrownstarboardwing, and so on. As it happened, we didn't. Why? The answer seems to be that for most purposes, one robin is pretty much as good as any other. If the kids rush in saying "The cat's caught a robin!" we don't stop to ask "Was one of its wings lighter brown than the other?" We either rush out in the hope of chasing the cat away, or we remark, "Yes, cats do that kind of thing; it's nature." We don't really care about the fine details of the bird's markings. Lawyers call this property fungibility. Each can of baked beans differs from every other can—in number of beans, or in their arrangement—but you would never win a lawsuit on the grounds that the can contained 753 beans instead of 754. Cans of beans are fungible. Birds within a given species are fungible.

There's no particular reason why the labels we assign should match any biological concept; but the curious thing is that on the whole they do. Jungle tribespeople, asked to name different birds, and scientists asked to assign them to species, tend to divide the birds up in exactly the same manner. The tribespeople and the scientists have the same concept of fungibility. In this view, creatures belong to the same species if they resemble each other sufficiently strongly and in sufficiently many respects that they are interchangeable for all practical purposes. That is, they interact with the outside world in pretty much the same manner.

It is of course possible that both scientists and tribespeople are imposing upon nature patterns that do not really exist and that "species" is just a brain pun, that robins may not be fungible to nature. However, the gay abandon with which nature flings robins around, the way it permits arbitrary robins to be eaten by arbitrary cats, tends to convey a strong impression that any robin will suffice, as long as it's got a red breast and yanks worms out of the lawn. It is certainly surprising to get the same brain pun from two cultures as different as jungle dwellers and white-coated technicians; but it's not out of the question, since their brains have similar structure, even if their cultures do not. It therefore makes sense to found the concept of species on something more solid than human opinion.

Fungibility is an external, contextual concept. It's not what's inside that matters; it's what a given item looks like to the outside world, and more generally how it interacts with its surroundings. The number of beans in the can doesn't matter; the fact that you expect to be able to make baked beans and frankfurters with it does. The important feature of a species is that its members interact in much the same way with everything around them. To a cat, robins are fungible, and robins have much the same opinion of cats.

BARKING UP THE EVOLUTIONARY TREE

The reductionist instinct is never satisfied with the looseness of a term such as "fungible." It wants to draw lines, to define just how many beans there must be in a can, to assign sell-by dates instead of just requiring the food on supermarket shelves to be of good edible quality. It finds it impossible to accept that boundaries may be fuzzy, fractal, or nonexistent. The approach of assigning species in terms of specific characters has been systematized into a quantitative mathematical technique. Because there is a quantitative measure, it becomes possible not only to say that two creatures belong to different species but also to assign a numerical measure of how different they are. This technique, favored by "quantitative systematists," in turn permits the reconstruction of an evolutionary tree showing where different species branched off from a common ancestor. The technique suffers from two major disadvantages: The choice of characters is essentially arbitrary and so is the quantity that measures the degree of resemblance. The evolutionary tree that is derived from the mathematical analysis can change if different choices are made.

Again, this approach can be refined by counting resemblances in DNA strings. But the problem is that the quantitative systematists are barking up the wrong evolutionary tree. They are using quantitative differences to detect a qualitative phenomenon. Recall our image of creature space—the space not just of all creatures that exist, but all that might conceivably exist. When we suggested you think of a chain of animals linking hummingbirds to horses, that was a chain in creature space. Only its two ends correspond to actual creatures. Indeed, that's to be expected. Creature space possesses its own geography, and large parts of the landscape are uninhabited. Viable creatures concentrate around particular areas; one is labeled "hummingbirds," another "horses." In between come portions of creature space that correspond to hummingbirds with tiny hooves, or horses that hover in midair and suck nectar. Assuming, as is reasonable, that those are less viable creatures, they will be eliminated by the evolutionary dynamic. Evolution causes creatures to cluster around fitness peaks in the landscape of creature space. Such a cluster we call a species. The cluster maintains itself by reproduction.

But why does a species, in this sense, form a single breeding group? Why couldn't several different, independent breeding groups of almost identical creatures occupy the same peak? The answer is that the strongest competition for resources always comes from nearby creatures in the landscape: Your competitors are the things that are after the same resources as you, and the more closely they resemble you, the more this is the case. Recall our fungibility approach: A species is determined by what it looks like to the outside world, *including other members of the same ecological niche*. Whether or not you can breed with something is a major feature of what it "looks like" to you.

Let's pursue that thought and nail it down. Suppose that two different groups occupy the same peak, but that they cannot interbreed. If one group is slightly better adapted to the available resources than the other, if it slightly outnumbers the other, or if it just gets lucky, it will win the competition and eliminate its "opponent." A single interbreeding group, however, is stable; although there is still competition between individuals of the group, the winners always belong to the same breeding group. Individuals may win or lose, but the group thrives. It's the difference between a single football team and two, a difference that does not depend upon the individual team members, but on how they interact. The reasoning here is often called the principle of competitive exclusion, and it's a Panda Principle with feedback.

The same argument holds for any connected grouping of organisms in creature space. It implies that the viable organisms in creature space will become segregated into relatively well-defined and distinct groups, each occupying a peak of the evolutionary landscape, and that, barring incidental obstacles such as infertility or the unavailability of enough mates, all the creatures occupying a given peak will be able to interbreed. It is in the nature of peaks that they are separated from each other, for a peak is a point that is higher than anything sufficiently close by. So the multidimensional continuum of creature space becomes separated, by its geography, into isolated clusters of distinct species.

TWO GEOGRAPHIES

This picture explains why we see variation within a species—not all robins are identical—but generally much less than the variation between one species and another. It emphasizes that reproduction happens on the level of species, not individuals. It also explains why the "interbreeding" criterion usually distinguishes species fairly effectively. Indeed, it illuminates the failures of that criterion, as we now explain.

Actual terrestrial geography introduces new dimensions into the picture, independent of those that determine the geography of creature space. As the habitat of a given species varies in space, there can be an accompanying gradual variation of characters—a peak in the evolutionary landscape that moves through creature space depending on the location of the corresponding creatures on the face of the earth. No single small movement of that peak will prevent interbreeding, for the differences are no greater than those within the species at a fixed location. But enough small movements can combine to produce a big one, and—just as for the gulls mentioned earlier—creatures at opposite ends of the chain are sufficiently different not to be able to interbreed. As you move along the chain the gulls move apart genetically, not together. Most species occupy isolated hills in the landscape of creature space, but there are also long ridges. Do they count as hills or not? Where do you draw the line? You don't. The real world has isolated hills *and* longer ridges, with no clear-cut distinction between the two. So does creature space.

Mountain ranges have features other than peaks—for example, ridges and passes. So do mathematical models of this general type. The problem of distinguishing between species in unusual circumstances is closely related to the problem of distinguishing between mountain peaks in unusual circumstances. Does a slowly rising spiral ramp count as a peak, a ridge, or what? Whatever it is, it's a vivid image for the part of creature space inhabited by the black-backed gulls and the herring gulls.

This metaphorical description of creature space should not be taken too literally. We've already emphasized that the evolutionary landscape is itself changing, indeed is changed by the events that take place within it. The mountain peaks of viable species are not fixed; like real mountains, they can be thrust up or eroded away as time passes. Or, in the upside-down version of that metaphor, with lakes in place of mountains, there exist rivers that erode the landscape, carrying away sediment to deposit it elsewhere. The point is that, even ignoring these complications, fixed features moving around a landscape can lead to the apparently paradoxical result that continuous changes along one route may appear discontinuous along another. This seems paradoxical because we want to label a given feature with a name ("herring gull") and to retain the same name if the feature changes continuously. We never step into the same river twice, but we give "that" river the same name at different times. But what if the rivers Exe and Wye move around, continuously, and change places completely after a hundred thousand years? Continuity has now interchanged the names, as far as geographical location goes. No wonder the dynamic real world sometimes fails to fit our mental pseudo-world of fixed verbal labels!

This same image resolves the equally difficult question "What is a character?" If "red breast" is a character and "yellow breast" is a character, what about all the shades of orange in between? Again the crucial test is what things are fungible. How is the organism seen by the outside world? The distinction to be made is between what is possible in principle—in character space—and what occurs in practice, which is determined by a dynamic operating within the geography of character space. Characters are the coordinate axes of creature space, the variables that define its dimensionality. Since actual creatures are clustered around the peaks of creature space in discrete lumps, the same holds for measurements of those creatures made along some particular axis—characters. What we call characters are just the convenient and important distinctions among those creatures that are actually observed. Characters, peaks, attractors, lakes: We see that they are all aspects of the same universal pattern.

■ ECO-SPACE

The context for species—types of creature—is the geography of creature space. Not just all the creatures there are, but all those that might have been as well. The context for ecologies, then, is eco-space: the space of all ecosystems that exist and might have existed. Just as the evolutionary dynamic in creature space provides it with a geography of hills and valleys, so the evolutionary dynamic in eco-space provides it with its own kind of geography.

Ecologists talk of "succession," regular rolling changes in a local ecology as the growth of one set of species creates suitable conditions for others to replace them. One such sequence begins with a pond inhabited by algae, reeds, and rushes. As decaying organic matter builds up, it can reach the surface of the water, where mosses, grasses, and shrubs begin to grow. These add to the layers of organic matter, and slowly the pond fills in. The aquatic plants disappear. But now the grass and shrubs begin to be replaced by birch and alder trees, forming woods and forests; then those in turn give way to oak forests. Succession is not quite the evolutionary dynamic of which we are talking; that operates on much larger time scales. But it works in a very similar way, and it shows that there is a natural dynamic even in the space of current ecosystems. Indeed, it makes a point that somewhat complicates our mental image of eco-space: An ecosystem is a dynamic succession of organisms, not a static one. Mathematicians would describe the whole picture as a system with two time scales: the fast dynamic of short-term local changes, and the slow dynamic of long-term evolutionary changes. The fast dynamic works only with existing organisms, bringing them in from elsewhere in the general locality; the slow dynamic leads to the evolution of totally new forms.

■ BALLOON ISLAND

Eco-space contains far more than those species that exist or ever have existed. The particular track through eco-space that our own planet has followed explored various territories: the blue-green algae; the soft-bodied animals of the Burgess shale; insects; fishes; dinosaurs; mammals; birds . . . At various stages along the way, the space of possible ecosystems itself expanded. Oxygen-breathing creatures could not exist, not even in principle, until bacteria swamped the atmosphere with oxygen, but from that moment on the whole evolutionary game was irrevocably different. The development of wings introduced a metaphorical new dimension into eco-space—paralleled by a literal new dimension, the vertical one, in the geographical space the organisms inhabited. We have likened these major changes in the space of the possible to the appearance of new bridges leading to entire new islands. What other islands might our planet have explored had other bridges been crossed? What else lurks unexplored in ecospace?

It's a question for imagination rather than observation or experiment, the kind of scientifically consistent thought experiment that characterizes "hard" science fiction. Here's one example. Suppose that on earth flight had first been invented by creatures that evolved tiny sacs filled with hydrogen gas, and not by the development of wings. These creatures could, for instance, have produced hydrogen from water or methane by enzyme action, or even electrochemistry. They would have taken to the air, at first borne randomly on the breeze but gradually evolving more sophisticated controls that would at least let them decide when to descend to the ground again. With—to begin with—no competition up there, they would thrive, just as the winged creatures did along *our* evolutionary track. Then the usual story of parasites and predators would unfold, all carried out using balloonists. Some balloonists might develop the trick of expelling their hydrogen in a stream, igniting it, and directing it at unsuspecting prey. **Here Be Dragons!** As the prey fell to the ground, the dragon would fall with it, its hydrogen having been expelled. Back on the ground it would consume the remaining parts of its prey and so recharge its hydrogen batteries. This kind of exploration of Balloon Island would soon lead to a teeming airborne ecology.

If balloonism got going first, then conventional winged flight probably wouldn't have gotten going at all, because of the Panda Principle. The moment some ground-based animal developed a rudimentary wing and flopped an inch or two skyward, a finely honed dragon would have roasted it to a crisp. For a similar reason, balloonism probably can't evolve in our present ecosystem. Incidentally, this kind of exclusivity is the main reason why we suggest a mental picture of a changing space of the possible for evolutionary systems, rather than building everything that ever did, does, will, or can happen into a single superspace of the possible/impossible (delete whichever is inapplicable).

The different things going on overhead would lead to all kinds of changes back down on the ground. Instead of cows' behinds swarming with winged flies, they would swarm with balloonist parasites stealing hydrogen from the cows' prodigious output of methane. Instead of the current symbiosis between plants and insects, there might have been one between plants and "bee-loons," with the plants providing hydrogen "filling stations" to tempt bee-loons into their pollination forecourts. An intelligent race evolved from apes, birds, or something totally absent from our evolutionary track would have learned to tame the great blimps. Air travel would be commonplace, and not a horse-drawn buggy to be seen.

You get the idea. If you want to have fun, rethink it all on Jupiter.

"Ah, yes, but that's just science fiction," we hear some of you saying. Of course it is; that's what the unrealized parts of eco-space have to be. But actually, our own evolutionary track did encounter something very similar to Balloon Island. However, it happened not in the air, but underwater. The true fishes, the teleosts, were the balloonists, and their invention was not the hydrogen sac, but the air bladder. Without this invention, the seas might instead have contained only bottom-dwelling fish and fish that "flew" above the bottom but generally returned to it, as most of the selachians—skates, rays, and sharks—do today.

■ WHATEVER GAIA WANTS . . .

Phrases such as "the global ecosystem" encourage us to think of the entire planet as a single thing. And that encourages us to come up with new images of global changes. "The dinosaurs were wiped out by a meteorite. Mother Earth was disappointed to lose the great creatures, so she set about developing a spacefaring ape, capable of defending her from meteorites." It's a heartwarming tale, but does it really make sense to say that the earth wants something for its creatures? That the entire planet has purposes of its own? Does it make any more sense to say this kind of thing of a planet than it does to say ocean waves "want" to break on the shore? Does it perhaps make sense at the level of a species? When people say that the human race wants something, they're using that kind of image; is there any meaning to it beyond metaphor? And even if it's not a matter of wants, of purposes, but just of evolution toward increasing organization and sophistication, does it make sense to talk of a planet as if it were a single vast organism? To discuss its health? To invent a new science—planetary medicine?

In 1982 James Lovelock did just these things when he invented the concept of Gaia, earth-as-organism. He started with a theme that we have been hammering home for several chapters: Evolution is not a matter of nature versus nurture, or a mixture of the two, but an integrated, interactive process in which both are intimately related. Lovelock's answer to the question "Where does evolution happen?" is that it happens to the entire planetary ecosystem. In much the same way that our bodies are the result of a process of evolutionary interaction between our cells, our food, our ancestry, and our culture, so the planet's ecosystem is the result of interaction between animals, plants, and environmental factors such as weather and geology. Taking this image to its logical conclusion, Lovelock was led to a concept that he named Gaia, after the ancient earth goddess: the planetary ecology as a single living creature, a super-organism.

This image had immense appeal to the growing "green" movement for many reasons, some admirable, some not. Many green activists have a romanticized view of the lives of animals "in the wild," which in reality are mostly nasty, brutish, and short. An earth animal obviously deserves to be protected; an earth mother would watch over us and protect us. In Gaia the "romantic greens" got both animal and mother. Lovelock has himself repudiated much of the greens' enthusiasm; in his view, they identified what was best for Gaia with what was best for themselves. A real Gaia might have an entirely different agenda. A Gaia that could cheerfully tolerate the loss of the dinosaurs might have no special affection for humanity.

A lot of people, especially the more orthodox scientists, had a simpler view still: They thought Lovelock was talking nonsense. In fact what he was talking about was a paradigm shift. It makes little if any sense to ask whether the earth is "really" a single living creature. It's like asking what it would be like to be a mouse, or a Zarathustran. Only the mouse, or the alien, knows—or not, if having knowledge is not one of their abilities. But it makes perfectly good sense to ask whether we gain useful new insights into the global ecosystem by thinking of it as a single creature. Does this viewpoint make us ask new, interesting questions? Does it give us a new perspective on the world and our place within it? Arguably it does—planetary medicine being an example—and that means that it can't be all bad.

On the other hand, metaphors and analogies should never be taken too far. The same kind of argument could be applied to, say, a football team. A good football team must function as a well-tuned unit; players must anticipate the actions of their fellows. Does this mean that a football team is a creature in its own right? Does it have purposes over and above the collective purposes of its members and its manager? One of the hallmarks of such a superconsciousness would be its willingness to sacrifice individuals for the sake of the whole. By those terms the Communist party would be an organism. Gaia also passes this test: We can tell stories of how the early bacteria were sacrificed because oxygen-based life would be superior, or how humanity may yet be sacrificed, killed by its own waste products, in order to save the planet. It's less clear with a football team. Although players have been known to risk serious injury for the sake of the team, they tend to do so as part of an individual decision based on loyalty or altruism. (Though sometimes it maybe a calculated risk based on their own individual success.)

To complicate the problem further, there is one sense in which it may be more respectable to think of a football team as forming a collective organism than it is to claim the same of a planet. Football teams evolve, through competition: Success brings in money, which brings in better players, which makes teams better. Changes to players, or to managers, tactics, or even field layout, are analogous to mutations. The game provides for natural selection. Teams, and their styles, evolve *as teams:* "Green Bay's not playing too well this year." Players are the team's physical manifestation, just as DNA is manifested as molecules. The managerial strategy plays the role of genotype, and the team itself is the phenotype. It is upon the phenotype that selection acts. If a team were to be replaced by one that is outwardly identical, that plays in exactly the same way under all circumstances, then the new version of the team would win precisely the same matches as the old. Only the phenotype matters when it comes to winning matches. This fact invests the phenotype, the team, with a kind of independent existence. Teams live in team space and within that space it may be sensible to speak of them as organisms.

The concept of Earth as Gaia, as a single creature, has one basic defect, which has been pointed out by Richard Dawkins: Earth is not in competition with anything. Yes, its creatures evolve in competition with each other, and that must lead to a kind of sophisticated organization of the entire ecology, but nothing is selecting for an overall good ecology. Dawkins's argument is that there is no geography in Gaia space, just a single rock at rest in a timeless sea. The Gaia organism has no place to live, no place in which to develop, nothing to fight with, nothing to hide from, and no way to reproduce itself. It can succeed as a whole, or fail as a whole, but it can't evolve as a whole because it can't compete as a whole.

That could change. Planetary ecologies that could use apes to protect themselves against meteorites would tend to survive better. However, they still wouldn't be able to reproduce. They'd need to develop terraforming apes to do that. Once the galaxy fills with world-forming species, sprung from a thousand alien ecologies, the scene will be set for genuine evolution, by competition, selection, and reproduction of those "planetary systems"—planets plus creatures that actively replicate those planets—that survive. However, we don't seem to be living in such a galaxy right now, and that robs the Gaia image of much of its appeal. As a metaphor, it's rather limited. Gaia is like a football team that has never played a game; the principles upon which she is organized are unlikely to be those that apply to a trained winner.

■ . . . GAIA GETS

On the other hand, there are large-scale patterns in the global ecology. Gaia does behave in some respects like an organism, and may have purposes of her own in the sense that the collective behavior of individual creatures may produce systematic effects that the creatures themselves never intended, such as the way in which tiny sea creatures lock up excess carbon in their shells and deposit it on the seabed to become limestone. Where do these collective patterns, these stabilizing mechanisms, Gaia's coherent dynamics, come from? As Dawkins points out, they haven't evolved from competition with other planetary-scale organisms.

There's another more specific problem, too. Suppose that the evolution of ordinary creatures, the bits and pieces from which unified Gaia emerges, develops in a direction that is "bad for" Gaia. In more conventional parlance, suppose that some species develops some destabilizing behavioral characteristics that threaten the survival of the planet. Is there any natural restabilizing dynamic that will necessarily react against such a species, to restore a viable balance? The Gaia concept leads us to expect such a reaction, but the record suggests not. An example of just such a process is the production of oxygen by the early bacteria, which totally destabilized the prevailing ecology. Gaia—as she then was—did not react by producing some countervailing force to restabilize it. Instead, she adapted to the new conditions. That this adaptation led to a viable ecosystem would seem to have been largely chance; there's no very clear sign of any overall pattern of behavior that would necessarily ensure Gaia's survival. The oxygen disease may well have been terminal.

There are plenty of documented cases of more local ecological dynamics leading to extinction, cases nearer to home than oxygen pollution billions of years behind us. Some time toward the end of the Cretaceous era, just before the time when the dinosaurs were wiped out by the K/T meteorite, a new kind of plant appeared. Until then, herbaceous plants had been small, weedy things or long creepers; the new grasses had tiny silica needles in their leaves. Herbivores suddenly needed continually growing grinding teeth. Grasses were very aggressive and spread very rapidly; ponds and lakes suddenly became a lot less permanent. If the K/T meteorite hadn't done the dinosaurs in, the grass might well have got them anyway; grass changed all the rules for land life.

Gaia didn't do anything about that either. Unless you think that grass was Gaia's way to get rid of the dinosaurs and let the mammals have a better chance to get going . . . and that the meteorite was just a lucky extra piece of assistance.

There's another example, which we humans ought to feel very bad about, because our ancestors were the culprits. We used to think that the dodo was a remarkable bird, a dopey flightless turkey. Then we discovered that there had been many flightless birds in New Zealand, Madagascar, and Sri Lanka; many were more spectacular than dodoes. It certainly looked as if they'd been finished off by the people who colonized those islands. Then it turned out that all of the big mammals throughout the Americas—the mammoths and mastodons and horses in the north, the giant sloths at the southern tip—all died out at about the time the human species colonized the continent. Humans killed them all. We are just as bad as grass, and nearly as bad as the oxygen disease. Even with a very optimistic reading of our own history and our inherent nature, it looks as though we've only just got started. We might easily destroy about 99.99 percent of the eukaryote species on this planet in the next five hundred years.

LEGAL LOOPHOLE

There's a loophole in Dawkins's argument that Gaia's (alleged) coherent dynamic couldn't have arisen through evolution—and, as we shall see, it's a rather interesting one. Gaia could be competing with herself. Not on the level of odd problem species, which as we've just seen tend to be ignored, but on the level of big ecosystems. Although Gaia can't evolve as a whole, as an integrated planetary system, she might get put together from subsystems that have indeed evolved. Ecological systems are strongly affected by their physical and chemical environments. Ponds, hillsides, and clifftops have different ecological patterns. These differences have, throughout history, permitted the coexistence on the planet of many diverse local ecosystems, linked only loosely to their neighbors. Ecological patches of this kind compete and evolve. Successful ecological mixes—ones that survive stably—tend to colonize any suitable areas nearby, and because of their success, they displace any less successful ecological mixes already present. Corals and brambles are examples. So, like individual organisms, ecologies can reproduce and spread, subject to the same general limitations of resources and space. The analogy isn't perfect, because ecologies interact differently from organisms, but it's reasonable.

From this viewpoint the environment is the site of an ongoing battle between neighboring patches of different ecologies. The woodlands are trying to invade the meadows; the rabbits that live on the grassy hillocks are taking to the woods and destroying them. In Africa's Okavango swamp, a water weed that was introduced from outside is spreading widely, and the creature responsible is the hippopotamus, which innocently and unwittingly carries odd bits of weed around draped over its body. To control the weed, the rivers have to be de-hippo'd, using barriers to confine the beasts to regions already infested with weed. The desert is encroaching on the savannahs, the forests are creeping into the wetlands, and the swamps are eating away at the edge of the jungle. . . . Over geological time, these patches have evolved, fine-tuned their responses, learned to protect themselves against a dry summer or a cold winter. The grasses have "symbiotic" insects such as ladybugs and mantises that eat the insects that want to eat the grasses. Small ecologies such as a coral reef or an area of rain forest, in which there is huge competition for resources, possess enormous diversity—and also an enormous amount of symbiosis.

This line of thought does suggest a possible stabilizing mechanism, of an evolutionary nature, that might apply to any sufficiently localized environmental change. Suppose that a "bad" development—one that lessens the chance of survival—occurs in a limited region. An example might be a superweed that spreads so fast and competes so successfully that it wipes out most other vegetable matter. The test of whether this *is* bad does not rest with human prejudices about biological diversity; it rests with Gaia herself. If the superweed is bad for the whole planet, then it is probably bad for the limited region in which it resides. For example, the birds that help spread the superweed may eat insects that in turn eat the plants that the superweed has ousted. Those insects then die out, so the birds die too, and the superweed destroys itself because it doesn't fit into the existing pattern well enough to survive. The devastated area can then—slowly—be recolonized from outside by a more viable mix of organisms. This mechanism lets local patches compete, therefore evolve, therefore fine-tune their responses and defenses. So local ecosystem patches can evolve finely tuned defense mechanisms. These mini-Gaias can reasonably be viewed as analogues of organisms, in the sense that they are highly structured, complex, and replete with protective mechanisms against most of the eventualities that they have encountered in the past. They also arguably have their own agenda: to occupy as much territory as possible.

■ SYMBIOSIS

The evolution of ecosystems differs from that of species in one crucial respect. Most of the time, competition drives evolving species to occupy distinct niches in the environment. One of the results of this pressure to find a new niche, the more unusual the better, is symbiosis. Symbiosis is incredibly complicated and there are many kinds. There is a very strict biological definition, but the word can be used in a much looser way, to indicate any kind of serious cooperation between distinct organisms for their mutual benefit; that's the meaning we will adopt here.

Although symbiosis is fairly common, it's not the commonest line of development. Ecosystem patches, however, tend to overlap; or, more accurately, the edges between them are blurred; more accurately still, there aren't any well-defined edges. The result of border skirmishes between neighboring ecosystem patches might be just that one wins and one loses—the desert spreading into the grasslands—but, as often as not, some bits of both get incorporated into something a bit different from either. When the forest takes over from the pond, the mosses migrate to the trees. For ecosystem evolution, symbiosis is the rule. And that means that as the micro-Gaias grow into mega-Gaias, becoming both more sophisticated and more geographically extensive, they fit together as a smooth, integrated whole. Gaia, a global "organism" with her own very well-developed set of stabilizing mechanisms, could very well have evolved piecemeal.

Let's not get carried away. The result of such a process is more like a committee than an organism. A truly global organism would evolve methods for preventing the superweed from ever getting started, just as our immune system has evolved to keep most disease organisms at bay. It would get in there and tinker with the superweed, not wait passively for it to destroy itself. It is hard to see how such capabilities could evolve except through competition with other such organisms for shared resources. There is nothing to suggest that the "earth organism" is doing this.

We may be tempted to think that "higher" organisms are in fact committees, whose members are individual cells. They aren't. They have evolved decision-making systems that operate on the level of the whole organism. These structures have emerged from the population of component cells, but they have transcended their origins because they have evolved within a framework that affects the whole creature. When complicated systems of human beings, such as cities, arise, they don't do so by first putting a huge number of people together without any infrastructure, and then developing City Hall, police, hospitals, and street cleaners. The city and its population evolve together. And so it is with many-celled animals: The multiplicity of cells and their infrastructure coevolve. A cow and the symbiotic bacteria in its digestive system evolve together because the cow eats grass that provides nutrients to the bacteria and the bacteria help digest the grass and provide nutrients to the cow. Symbiosis between organisms has a resonance with "symbiosis" of organs within an organism. In those terms the rain forest is an organ of Gaia—the usual image is that of lungs, but that's something of a brain pun—and it has evolved into an incredibly intricate network of symbionts. On the level of real organisms, think of a cow's relation to its eyes. There is a sense in which a cow and its eyes evolve together because the eyesight benefits the whole cow, and the cow benefits its eyes. The eyes help the cow find food and survive attacks by predators—and if the cow thrives, so do its eyes. Similarly a jaguar's speed benefits the whole jaguar. And everything evolved in a much larger ecosystem, where cow and jaguar, not just eyesight and speed, were in competition with each other and with alternatives. No wonder it's finely tuned.

■ THE ARCHETYPAL SUPERWEED

The archetypal superweed, of course, is humanity. We have spread like vermin across the planet, tearing it apart to make short-term gains for ourselves. Does our discussion of Gaia's piecemeal evolution imply that she automatically possesses defenses against us?

While our population stayed low and our powers stayed small, our "antisocial" activities did relatively little harm—which is not to subscribe to the romantic view of "primitive" humans "in tune with nature." Many "primitive" societies have adopted lifestyles that cause considerable damage to their environment; they have survived by moving on, to cause more damage elsewhere. Slash-and-burn agriculture is an example. In certain desert areas of the United States are ruined cliff dwellings, built by Indians. Now, you don't go to the enormous effort of building dwellings in order to live in a desert. And indeed those areas were once thriving forests and woodlands. Archaeologists have belatedly seen the obvious: When the Indians cut down the forests to provide wood for building and cleared the ground for agriculture, they also triggered the formation, over a longer period, of the deserts we now see.

In the past the planet survived such depredations because we couldn't damage it enough to matter, not because "in the wild" we instinctively respected it. But now there are multitudes of human beings and the damage that we do is systematic and widespread. It is also fair to say that our self-styled civilized society does not possess much respect for the planet at all. If the Gaia image of earth as organism is valid, then Gaia should be taking steps to eliminate the threat. Is she?

Perhaps. When we damage the planet, we also damage ourselves, and we are considerably less robust than the earth. We are heading for the great Twenty-first Century Water Crisis. And overbreeding, cramped conditions, and poor sanitation provide ideal breeding grounds for disease. Our medical abilities are more limited than we tend to think. Diseases that we had imagined to be beaten, such as tuberculosis, are on the way back, only now they are resistant to the drugs that—for a very short time, a few decades—seemed to have defeated them. We got rid of smallpox, and up popped AIDS. Medicine is an ongoing battle, with few final victories.

Gaia could indeed "decide" to get rid of the vermin.

However, if she did, it would seem to be as an accidental by-product of mechanisms that were never "intended" to deal with us specifically, just with any outbreak. We don't mean that Gaia has wishes and intentions, but we want to state our point in that language because it is directed against precisely such a hypothesis. What we mean is that, when Gaia evolved her defense mechanisms, we weren't present. We've evolved very recently. That wouldn't matter so much if we were just another example of something that Gaia learned to handle long ago, but we seem to have brought an entirely new trick to bear: intelligence, and its associated effects such as culture. Moreover, it is those features of humanity that are responsible for the devastating effect we are having on our planet. Gaia, however, may be very good at dealing with the consequences of intelligence, such as overexploitation of resources and uncontrolled geographical spread, even if she is "unaware" of intelligence itself.

Organism or not, the earth is a very complex system, with all sorts of feedback loops and hidden capabilities. Mathematically we can picture it as a complicated, high-dimensional dynamical system, and like all such it settles down onto an attractor. Attractors, by definition, are stable; small disturbances that take the system off its attractor die away automatically as it heads back toward the attractor again. Such temporary disturbances are called "transient." It may be that all complex systems display inherent stability of this type. In these terms, the big problem facing the planet is the possibility that human activity may tip it over the borderline toward a totally new attractor. The stability of attractors is not unlimited; each attracts only those system states inside its "basin of attraction," its sphere of influence. Big disturbances may not die down. An outside observer may have considerable trouble distinguishing between chaotic motion on the attractor and erratic transients away from it, but the distinction is a consequence of the system's structure; it follows from the system's rules, and the rules kill the transients. It isn't necessary for the system to "know" where the attractor is, any more than a stone bouncing down a hill has to "know"

Newton's law of motion. Attractors are the way *we* understand dynamical systems.

As we have said earlier, motion toward an attractor may give human observers a misleading sense of collective purpose. For example we speak, rather loosely, of water "wanting" to flow downhill. Not so many centuries ago the fact that objects fall was explained by their "seeking out their natural place." Objects should live on the ground; an object hovering in midair is out of its natural position and so must return to it. Given a dynamic, it is tempting to see a purpose, and we warned you earlier not to make that assumption. Gaia as integrated dynamical system, replete with feedback loops and stabilizing subsystems, is an entirely respectable concept. But the image of Gaia as an organism, with self-interest and powerful defenses, is stretching a point. By thinking of Gaia as an organism, the greens, ironically, credit her with more power to defend herself than she may actually have. The irony is that by doing this they may inadvertently be increasing human complacency and thus making it easier for us to harm her. At any rate, Gaia's defenses have not evolved to deal with intelligence and the cultural transmission of ideas and technology. We are a huge wrench flung into Gaia's fine-tuned works. The planetary ecosystem's machinery is so complex and adaptable that it's not at all clear what effect the wrench will have, except that it will be unpredictable. Intelligence is a new gimmick in evolutionary space; Gaia has blundered across a new bridge. Together with the human race, she is now starting to explore the new island.

What will happen next? This is an opportunity for us to apply our theories about large-scale patterns being common to many different systems. There is a typical pattern when evolutionary spaces expand in this dramatic manner: Before the geography of the new island freezes into solid existence, contingency and diversity rule. Then, slowly, the new geography becomes established and new patterns settle in. What this analogy predicts for the future of humanity is that it will be—unpredictable. But the general outlines *are* predictable. There will be a lengthy period (in human terms; a short period in geological ones) of cultural contingency and diversity. Many distinct cultures will compete, fight, absorb each other symbiotically. A genuinely multicultural world, but one still in the process of evolving, and whose ground rules are uncertain. Something will eventually settle out—but what? It could be a single "winning" culture, or it could be a "cultural rain forest" with enormous diversity and a lot of symbiosis. Paradoxically, if the culture is short of resources, then—like a rain forest—it will evolve deeply interdependent symbiotic subcultures; if the culture is affluent then its diversity will be far less.

Alternatively, Gaia might simply eradicate us. But one ray of hope is the existence of meta-rules, patterns on an even larger scale. Back to Balloon Island. Although the substitution of balloons for wings changes the details of the global ecology completely, a lot of the general structure remains the same. There are land animals and airborne ones, parasites and predators, and so on. The meta-geography has not changed nearly as much as the fine detail. As well as contingency, there is convergence.

Applying the analogies that are expressions of these meta-rules, we find precedents for our current predicament that give us a degree of hope for its long-term resolution. When rats invaded New Zealand and exterminated most of the ground birds, they had a whale of a time to begin with, but eventually they ran out of ground birds to exterminate. They changed their lifestyle accordingly, and went back to doing what rats usually do. The rats survived—though the ground birds unfortunately did not. The rats survived because they were generalists, far better adapted to a much less survivable environment than the ground birds, who never knew what hit them. We, too, are generalists, and we may well survive the disruption that we are causing to the planet. The result will be a global ecology into which we are fully integrated, one that has collapsed the chaos that we are currently causing. However, it may well be an impoverished ecology compared to what exists today, just as New Zealand's ecology without ground birds is impoverished. This impoverishment will last a long time on a human scale, but it will be temporary on a geological time scale, for to Gaia individual species, even entire food webs, are fungible.

Moreover, the ecology might become so impoverished that Gaia herself ceases to function. You can't have a complex fine-tuned dynamical system when most of the pieces are missing. Earlier we mentioned the Nile perch, a generalized predator recently (and stupidly) introduced into the lakes of the African Rift Valley to provide a fishing industry. Like the catfish, also introduced into the same lakes, it is wiping out its cichlid competitors. But as a domino effect, it is wiping out the deep-water ecology altogether, and the lakes are starting to die. If they complete that process, the Nile perch will die with them. There is a moral for us.

If the outcome were determined by the contingency of the evolutionary dynamic, then anything might happen just after a new island opened up. You pays your money and you takes your chances. But there is a further ray of hope. Intelligence, the new evolutionary gimmick that got us all into this mess, also opens up new ways to get both us and the planet out of it. Until recently we have been acting as if the global ecology were a context for humanity. We saw our task as a narrow one: to extract what was best for us, and to hell with everything else. We are now in the process of discovering, painfully, that we are also a context for the global ecology. Everything we do affects it. If we can understand how to be a context, we will understand how to coexist with the rest of our planet.

■ HIGHER GAIA

Gaia is a big concept, but she pales into insignificance in comparison to a theory recently announced by Lee Smolin: an evolutionary principle for universes. Current cosmology provides some general rules for the behavior of universes, but various arbitrary choices must be made before the rules can be applied. These include the values of fundamental constants, such as the speed of light or Planck's constant. No existing physical theory explains why these constants take the values they do; instead, the constants are built in. The same rules that cosmologists believe govern our own universe can be applied, unchanged, to a hypothetical universe in which the speed of light is six hundred miles per hour.

Cosmologists are fascinated by black holes. These occur when a sufficiently large mass starts to contract under its own gravitational attraction. Close enough to such a mass, light cannot travel fast enough to escape. An "event horizon" forms, and nothing that happens beyond the event horizon can be detected from outside. Effectively, a separate mini-universe is pinched off from ours. According to general relativity, our universe should make black holes rather easily. We're admittedly having trouble finding conclusive observational evidence for their existence, but we're trying to spot them from a long way off, through a lot of junk. It seems generally accepted that at least some galaxies, ours included, contain black holes at their cores, and the evidence in favor is growing rapidly. At least four black holes are considered "known" by the astronomical orthodoxy.

The basic reason why our universe is theoretically so good at making black holes is that its fundamental constants happen to be just right for that task. From the point of view of efficient black-hole manufacture, our universe made a brilliant choice of the values for Planck's constant, the speed of light, the charge on an electron, and so on. Smolin's idea is that this is no coincidence. Suppose that whenever a new universe pinches off via a black hole, its fundamental constants change a little, randomly. Universes effectively reproduce, with small random mutations. Those that are more successful at reproducing—those that create the most black holes—become far more common than those that do not. The evolutionary process will fine-tune the constants in just the right direction to encourage the formation of black holes. The evidence for Smolin's theory is at best circumstantial. Because we don't know what happens on the far side of an event horizon, we can't find out whether the fundamental constants can change. Alternative explanations of the values of our universe's fundamental constants generally aim at a more extensive theory that makes those values unique, so that all universes would be just like ours.

Smolin's theory possesses the same defect that Dawkins detected in Gaia. There's no reason to suppose that the separate universes are competing with each other for resources, so no evolution in the true sense can take place. This time there's no loophole comparable to local ecosystem patches. Universes work as a whole or not at all; they don't have flexible overlaps with adjacent universes that operate under slightly different versions of natural law. Smolin's theory also fails to explain one further coincidence: why the values of the natural constants that are good for producing black holes are also those that are good for making carbon-based life like us. And it ignores the possibility that there are more regions of universe-space than you can explore just by varying universal constants. What about universes in which light does not exist, and so *has* no speed? But it does illustrate, in perhaps an extreme fashion, the unexpected directions in which contextual thinking can lead, and it dramatizes the questions that reductionism leaves not just unanswered but unasked.

■ WHERE DO PATTERNS LIVE?

We've now seen that ecosystems do have outsides. *Everything* has outsides; the more imaginative you are, the more outsides you can think of. And the outside view offers a different emphasis on scientific problems, raises new questions, suggests new answers. The contextual philosophy focuses attention not on the fine details of particular systems, so beloved of reductionists, but upon large-scale patterns, rules, meta-rules—analogies, metaphors.

Where do these "big picture" patterns and regularities come from? Are they just "there," or do our minds invent them?

Mathematics is the science of pattern; we can gain some insight into the source of patterns in our mental models of nature by dissecting a few mathematical patterns and trying to see how they arise. We've mentioned the famous Feigenbaum number, 4.669... and its universal manifestations in chaos theory. Whenever a dynamical system exhibits a period-doubling cascade—an ever more rapid sequence of events, in which it goes through cycles whose periods continually double, piling up into a chaotic motion with no periodicity at all—the steps in the cascade are multiplied by a factor that gets closer and closer to $1/4.669$. The Feigenbaum number is a documented instance of an emergent phenomenon, an unexpected collapse of chaos.

When Feigenbaum first discovered his number, its universality was a complete mystery. However, mathematicians and physicists have now discovered a full explanation of the reasons why the Feigenbaum number is ubiquitous in chaos, so we are in the happy position of being able to dig beneath the surface and observe what makes the mystery tick. To make the arguments easier to understand, we first develop a parallel treatment of the equally special number $\pi = 3.14159...$, which arises wherever there are circles. The twist is that we're not going to assume any knowledge of circles; the story takes place in a world that has not yet discovered the circle. How can you stumble across π without knowing about circles? It seems only appropriate to let the Zarathustrans take up the tale.

NEEPLPHUT: You will/will not (delete whichever is inapplicable) be interested to hear about a famous historical debate between the octimists and the septimists, which also involved repeated doubling and a limiting constant.

CAPTAIN ARTHUR [*in a resigned tone of voice*]: I find alien ideological arguments totally fascinating, Neeplphut.

NEEPLPHUT: As I recall, it was the philosopher Octimedes who began the debate, in his seminal analysis of the regular octagon. He had proved that the square of the perimeter of an octagon is proportional to its area, and found an elegant new way to calculate the ratio of the two, which he found to be 13.25. One of his research students—her name was Baugenphyme—invented a process of edge doubling that converted the octagon into a hexadecagon, a regular polygon with sixteen sides. Again it transpired that the square of the perimeter is proportional to the area, but this time the constant of proportionality is slightly different, namely 12.73. Baugenphyme's edge-doubling process could be repeated to yield polygons with 32, 64, 128, or more sides, the number doubling each time: an edge-doubling cascade.

CAPTAIN ARTHUR: Remarkable.

NEEPLPHUT: Baugenphyme noticed that in all cases the square of the perimeter is proportional to the area, but the ratio decreases as more and more sides are included. However, it does so in a regular and predictable way; and as the number of sides grows without limit the ratio becomes ever closer to the strange constant 12.56637..., which of course became known as the Octimedes number.

CAPTAIN ARTHUR: Why not the Baugenphyme number?

STANLEY: She was the research student, Arthur.

CAPTAIN ARTHUR: Oh, yeah.

NEEPLPHUT: This was a celebrated discovery. Octimedes wrote a treatise explaining that the existence of his number proved the correctness of the Principle of Octimality. But the septimists disagreed. Their leader Sepcrates proved that the square of the perimeter of a regular heptagon—a seven-sided polygon—is proportional to its area, with a constant of proportionality 13.48. This failed to impress the octimists, since it bore no relation to any of the constants derived by Octimedes and Baugenphyme.

But Baugenphyme was something of a maverick. In an unprecedented act of ideological betrayal, she applied a heretical edge-septupling variant of her process to the heptagon, getting polygons of 7, 49, and 343 sides, and so on. She proved that in all cases the square of the perimeter is proportional to the area, calculated the constants of proportionality, and observed that again they decrease, tending to a fixed limit as the number of septuplings becomes arbitrarily large.

CAPTAIN ARTHUR: That's not terribly surprising, is it?

STANLEY: Since they decrease and are positive they've got to tend to *something*!

NEEPLPHUT: Yes. But the limit was 12.56637... again. Baugenphyme had derived the Octimedes number from a purely septimist procedure. It very nearly knocked the bottom out of the Principle of Octimality. I shudder to think how close Zarathustra came to rampant septimism. However, in a masterstroke Octimedes claimed that the agreement was superficial and that it would only be necessary to carry the calculation to a few more octimal places for discrepancies between the two constants to become apparent.

STANLEY: That's nonsense. Octimedes was wrong.

NEEPLPHUT: Why?

STANLEY: In either case the sequence of polygons is getting closer and closer to a circle. For a circle of radius r the ratio of the square of the perimeter to the area is $(2\pi r)^2/\pi r^2$, which is 4π. And that's 12.56637..., of course.

NEEPLPHUT: Excellent.

STANLEY: So it wouldn't matter how many sides you started with: 8, 7, 153, anything. Baugenphyme's polygon-doubling or polygon-septupling cascades would always lead to the identical number, 4π.

NEEPLPHUT: You are right. However, in those days the circle was not recognized as a legitimate geometrical object, because it was not composed of straight lines. It took several eight-by-eights of years before the full truth emerged. By then the Principle of Octimality was thoroughly accepted by all but the most die-hard septimists—but on quite different and logically impeccable grounds, you understand.

Baugenphyme's process, then, produces the "universal" number 4π from the edge-doubling cascade. It is universal because the same number appears from that process, whatever the number of sides of the initial polygon. You can check that the numbers all look the same, purely by calculating them, and the only objects you ever use are polygons. Circles don't appear. But the explanation of the universality lies outside polygon space; indeed in a reasonable sense it lies just off the edge. It resides in a nonpolygon, the circle, which has special and unusual properties that single it out. Baugenphyme's ratios tended toward the universal number 4π because her polygons tended toward the circle. But nobody noticed this to begin with because "circle" wasn't in their mathematical vocabulary.

It took a while for us humans to find the trick behind Feigenbaum's number, too. The secret is that as you get nearer and nearer to the end of the period-doubling cascade, each new period-doubling looks more and more like a tiny shrunken copy of the previous one. Indeed there is a special, essentially unique period-doubling cascade—let us call it a furcle—for which each new period-doubling is precisely a shrunken copy of the previous one. As you travel along the cascade, everything looks more and more like a furcle, just as Baugenphyme's polygons look more and more like a circle. But, of course, you don't notice that unless you know that furcles exist—and for a long time people didn't, except for Feigenbaum, who conjectured just that, right at the beginning. They're not "off the edge"; they just don't stand out unless you know what to look for, and then you have to work hard to find them. What you can easily observe is not furcles, but a puzzling numerical convergence—just as in Baugenphyme's edge-doubling cascade—that always leads to the same answer. That answer is the one you get for the special object on which everything works unusually nicely. In Baugenphyme's case it's the circle, with magic constant 4π; in Feigenbaum's case it's the furcle, with magic constant 4.669. In both cases there's only one number because there's only one special object. The processes used extract that number from nonspecial objects by (cryptically) making them resemble the special one more and more closely.

■ THE GEOGRAPHY OF PATTERN

Let's interpret these two tales in terms of spaces of the possible. Baugenphyme first. Her process takes place in polygon space. It converts any given polygon into one with twice as many sides and labels all of the polygons with a number, the ratio of the square of the perimeter to the area. That process, repeated indefinitely, gives rise to a dynamic that pushes any initial polygon off through polygon space: 8, 16, 32, 64 sides, and so on. Sitting just off the edge of polygon space is the circle, a polygon with infinitely many sides. It must have *some* label, and as it happens its label is "12.56637" or, in symbols, "4π." The Baugenphyme dynamic on polygon space drives any initial polygon toward the circle. The dynamic creates a geography of polygon space, with the circle as an attractor, a peak in the landscape. And that's where the pattern "4π" comes from; it is the choice of dynamic that causes it to emerge.

The same is true of Feigenbaum's number, which arises in period-doubling-cascade space, PDC space for short. The overt process of measuring the lengths of the branches in the cascade is paralleled by a cryptic one of chopping out small bits of the cascade and magnifying them. Both generate the same sequence of numbers. The cryptic process creates a dynamic that drives all initial cascades in PDC space toward the furcle. This has to have *some* label, and it so happens that the label is "4.669." So again we have a space of the possible, a process within that space that creates a dynamic, and an emergent pattern associated with the attractor of that dynamic.

Emergent simplicities are the peaks in landscapes of the possible. The bigger the peak, the more important that simplicity tends to be. No self-respecting mountaineer would write a book called *The Ascent of the Insignificant Anthill Just Outside the Post Office,* but they do get excited about Nanga Parbat. In mathematics the range of peaks available is also determined by the local geography: the system of those things that are likely to interest mathematicians. They ought to write exciting books like *The North Face of Feigenbaum's Constant,* but the titles usually come out more like *Renormalization Techniques for Discrete Dynamical Systems.*

In the same way, the "big questions" in science are about the big peaks. Reductionism tries to understand them by digging deep down inside the peak, to see what lies beneath it, what it's built upon. But mountain peaks aren't built by piling up long thin tubes of rock; they come from the overall folding of the total landscape. Reductionism, as we have repeatedly stated, is highly successful within certain limits. In particular, it leads to powerful technologies. There is more, however, to understanding nature than reducing it to "underlying" laws. Nature's geography is not the geography of laws, but of the landscapes that emerge from those laws. We also need a description that makes sense on the level of the landscape itself.

Human cultures have many views of the deity. One is a human-centered view, the vision of God as content. An example is Christianity, in which God is to be understood by looking inside human beings, at humanity's interaction with Jesus, and so on. But there is also a pantheistic view, the Spinozan vision of God as context, built into the very warp and woof of the universe. Reductionist science, with its inward-seeking obsessions, adopts a rationalist parallel to Christianity. What is lacking is a rationalist parallel to pantheism—a panscientific vision of the universe.

13 COMPLICITY AND SIMPLEXITY

A Texan was driving in the Israeli desert. He was feeling thirsty, and he noticed a tiny house in the distance, so he drove up and knocked on the door.

The owner, a wrinkled old man, let him in and gave him a glass of water.

"Do you own this place?" asked the Texan, making polite conversation.

"Yes."

"Why do you live out here in the desert?"

"I raise chickens."

The Texan looked around him. "How big is this place, anyway?"

"Hmmm—it must be about twenty yards at the front. And at the back, it's maybe sixty! Well, fifty, at least."

The Texan grinned. "Back in the States," he said, "I own a ranch just outside Dallas. I get up at dawn, and I get in my car and drive. I keep driving all day, right into the evening. And I still don't reach the boundary of my ranch."

"Oh dear," said the old man, "I once had a car like that."

We have been looking at the relationship between a complex universe and the simple laws concealed within it. The first part of this book has a coherent theme: the reductionist approach to the understanding of nature, which explains complexity on one level of description in terms of simple structure on a lower level. In chapter 7 we argued that this approach ignores an important question: Why do high-level simplicities exist? How does nature collapse the chaos of highly complicated internal substructures to produce stable large-scale phenomena? We developed this theme in the second half of the book. In this final chapter we attempt to take stock, and ask whether the second half also has a coherent theme comparable to the elegant certainties of reductionism.

One running theme in the second half is the role of context, rather than content. We view reductionism as seeking explanations by looking at the insides of things, in the sense that atoms are "inside" crystals, or DNA is "inside" a cell nucleus. "Inside" is a convenient word, but it mustn't be taken too literally. By the inside of a system we mean the things that you find when you break it down into pieces. By this definition the inverse-square law of gravitation is "inside" the solar system. Everything that is not inside, in this sense, will be considered as "outside," and that's what we mean by context. The content/context alternative is a neat piece of wordplay, but it's not exactly profound to suggest that internal influences should be complemented by external ones. The common theme of outsides is not strong enough, on its own, to give unity to the second half of the book.

But there are, we believe, deeper themes. The main one is the story of emergence, of high-level patterns arising from the indescribably complex interaction of lower-level subsystems. We distinguish two kinds of emergence, "regular" and "super." We call regular emergence "simplexity" and the super version "complicity." The archetypal examples of complicity are evolution and consciousness, and the complicit view gives interesting insights into the questions of mind and the human spirit. We are then, finally, ready to return to the central question of the whole book: What is a "law of nature"? We give the beginnings of a formalization of the phenomenon of emergence, and suggest how science can break out of the reductionist mold without losing the valuable insights that it has discovered there.

We leave the last word to the Zarathustrans.

A RULE FOR FINDING RULES

A deeper view of reductionism is that it is about how rules and regularities (for insides) generate behavior, and how meta-rules (for insides of insides) generate rules, and so on. As theories destroy facts, so do meta-rules destroy rules: More and more rules are condensed into fewer—but more abstract—general principles. The ultimate goal of reductionist science, a Theory of Everything, aims to condense the entire universe into a single system of meta-meta-. . .-meta-rules. The mental image of nature's laws is a

Tree of Everything, ultimately rooted in a single universal mental funnel (Fig.44).

There doesn't seem to be a convenient word for the process complementary to reductionism, for the description of a system in terms of rules for its outsides. "Holism" isn't quite the word we want; holism is certainly an alternative to reductionism, but it considers a system as a unit and often ignores its context. "Contextualism," perhaps? Whatever we call it, the external view also has a hierarchy of rules, meta-rules, meta-meta-. . .-rules—but this time for outsides. Darwin's principle of natural selection seems to be some kind of external meta-rule (meta because it is one step more abstract than the "rules of competition," whose consequences it attempts to encapsulate in a simple statement). It explains the complex form of living

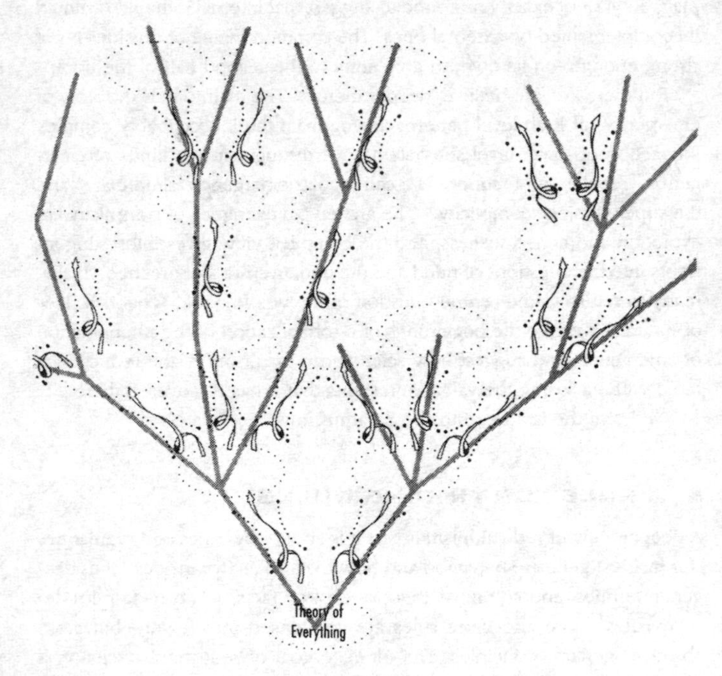

FIGURE 44

The Tree of Everything

creatures, not in terms of their DNA but in terms of their historical development and evolution within a changing external environment. That's one respect in which Darwinism is different from, say, Newton's laws of motion, and it's one cause of the persistent arguments about whether or not Darwinism is genuinely scientific. It is, but strictly speaking it's not reductionist science.

It may seem that the second half of this book explains simplicity in terms of internal complexity: an approach complementary to reductionism, but still internal. However, we have argued that the real cause of large-scale simplicities is not internal complexity at all, but external constraints, which collapse chaos and render systems independent of much of their own internal complexity.

Reductionism, as a scientific-philosophical principle, is a grand meta-meta-. . .-rule about how to understand systems in terms of ever finer internal decompositions. The complementary principle that we are seeking must be an equally grand meta-meta-. . .-rule about ever expanding sets of outsides. However, we are not obliged to mirror reductionism, replacing every "in" by "out." The principle or principles that we seek may well have their own special features, lacking any parallels in conventional reductionism.

And they do.

■ SIMPLEXITY AND COMPLICITY

In this closing chapter we argue the case for two unifying principles that between them govern not just the second half of this book, but the first half as well, and the relation between the two. These principles must relate to the fundamental issues of simplicity and complexity; moreover, our format of paired chapters implies that chapters 1, 7, and 13 have to have titles that rearrange one another. We have therefore played another word game: We call these principles simplexity and complicity.

The word "simplexity" is an obsolete form of "simplicity"; the related word "simplex" means "composed of a single part," which is one meaning of the word "simple." Our notion of simplexity is itself a simple concept, easy to describe and exemplify, and we don't think it will be very contentious to suggest that something like it exists.

In contrast, the word "complicity" already exists in common parlance, and its conventional meaning is remarkably close to what we have in mind: "the state or condition of being an accomplice." It has an older, now obsolete meaning: "the state of being complex." The concept to which we apply it has elements of both meanings. It's much harder to explain than simplexity, and we wouldn't be surprised if many of our (reductionist) colleagues were to deny that complicity—in our sense—can possibly occur. We think it does; in fact, we think it is an inextricable part of the way in which nature functions and an essential, though neglected, aspect of scientific understanding.

We must add that we don't for a moment imagine we've found the ultimate answer to life, the universe, and everything. Just a point of view. We're aware that our ideas are themselves a result of complicity, and that among other things we are attracted to linguistic tricks. Given a different scientific upbringing, we might well be trying to sell you the virtues of "conspansion" and "extraction," say. But we think that what we're saying isn't *just* a matter of linguistics, and we hope to convince you of that as the chapter unfolds.

■ CAUSALITY AND CONTINGENCY

What we are now discussing is not physics, the laws of nature, but metaphysics—the philosophical basis of natural law. People who look for philosophical frameworks in which to think about rule-based explanations of nature are attracted by apparently contradictory styles of explanation. Some of them like the connectivity-contingency model, the world of Francis Thompson's "The Mistress of Vision":

All things by immortal power
Near or far
Hiddenly
To each other linked are
That thou canst not stir a flower
Without troubling of a star.

In this view, the entire universe is an intricate web of step-by-step causality, but with so many steps that on a large scale we seem to see only contingency. The "importance" of any cause then has no meaning, because momentous events are always initiated in some sense by trivial, indeed immeasurably small, causes. People who like this worldview tend to be attracted to chaos theory, in which the flapping of a butterfly's wings can change the weather—although they don't always sympathize with its underlying determinism. In this philosophy there is usually a demotion of single causes; instead, everything is caused by everything that preceded it. The view of human evolution presented in Stephen Jay Gould's *Wonderful Life* is a recent example of this style of explanation.

The alternative view looks for patterns, rather than isolated steps of causality. Its proponents are impressed by how many animal groups independently invented flight (insects, birds, bats, pterosaurs, several different families of fish); by how many different plants independently invented photosynthesis; by how many animal groups—such as mammalian primates, carnivores, and dolphins; molluscan octopi and squid; perhaps even mantis shrimps—have independently produced brains as a general solution to nature's waywardness. They are more interested in the collapse of chaos, in the way that large-scale systems protect themselves against underlying randomness or intricate fine structure, than in the fine structure itself. They argue, as we do, that whatever the details of the route had been, something like Stephen Jay Gould could easily have evolved; and if it was sufficiently like him in the way it thought about evolution, then it would have produced a similar theory when exposed to a Burgess shale. They concede that fluttering butterflies may contribute to the trillions of microeffects that provide the chaotic background from which the weather emerges, but they prefer the dynamics of large masses of warm, humid air as an explanation of hurricanes. They point to the contrast between the erratic variability of weather and the relative stability of climate.

Embryologists of this persuasion emphasize not contingency but convergence; they are impressed by the way in which the same frog can be produced from a frog's egg by very different developmental routes and under very different environmental conditions. The same goes for ecologists who find that on islands or in lakes, similar ecologies result from very different initial combinations of creatures and plants.

But convergence also has its problems. In particular, any attempt to elucidate the manner by which convergence is attained is likely to expose a complex web of control mechanisms whose detailed function rapidly subverts all lucidity in an explosive orgy of reductionist analysis. Instead of the

Tree of Everything, whose stock of mental funnels decreases as we penetrate into the reductionist depths, we have the opposite: the Reductionist Nightmare, branching forever, with nothing solid at the bottom at all (Fig. 45).

However, the distinction between the contingent and the convergent types of explanation is not simply that between reductionism and its alternatives. Whether you are reductionist or antireductionist, you might espouse either of the above philosophies or methodologies of science. The difference is not one of scientific methodology but of attitude. Different attitudes can generate different, often conflicting explanations.

Step-by-step causality is appropriate to problems in which the explanation being sought—the answer to some question about nature—is stated in

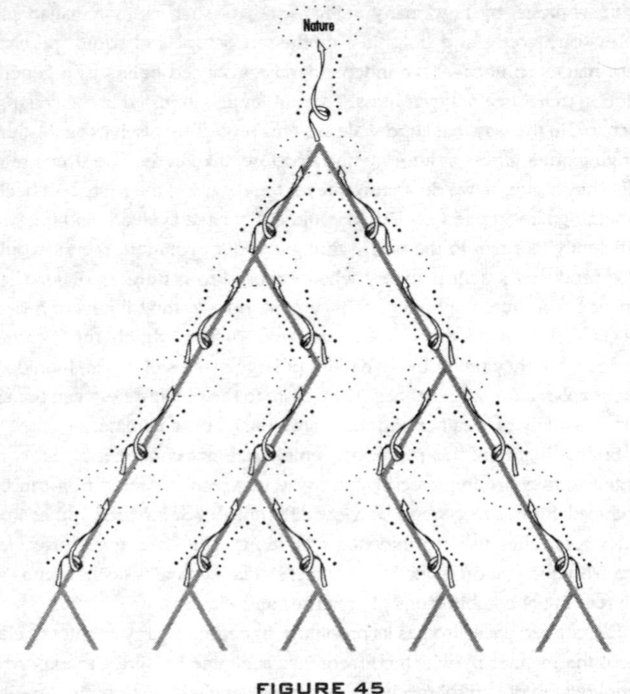

FIGURE 45

The Reductionist Nightmare

the same terms as the question. "Why does carbon have atomic weight 12?" "Because it has six protons and six neutrons." On the other hand, pattern seekers want to compare the question with other similar ones; their "explanations" are assemblies of disparate but homologous questions. "Why are there so many carbon compounds in living creatures?" "Because carbon is the only element that can form the backbone of enormous, complex molecules, capable of performing the molecular computations needed for life to occur."

NEEPLPHUT: Hello, Captain. Stanley and I have an interesting discovery to report.

CAPTAIN ARTHUR: Really?

STANLEY: More of a nondiscovery. We used the ship's computer to examine ZNA sequences for half a million organisms, and found no further resemblances to terrestrial DNA sequences.

CAPTAIN ARTHUR: Fascinating. Neeplphut, that reminds me of something that's been puzzling me.

NEEPLPHUT: What, Captain?

CAPTAIN ARTHUR: Well, ZNA is chemically quite different from terrestrial DNA. It's true that the human genome and the Zarathustran zenome do have equivalent code sequences, but that implies no resemblance between humans and Zarathustrans.

STANLEY: Why not?

CAPTAIN ARTHUR: Look at Neeplphut, Stanley.

STANLEY: Oh, yeah.

CAPTAIN ARTHUR: And now you tell me that no other Zarathustran or terrestrial organisms have equivalent code sequences. Yet the Zarathustran ecology has a lot in common with ours: flying forms, streamlined aquatic forms, even intelligent life-forms such as yourself. It must be some kind of convergent evolution.

NEEPLPHUT: Yes, there are striking parallels. But there are also differences. On Zarathustra no creature has its foodway crossing its airway. And they all have four fingers, not five—so that a pair of limbs provides the perfect octimal number of digits, of course. And we most certainly do not get our reproductive and excretory apparatus mixed up, as yours—and all other terrestrial vertebrates'—are.

CAPTAIN ARTHUR: That's what bothers me. If you're allowed to select the parallels and ignore the differences, in what sense can you talk of universal patterns?

STANLEY: The point must surely be, Captain, that you can distinguish the two types in advance. Some are "parochials," obviously dependent upon historical accidents and of no great importance in themselves. All of the features that Neeplphut mentioned are parochials—they were inherited from *Eusthenopteron* when it came out of the water and invaded the dry land. The big patterns of flight, swimming, and so on, though, those are universals. Whether or not you can fly doesn't depend on where you've put your sex organs.

NEEPLPHUT: Yes, but where you put your sex organs *may* depend upon whether you can fly. Think about landing.

CAPTAIN ARTHUR: Touché. Stanley, you must explain how to distinguish universals from parochials *in advance*.

STANLEY: Hmmm, tricky. Well, the distinction has to be phrased in terms of large-scale patterns. I've just been watching a weird old videobook about laws of nature, called *The Collapse of Chaos*. It points out that universals must be immune to differences of internal, small-scale structure. Otherwise they wouldn't be universals! Universals are structural features of the geography of the space of the possible—like the islands that correspond to big new evolutionary gimmicks and the bridges that connect them.

NEEPLPHUT: I think I see what you mean. In order to observe a universal, you must think contextually. You can never notice its universality if you contemplate only its internal structure. Take the concept "bridge," for example. Different civilizations see it in very different ways, as we Zarathustrans discovered when we wanted to honor Baugenphyme by naming a small island after her. The island was not connected to the mainland, so it was decided that a bridge should be built. The city authorities decided that the project was so important that it warranted an interplanetary competition.

The proposal from the Weezlies of Argyris ii3 came as a freighter load of computer disks, all of which read something like this: "A bridge starts with grain of sand #1, 24 facets as follows: triangles 18, pentagons 4, hexagons 2, impurities as follows: iron 0.000345, aluminum 0.014673, cadmium 0.000022, magnesium 0.009756; . . . ; grain of sand #107,895,674,593, 493 facets as follows: triangles 239, heptagons 11, dodecag—"

STANLEY: Okay, Neeplphut, I think we've got the point.

NEEPLPHUT: That kind of approach requires a huge amount of information, but none of it is informative to anyone other than a Weezly. The second proposal, from the Blakboxians of Boolywood-Holovard, seemed more promising. It was a large crate, labeled "Bridge—No User-Serviceable Parts Inside."

STANLEY: Economical with the information requirement, certainly.

NEEPLPHUT: Yes. The problem emerged when we opened the crate and discovered that the bridge was too long.

CAPTAIN ARTHUR: Couldn't you shorten it?

NEEPLPHUT: "No user-serviceable parts inside." It was a single, indivisible unit.

CAPTAIN ARTHUR: Ah.

NEEPLPHUT: Both competitors seemed to have missed the essential feature of a bridge—that it is a fixed device that lets you transit a discontinuity without getting knurfled. It could have been constructed from woven bamboozle, prestressed clouds, or molded flaum droppings, for all we cared. The concept is conveyed in a few words; the information requirement is tiny because the object is defined in terms of its interaction with its surroundings.

STANLEY: I suppose the Weezlies would object that specifying those surroundings requires even more incredibly detailed information.

NEEPLPHUT: No doubt. But the surroundings were there already; they did not have to be specified.

CAPTAIN ARTHUR: So who won the competition?

NEEPLPHUT: Nobody. We dug a tunnel.

STANLEY: Well, that would fit your definition of "bridge" equally well.

NEEPLPHUT: Precisely. A tunnel *is* a bridge, for that purpose. But not according to the fine print of the contract. So we saved a fortune in bridge-design fees . . .

FROM RULES TO NATURE

Having examined these different styles of explanation—or nonexplanation—we can now move toward our twin concepts of simplexity and complicity. To that end we shall examine two key examples of rule-based explanations of nature: Newton's laws of motion and gravitation, and Darwin's concept of natural selection. One mathematical rule, one biological; this is a very democratic book.

Newton's laws of motion and gravity, as applied to the motion of the solar system, seem to be a clear-cut case. On the one hand, there is a huge body of experimental and observational evidence on the movements of the planets, their moons, comets, and so forth; on the other, there is a system of mathematical rules compact enough to be written on one sheet of paper. We're not very fond of the word "law," and while we're not prepared to counter history and speak of "Newton's rule of gravity," we'd like to. We'll avoid "law" whenever possible, at least until we can explain what we think a law of nature *is*. Together with one relatively small set of data (the current positions and velocities of the planets), Newton's rules—in principle, and very nearly in practice—imply all subsequent behavior of the solar system. If ever a theory destroyed facts, Newton's did. Astronomers have recently used Newton's laws to calculate the future motion of all the planets of the solar system for the next million years.

The relation between Newton's mathematical rules and nature is straightforward. The rules condense into a few formulas all the interacting forces that determine motion. Their consequences, derived by mathematical calculations, have a direct interpretation in the real world. They can be compared with observations, and they work. "I have found the System of the World," said Newton, with some considerable justification. In this sense, they explain the observations by reducing them to something simpler and humanly graspable. Schematically, the relationship seems to be something like Figure 46.

FIGURE 46

The relation between laws and nature—conventional view

Exactly the same structure occurred in Kepler's prototypes for Newton's laws, his three laws of planetary motion, such as "Planets move in ellipses." The main difference is that Newton's laws are more general. They apply not just to two-body systems (sun/planet or planet/moon) but to everything. They correctly describe the precession of the moon's axes; the Trojan asteroids that accompany Jupiter in orbit, leading or lagging by angles of sixty degrees; and an intricate dance between Jupiter and Saturn in which each alternately goes ahead of, and falls behind, the place where it would have been had the other not existed. They describe the orbits of binary stars, the newly discovered planets that are presumed to encircle pulsars, and the spiral form of galaxies.

Despite its different feel—verbal rather than mathematical, and descriptive of real things rather than constructed of mathematical fictions that model them—Darwin's principle of natural selection appears to bear a similar relation to reality. It too is a law, and although its implications are derived by verbal reasoning rather than mathematical calculations, it too appears to explain an enormous body of observed facts, including many counterintuitive ones such as the panda's "thumb."

Figure 46 is a neat, tidy, and entirely standard summary of these relationships.

And wrong in at least one crucial aspect.

■ FEATURES AND INSTANCES

In chapter 8 we asked what would happen if we were to take Newton's rules literally and use them to calculate Mars's orbit, taking into account every single atom of Mars on the grounds that if Newton's rules are *nature's* laws, then that's what nature has to do. You have to do the same thing if you take the visual rhetoric of Figure 46 literally. Remember the bit about "atom of iron, longitude 123.777 . . . 79 degrees west, latitude 47.883 . . . 66 degrees north, depth 132.755 . . . 43 kilometers. Atom of sulfur . . ."? Ring any bells? It's the Weezly approach to physics.

We then compared the rhetoric of reductionism with the (far more sensible) things that scientists actually do when they calculate Mars's orbit, such as pretending Mars is a rigid sphere. What we saw is typical of the human race's approach to every scientific problem. Behind it lie some broad, general principles. Those principles cannot be applied directly to the actual problem, because the real world is too complicated. So we fudge things. We make a few simplifications, which we hope don't alter the answers much; we apply some neat and tidy consequences of the general principles; and if we've done our job well, out pops an elegant answer. In this case we choose to replace the real, immensely complicated ever-changing body of atoms that we call Mars by a convenient fiction: a perfect, continuous, homogeneous sphere of matter. We do the same for the sun. And we ignore all other bodies altogether.

That's how science works, and it's not Figure 46, and it's not like the rhetoric of reductionism at all.

We need a third concept, in addition to "laws" and "nature," which we'll call a feature. Features are simplified general concepts, related to the mathematical or verbal world in which the laws are formulated. "Conservation of energy" is a feature of Newtonian mechanics; "sphere" is a feature of shapes. Features are large-scale simplicities. If you look back through previous chapters, you'll find that the word "feature" occurs quite often. We were using it in its normal sense, but in nearly every case it fits our more specialized meaning too.

We also need the idea of an instance, something in nature that corresponds to a feature. Mars is an instance of the feature "nearly spherical body." The relationship is more like Figure 47, with features emerging from the rules like tendrils of smoke and wafting toward corresponding instances.

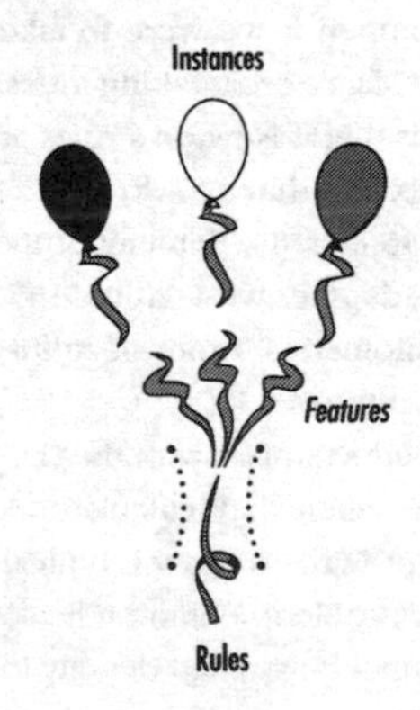

FIGURE 47
From rules to instances by way of features

Instances are just particular aspects of the natural world—a rather special subset of, or structure within, nature. They are the bits of nature that our brains can grasp as simple wholes. The experimental "verifications" and "falsifications" that are used to justify scientific theories all compare features with instances. Suppose, for example, that a mathematician carries out the analysis just described to show that a perfect uniform sphere (a feature of Newtonian mechanics) has an elliptical orbit (another feature). This deduction from the laws is then compared with observation (nature), but the appropriate part of nature is the *instance* "gravitating bodies that look pretty much like spheres"—or, more simply, "planets." Science tests its laws by deducing some of their features and trying to match them to approximately similar instances. As our discussion in chapter 8 of the atomic composition of Mars made clear, there is no genuine attempt to compare rules with nature—to test the rules as such on a realistic description of Mars. Everything is carried out using surrogates: features and instances.

This strategy is not confined to planetary orbits. Quantum mechanics is an established theory of the microstructure of matter, but a lot of the time we don't use it. The technique of X-ray crystallography, for instance, is central to molecular biology; it is how biologists work out the atomic structure of molecules such as hemoglobin. The idea is to shine a beam of X rays through a crystal made from the molecule under study, observe the directions in which the beam diffracts (bounces off the atoms), and deduce how the atoms are arranged in space. The theory behind this technique assumes that a crystal is a regular lattice of atoms, and proceeds from there using a fairly classical theory of wave diffraction. It does not derive the atomic lattice from quantum mechanics, because nobody knows how to. It uses a feature, "regular lattice," and tests it on instances—directions.

In the case of Mars, we have said that the gap in the reductionist story can in principle be filled fairly convincingly; it is possible to prove a mathematical theorem to the effect that the gravitational field of a solid sphere is a good approximation to that of a Mars-sized body composed of atoms. For crystal structure, no such theorem is known—though, as we have indicated, physicists expect such a theorem to be proved one day. Other gaps are more glaring: the gap between the DNA code and the organism that develops using it; the gap between individual plants and animals and the ecosystems they inhabit; the gap between a complex network of nerve cells and an intelligent mind. There are no theorems that bridge these gaps, nor are there ever likely to be. But scientists study such questions nonetheless, and the way they do it is through features and instances.

The major distinguishing aspect of both features and instances is that they should be simple. We do not test the law of gravitation by tracking the motion of a zillion Martian atoms. We test it through concepts such as "elliptical orbit"—a simple feature that is a consequence of the law of gravity. The human brain does not, probably cannot, comprehend complexity as such; it latches onto features and instances, and works with them instead.

BACK TO THE FEATURE

The word that scientists use for this "featuristic" approach is "modeling." Mathematical descriptions of nature are not fundamental truths about the world, but models. There are good models and bad models and indifferent models, and what model you use depends on the purposes for which you use it and the range of phenomena that you want to understand. It's here that reductionist rhetoric and practice diverge; as we've said, what's interesting and important is the practice. The rhetoric is an after-the-fact rationalization of the procedure actually used—and it's a flawed rationalization. It claims a degree of congruence between deep underlying rules and reality that is never justified by any actual calculation or experiment.

The actual practice of reductionism requires features and instances to serve as intermediaries. They fit between the rules that supposedly determine the behavior of the components into which reality has been dissected and the reality to which those rules are applied. Features concentrate our attention on particular instances, and these are used as a testing ground for the rules. Once mathematicians learned how to handle the feature "uniform sphere," by showing that it could be replaced by a point mass, they were in business. They could compare the predictions of the laws of gravitation with the motion of any natural object, such as a planet, that looks like a uniform sphere. The power of this rather cavalier approach must not be underestimated, for it works. Mars reacts to the sun's gravity just as a uniform sphere would; we can tell this because it follows an orbit astonishingly close to the theoretical orbit of a uniform sphere.

Nearly always, the instances that are considered to be important in some branch of science are derived from a corresponding system of theoretical features. In the terminology of Thomas Kuhn, a system of features is a paradigm, a worldview. Indeed a "paradigm" originally meant a standard instance, as in "rat," a paradigm of "rodent." The word "stereotype" applies when the paradigm metaphor has degenerated into cliché.

Only within a chosen worldview can the human mind interpret the world. But a system of features is more than just a worldview; the essence of a feature is simplicity. A highly complex worldview could, in principle, be a paradigm—for somebody smart enough—but only a worldview built from simplicities can be a system of features that we all can recognize and use. This is how science puts common sense back into the driver's seat: Features are *designed* for commonsense thinking; they allow us to ignore incomprehensibly intricate details of unimaginably complex processes.

The process can also proceed from instances to features. Mathematicians worried about the gravitational fields of spheres *because* Mars looks roughly spherical. When they got more ambitious they noticed that most planets are flattened at the poles, and they started to study the gravitational fields of ellipsoids. And so on: feedback between features and instances, each reinforcing and selecting the other in the mind of the thinker. Newton found his law of gravitation by working backward from a known feature, elliptical orbits, that seemed to match nature rather well. Kepler's "laws" of planetary motion are really no more than features: They apply to individual planets but not to the solar system as a whole. Darwin's principle of natural selection is a feature, not a law—but with a new twist. There is no agreed-upon—indeed no known—body of laws from which this feature can be deduced, no "Darwin's equation" from which natural selection can be derived. That is another reason why people argue whether Darwinism is "scientific." However, Darwinism is more subtle than just an unsupported feature, as we shall explain in the section after next. First, we summarize the story so far and invent a concept to capture its main . . . well, feature.

■ SIMPLEXITY

We shall give the name "simplexity" to the process whereby a system of rules can engender simple features. Simplexity is the emergence of large-scale simplicities as direct consequences of rules. Newton's laws—rules—of motion have direct mathematical implications about centers of mass, energy conservation, and so on. You can write them down in a few pages, grasp them in their entirety. Another word with a very similar meaning is Stuart Kaufmann's concept of antichaos: the occurrence of simple large-scale behavior in complicated systems. The Feigenbaum number—and its Zarathustran parallel, the number π—are examples of simplexity. So is Conway's game of Life, which we met in chapter 6; here we *know* the rules, and very simple they are. But what interests everybody who plays the game is its features: gliders, spaceships, glider guns, programmable computers. These are simplexities; their properties are direct and inescapable consequences of the rules.

An important point about simplexities is that their presence is guaranteed, once you have the rules. Any system with the same rules will necessarily exhibit exactly the same simplexities.

By extension, we also apply the word "simplexity" to a real-world concept that parallels this theoretical process. For example, a mathematical model of Mars can sensibly treat it as a uniform sphere because a mathematical simplexity operates. Any collection of particles sufficiently close to a uniform sphere has a distant gravitational field that is very close to that of a single point. That's not just a faint hope or an empirical observation; it's a theorem, which could be proved by making the right estimates. Theorems are mathematical simplexities, made explicit in their proofs. This theorem simplifies the problem enormously. Instead of an atom-by-atom calculation that would take longer than a human lifetime, you can get away with a few scribbles on a sheet of paper. The conversion of Newton's laws at the atomic level into very similar laws about planet-sized bodies is a case of simplexity, and without it we would never have been able to apply Newton's laws usefully to anything in our world.

The real-world parallel to simplexity is exploited to good effect by nature. Mars is indeed composed of atoms; these do—as far as we can tell—obey Newtonian dynamics to a good degree of approximation, and Mars conveniently moves in a near ellipse. This can't be coincidence. There must be a kind of natural simplexity in operation as well, otherwise Mars wouldn't behave in such a simple manner. The natural simplexity, indeed, should work for the same kind of reason that the theoretical one does: The net effect of all those interatomic forces, on another collection of atoms as distant as the sun, causes Mars to move just as it would if its mass were all concentrated at its center. This is no more than a restatement of the previous reductionist rhetoric, but one that makes the assumptions explicit. The features in the theoretical domain are congruent to the instances in the natural one.

Because features are simple, they "explain" the corresponding instances. The feature "parasite" explains instances such as "liver fluke," "bit string in Tierra," and "computer virus." By extension, people tend to claim that the rules from which the features were extracted therefore "explain" nature: *Because* Newton's law of gravitation implies elliptical orbits for spherical bodies, and *because* Mars has an elliptical orbit, *then* Newton's laws explain the motion of Mars. This is fair enough, as far as it goes. What is far less justified is the common assumption that the observed elliptical orbits confirm the truth of the gravitational rules themselves. They do confirm a particular consequence—a simplex feature—of those rules: elliptical orbits. But other, quite different rules might possess that very same feature. Lots of variants on Conway's rules for Life can also produce moving objects and programmable computers, for instance. As we've said earlier, every theory is surrounded, in theory space, by a cloud of virtually indistinguishable theories. Everything in that cloud possesses very similar features. By testing only features, we can confirm—or, as Karl Popper insists, fail to deny—only features. The truth or falsity of the rules continues to elude us.

Simplexity, then—either in nature or in theoretical models—is the emergence of simple features as a direct (though possibly highly intricate)

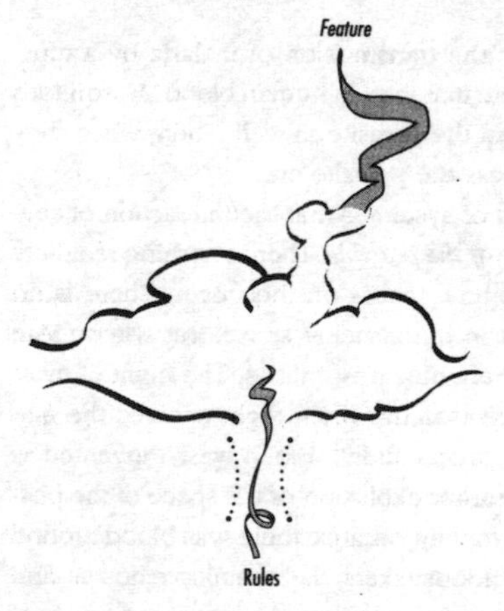

FIGURE 48
Mental image of simplexity

consequence of "deep down" rules. Simplexities are "theorems." We may be able to spot a simplexity without actually being able to write down its explicit deduction from those rules; often we just tell a plausible story without the hard details. Our mental picture of simplexity is not a funnel, but something more like rising smoke (Fig. 48). Every so often a neat tendril—a simplex feature—gets put together by the internal smoke dynamics.

As far as this feature goes, the complexity buried in the smoke or farther down the funnel doesn't matter, either to us or to anything in the natural world that sees it. To a cat, all robins are fungible, so cat-mediated natural selection acts on whatever features of robins are seen by cats. Only indirectly does this process affect internal structure, such as DNA. The structure that underlies any feature can be replaced by a different complexity if it yields a feature that is "seen" as being the same.

■ COMPLICITY

Simplexity is, appropriately, a relatively simple concept. It is the easy way for different rules to generate similar or even identical features; it works when the rules themselves are very similar. Now we consider something much more subtle, in which totally different rules converge to produce similar features, and so exhibit the same large-scale structural patterns. We call it complicity.

For example, let's think about the transmission of malaria by a mosquito. Malaria is caused by a parasite that lives in human blood. Mosquitoes suck blood and inadvertently ingest the parasite as well. Then, when they attack another human, they can pass the parasite on.

What is special about this kind of system is that the interaction of several subsystems *enlarges the space of the possible*. There's nothing remotely like malaria in any of the component spaces on their own. There is no cryptic malaria in blood space, or in bloodsucker space. But when those spaces interact, they open up entirely new possibilities. The flight of mosquitoes wasn't invented to transmit malaria; when flight evolved the malaria parasite didn't exist, because people didn't. Blood wasn't invented as food for mosquitoes; that was an earlier explosion of the space of the possible, with bloodsucking insects evolving because there was blood around to suck. Put all the bits together—bloodsuckers, flight, multiple hosts—and you *still* haven't caught any glimpse of parasites in the combined space of the possible. Wait, though, and they emerge from the new interactive dynamic.

And once a new feature of this kind appears, it leaves its history behind. You can understand how malaria is transmitted without going into all the evolutionary details.

Simplexity merely explores a fixed space of the possible.

Complicity enlarges it.

And both processes collapse the underlying chaos, producing stable features from a sea of complexity and randomness.

Complicity, by its nature, is so intricate and convoluted that any attempt to dissect out its internal workings and past history just leads to the Reductionist Nightmare. Despite this, there are patterns to complicity—patterns that let us recognize its presence. They are meta-rules, large-scale universals. You couldn't have predicted malaria, in all its gory detail, from the interaction of blood space and bloodsucker space. But with a bit of imagination you could have predicted the universal pattern "parasite" and guessed that the combined space opened up new niches for parasitism.

Another instance of the same meta-rule is "leech," the aquatic analogue of the mosquito. The analogy between mosquitoes in air and leeches in water closely parallels our earlier analogy between wings in air, and air bladders in the ocean. Maybe most airborne features have waterborne counterparts? If that's true, it's a meta-meta-rule—a universal pattern about universal patterns.

These meta-rules and meta-meta-rules work on the level of features, and only on that level. Their explanations do not lie inside the complexities of the component subsystems, which rapidly diverge as you progress to deeper reductionist levels. The meta-rules are emergent, not reductionist. Indeed they are so nonreductionist that it doesn't even matter whether a given feature arose through simplexity or complicity. As long as it looks the same to the outside world, that's all that matters. This fungibility or universality is what makes the patterns universal, and it's what lets them collapse chaos.

To go back to our example, malaria is special, but the feature "parasite" is everywhere: in Darwinian evolution (on several levels—viruses parasitizing bacteria, flukes parasitizing snails, tapeworms and malaria parasitizing humans); in the bit strings of Tierra; in social systems ("Such people are mere parasites on decent society!"); and in economic systems (where parasites take the form of speculators on the money markets). In no case—not even the simplistic world of Tierra—can we spell out a complete story of the link between rules and features; only in the case of Tierra do we even know what the rules are, because there Ray invented them. Nor does it seem very likely that the reductionist rules of economics (if such things exist) would in any way resemble the reductionist rules of evolution (if *they* exist). An appeal to a putative Theory of Everything cannot explain the parallel, even if it is true that the behavior of malaria parasites and market speculators can both be traced to the same underlying subatomic rules. How the molecules of a malaria parasite conspire to make it infest the human bloodstream is not likely to be similar to how the molecules of a speculator conspire to make him dump dollars in favor of deutsche marks. It is *only* on the level of large-scale features that we can even detect the analogy between tapeworms and speculators, and we must explain the analogy on the same level.

Another example is that of the elephant and the giraffe. Both have a common feature: "long thing for drinking water without kneeling down and tearing off leaves without hopping into the air." It so happens that for elephants we call this feature a trunk and for giraffes a neck, but it is a common feature nonetheless. It is also clear that, in terms of external constraints such as competition for food and water, the explanations of the evolution of trunks and necks probably parallel each other closely. Elephants evolved trunks for much the same reasons that giraffes evolved long necks. However, a reductionist analysis of elephant and giraffe DNA, even if it told us exactly how to make a trunk or a neck out of proteins and other chemicals, cannot present the analogy as anything other than an accident: two utterly distinct processes happening to produce similar results. In the absence of any overt link between rules and features, we cannot declare such things to be the result of simplexity. The developmental rules for elephants do not contain tiny bits of "trunkiness" which just get magnified and expressed in the adult creature. So, if not via simplexity, these resemblances must arise by some other, more subtle mechanism.

You could declare such resemblances between features of quite different systems with quite different rules to be mere curiosities, accidents of antichaos which attract our attention because human minds are like that. You could argue that the supposed analogies don't really hold water: Speculators aren't really parasites on economic systems, but just seem to share a few behavioral characteristics. However, we think that these resemblances deserve to be explained, not explained away; so we have coined the word "complicity" to describe the kind of process that we think must be occurring.

Our mental picture of complicity, comparable to Figure 48, is of smoke from two or more sources mingling, interacting, flowing into and through each other (Fig. 49). Again, the odd tendril of smoke, an identifiable feature arising from that interaction, becomes visible.

Complicity arises when simple systems interact in a way that changes both and erases their dependence on initial conditions. The hallmark of complicity is the occurrence of the same feature or features in systems whose rules are either known to be very different, or are expected to be very different if only we could find out what they are. This carries an important consequence: Complicity is a convergent process; it homes in on the same features regardless of fine detail in the rules. Another way to say this is that complicity leads to "replaceability" of (some) components. If

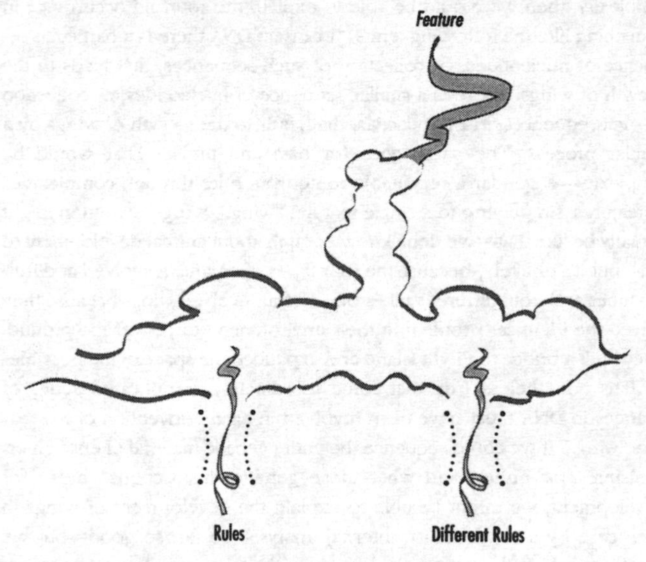

FIGURE 49
Mental image of complicity

some complicit process involves the construction of a bridge—real or metaphorical—then anything that performs the same function, such as a tunnel, might easily do the job equally well. And what a bridge is made of, or what is around a tunnel, is irrelevant. Once the requisite structure is present, and strong enough or otherwise well enough built to perform its intended function, then its materials become irrelevant to that function—in the mind of the thinker, the body of the traveler, the motor of the car, the person in the toll booth.

COMPLICITY IN EVOLUTION

The prime example of complicity is evolution, for which the two systems of rules are the chemistry of DNA and the systematic ways in which organisms interact with their environment. Consider the animal feature "wing." Such a feature has evolved several times, starting from entirely different circumstances, in insects, pterodactyls, bats, birds. If wings were a case of simplexity, then we would be able to explain this fourfold occurrence in something like the following terms: "In insect DNA there is a particular sequence of nucleotides, or collection of such sequences, that leads to the growth of wings. There is a similar sequence of nucleotides, or collection of such sequences, in pterodactyls, that leads to the growth of wings by a similar process. The same goes for bats and birds." That would be simplexity—a standard, repeatable route from rules through complexities to features, simplifying to a single feature, "wing." Wings' evolution might actually be like that—we don't know enough about animal development to tell—but it's unlikely, because the four types of organism evolved at different times and from different ancestors. All four evolved wings because they solved the identical problem in their environment: to get off the ground. They built a bridge to Flight Island and expanded the space of the possible.

It is nonetheless true that some inordinately complicated series of changes in DNA must have been involved in each reinvention of the feature "wing." If we could sequence the entire genetic material of enough organisms, *and* understand what those genetic "instructions" mean for development, we might be able to explain the development of wings in each case by a reductionist, internal analysis. So far, so good—but we would end up with four totally different explanations and no understanding at all of the source of the analogy. Similarly, if we started with the blueprints for a Boeing 747 and an Airbus, and counted the nuts and bolts and grommets and ashtrays, and wrote down just where each one goes, we would "explain" the structure of each, but without building them and seeing what they do, we'd probably be unable to tell that either of them flies and we'd certainly not be able to tell that they are both instances of the feature "aircraft."

Instead of disconnected and unrelated bits of reductionism, we prefer to emphasize the analogy; to capture it as a feature or a set of features; to describe it as an example of complicity; and to explain it in terms of external constraints. All four types of creature evolved wings because that's what you need if you want heavier-than-air flight. The rules of aerodynamics constrain the form of the gadget you develop. A miniature version of the Leaning Tower of Pisa wouldn't fly at all: You need flat surfaces, stretched out to the side. The two types of aircraft have similar wing designs—and much else in common too, such as wheels for landing—for the same reason.

GEOGRAPHICAL FEEDBACK

Because complicity, by its nature, is convergent—it leads to the same features independently of the underlying rules—it cannot be reduced to a particular system of rules in any useful manner. Even a complete set of such reductions—one for insects, another for pterodactyls, and so on—fails to explain how the analogy between the resulting features arises. It just explains each feature separately, without "noticing" any resemblance. It can't notice any resemblance because it's internal to a given feature. Therefore, we can't explain complicity within a purely reductionist framework. We can nevertheless explain roughly how it works, by appealing to the idea of spaces of the possible.

Continuing to focus on the DNA/organism example, we can describe the general nature of their interaction. DNA sequences live in DNA space, and in the absence of any other influences would wander around dynamically through the geography of DNA space, seeking attractors and settling on them. Similarly, organisms live in creature space, and in the absence of any other influences would wander around dynamically through the geography of creature space, seeking attractors and settling on them. However, there are other influences, which couple those two spaces. Biological development takes DNA strings and develops organisms with their aid. This happens in the context of already developed organisms; it's not just a simple mapping. The implication, however, is that changes occurring in DNA space affect the dynamic in creature space. Conversely, natural selection, which provides the dynamic in creature space, favors some organisms over others, and thereby favors the DNA that is contained in the favored organisms. Changes occurring in creature space affect the dynamic in DNA space. In short, there is a feedback loop between the two spaces. It causes them to coevolve toward a common dynamic.

However, the two separate spaces do not have identical geography. They don't even have similar geography. Small changes to DNA may produce a sequence that does not lead to any viable organism; in sexual species such changes typically lead to recessive genes that don't show up until at least one generation later, and then only if the mutation occurs elsewhere in the population. DNA changes have only a cryptic effect on organisms, and small movements through the geography of DNA space can have no meaningful creature space counterparts. The phenomenon of genetic assimilation shows that changes in organisms are effectively independent of changes in their DNA—or, more precisely, that any link is so indirect and cryptic that it is unobservable. And even if we can get around all of these difficulties—which is what the concept of the extended phenotype was invented for—it is still true that the "natural" ways to change DNA are very different from the "natural" ways to change organisms. The way to modify a bit of DNA is to change a few nucleotides, or snip a segment out, or paste one in. The way to modify an organism is to make its legs a little stronger, its chin a little weaker, its feathers bluish-green instead of greenish-blue, its tail a bit more stripy. The geography of the two spaces, the notion of what a small change can achieve, is simply *different*. Different, too, are the features of the two spaces. Features of DNA are things like "makes hemoglobin"; "junk DNA"; "high-level control gene." Features of creatures are things like "tendency to get stuck in trees" and "liking for honey." As we've already explained, even if you accept the Dawkins definition of "gene for liking honey" as "genetic variation in the liking for honey," there's still no reason to expect to be able to match it up to any particular feature of DNA. The same goes for getting stuck in trees.

Because the two spaces have very different geography, their individual attractors don't match up nicely, so the feedback between the spaces has a creative effect. It changes them both, usually in a rather unpredictable way. Feedback between spaces with different geographies tends to produce new types of behavior that are seriously different from anything that you find in either system alone. For example, suppose the combined system tries to sit on an attractor in creature space that does not match up with any attractor in DNA space. The DNA-space dynamic will try to change the state of the combined system by altering the DNA; the creature-space dynamic will try to preserve the state because in creature space we have a state that—were it not for feedback from DNA space—is an attractor. But because of the feedback, both dynamics are trying to operate, and each is also influencing the other. In place of compromise we find conflict, and conflict leads to unpredictable, complex interactions. Those interactions create a new, combined geography that in no sensible way can be thought of as a mixture of the two separate geographies.

By analogy, think of a child, living in child space, riding a horse that operates in horse space. The attractors in child space are things like ice cream vans. Those in horse space are things like bales of hay. If there is an ice cream van with a bale of hay outside, then a common attractor emerges. But when the rider wants the horse to stop next to the noisy engine of an ice cream van, while the horse has her eye on a bale of hay two fields away . . .

Don't be confused by this image. Several billion years of evolution have bound our present DNA space and creature space together, and a combined dynamic has settled out. As a result, we never see the individual dynamics of "naked" DNA space or "naked" creature space. But we do have mental images of what those spaces, and their geographies, should look like. For example, people have been enormously puzzled by the occurrence in DNA space of sequences to make really clever hemoglobin, with a mechanism for picking up oxygen in the lungs and releasing it in the tissues, because "random" assembly of nucleotides couldn't possibly produce such a thing. We are equally puzzled by the occurrence in creature space of the feature "heart," because a "random" assembly of cells would not be able to make such a high-tech structure. What we don't see, unless we think rather hard, is that each puzzle goes some way toward explaining the other. The DNA to make hemoglobin has arisen because the organism has evolved a circulatory system, including a heart; the heart has developed because the DNA provides the right proteins and controls to build it.

Neither has arisen on its own, in its own "naked" space, nor would it. The *joint* dynamic creates the attractors that will actually occur, not the individual dynamics. Those attractors define the macros that the monkeys need to type hemoglobin; and a heart comes too, as part of the same complicit package.

This mutual geography arising from feedback between different dynamical spaces is a metaphor for complicity. Because of our ignorance of the individual spaces, their individual dynamics, and the nature of the feedback, exacerbated by the complexity of all three, we have little hope of dissecting out the precise reductionist story of how the dynamic proceeds from DNA to organism via development, and even less so the entire historical tale whereby whatever was there a billion years ago became an eye today.

However, what we can work out is the kind of pressures that must result from the kinds of interaction that have just been described. The advantages of hearts to an organism are clear; through natural selection they feed back into DNA space and push the system toward those parts of it that lead to the development of better hearts . . . and around and around goes a self-reinforcing loop. So without knowing any details, we still have a good understanding of the general nature of the process. We can see why the complicity should occur.

But not how it does. "How" questions are those with reductionist answers. "The wing of the lesser spotted kinki-bird has 435 feathers. Feather number 1 is 2.6 inches long, and when the bird flies horizontally at a speed of 15 m.p.h., it vibrates at a frequency of 16.77 Hz. When the bird flies at 16 m.p.h., this frequency changes to . . ." This is the Weezly bridge again. In contrast, "why" questions have external answers that often operate via complicity. "The lesser spotted kinki-bird has wings because it has to escape from the claws of the great rumpuscat." But such answers can never be single, because of complicity. "It also needs them to feed on goolifruit high in the branches of the wangi tree." The distinction between genuine explanations and Just So Stories is a subtle one.

■ STRANGE LOOPS

Our discussion of scientific theories in terms of levels of description includes a simplification that must sometimes be borne in mind, and now is the time to point it out. Most of the time it may make sense to say that a given theory is "at a deeper level" than another—to say, for example, that quantum mechanics is deeper than crystallography. But sometimes phenomena that seem to belong to one level show up on an entirely different one. Similarly, phenomena that seem to be peculiar to one system may unexpectedly influence another. We call such interactions strange loops or strange links and picture them as in Figure 50. We'll give a few examples.

Our first strange loop involves eels. European eels look quite different from American ones, but both types of eel breed underneath the Sargasso Sea, a region in the Atlantic Ocean where large amounts of floating seaweed and other junk collect. The Sargasso Sea is roughly the size of the Gulf of Mexico and lies to the east of Florida. In Europe the eels grow to an age of eleven to twelve years before they head off into the Atlantic Ocean; you also see elvers, young eels, coming back from the ocean to Europe. In America the eels also head out into the Atlantic, and you can find larvae returning, drifting on the tides.

The problem is, the adult European eels never make it to the Sargasso. They can't eat along the way, and they simply don't have enough food or energy to get there alive. So where do the European eels, "quite different" from the American ones, come from? Spontaneous biogenesis?

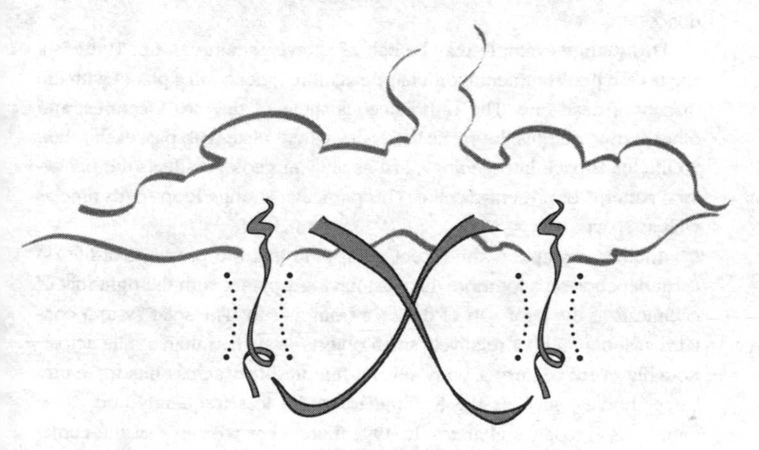

FIGURE 50
Strange loops

The answer is more prosaic: They come from American eels. Those *do* make it to the Sargasso. Their planktonic larvae drift into American rivers. They also drift across to Europe in the Gulf Stream, taking two or three years to get there. Because they're drifting for much longer, they grow larger than their American cousins, so they develop into adult eels that look different. Remember anglerfish and *Bonellia*? There the same genetics led to different sexes, but here the same genetics leads to different forms of adult eels.

This is all rather strange, and while it's clear that conditions in the Sargasso must be unusually suitable for eels to breed, one still wonders why the European eels can't find a better place nearer to home, and how the whole thing got started. The answer seems to be geological, not biological: continental drift. The continents float on top of huge plates on the earth's molten magma, and over periods of geological time the plates move, taking the continents with them. A plate boundary runs roughly down the middle of the Atlantic, and new material wells up through it, so that in the long run the ocean widens and the continents move apart. It looks as though the eels' breeding strategy got going when the Atlantic was much narrower and adults could survive the return trip from both sides. As the continents gradually separated, the eels' geographic programming didn't notice.

This strange example may be half of an even stranger loop. There is a suggestion that continental drift happens more quickly on a planet with life than on a dead one. The carbonaceous shells of tiny sea creatures, and other marine detritus, lubricate the zones where plates rub past each other. So life feeds back into geology, just as ancient geology affects the behavioral patterns of present-day life. This particular strange loop spans time as well as space.

Another example is the recent realization that the presence of life on earth depends on a lot more than just having a planet with the right mix of chemicals at the right sort of distance from its sun. The solar system contains vast numbers of relatively small objects—say, less than a mile across. Roughly every century a body several hundred feet across hits the earth. Larger bodies, such as the K/T meteorite, hit less frequently and cause enormous ecological changes. In 1992 there were worries that the comet Swift-Tuttle would hit the earth about 130 years from now—although more accurate calculations suggested it would miss by a comfortable margin.

Fragments of another comet, Shoemaker-Levy, are predicted to hit Jupiter on July 25, 1994. Indeed, it has been suggested that the gas giants of the solar system—planets such as Jupiter and Saturn—prevent most such bodies from hitting the earth because their gravitational fields effectively sweep the solar system clean of most of this rubbish. So your presence on this planet depends upon the existence of Jupiter and Saturn, a strange link from the top of the astronomy smokestack into the bottom of the biology one.

THE I'S MIND

Our brains have evolved an impressive ability to detect features. Features are generated in nature by the collapse of chaos, and they provide a quick-and-dirty method for anticipating events in our environment so that we can respond more rapidly to possible threats. There is thus a lot of evolutionary pressure for brains to evolve feature detectors. But feature detectors are themselves features, so a generalized feature detector will be self-referential. We give the label "consciousness" to our feature-detection system: We become conscious of a feature of the world when our brain detects it. Therefore consciousness is self-referential—that is, we are conscious that we possess consciousness. We tend to think that this property of consciousness is the most surprising one, but actually, it is a simple consequence of the generality of our feature-detection systems. What is surprising about consciousness is not that we are conscious we are conscious, but simply that we are conscious. What needs explaining is not the self-referential nature of consciousness, but how it can take a huge quantity of partially structured sensory data and extract important features—the same features that "natural" interactions see.

Complicity helps us to see some deep philosophical issues, such as consciousness and intelligence, from a new point of view. Many people view consciousness as an irreducible mystery—perhaps a gift from God, or a manifestation of some supernatural principle. To them we have nothing to say except that we personally don't find that kind of answer satisfying. Others try to reduce consciousness to something simpler—for example, quantum mechanics. The argument goes something like this. The big problem of consciousness is not that creatures seem to have some concept of self, but that conscious creatures make conscious *choices*. There is a strong indeterminacy about consciousness. Now, whereabouts in the physical world do we find indeterminacy? In classical Newtonian mechanics? No. In quantum mechanics? Yes. Aha! The indeterminacy of consciousness must arise from the indeterminacy of quantum mechanics!

To advocates of this kind of argument we have rather more to say. We think that their reasoning is, frankly, rather silly. It's as if the question were "Chartres cathedral?" and the proposed answer were "Bricks": Chartres cathedral has beauty and form and structure because each brick that goes to make it up has a *very tiny amount* of beauty and form and structure. This is a preformationist argument of the worst kind: "everything is as it is now because it grew from something in the past that contained, in microscopic form, the entire thing that it now is." Similarly, the homunculus theories of human reproduction: Inside each sperm is a very tiny but complete man or woman. Or the theory that DNA contains a tiny coded version of a complete man or woman. Or the idea that every piece of human behavior must come from an animal precursor.

To answer the question "Consciousness?" with "Quantum mechanics" is to look in the wrong place, at the wrong level, and in the wrong terms. Preformationist theories just lead to infinite regressions; everything has to have come from tiny copies of itself, so nothing can ever get started. The indeterminacy of consciousness must come from the indeterminacy of electrons; then the indeterminacy of electrons must come from the indeterminacy of quarks; then . . . Even if you stop when you get down to a Theory of Everything, you haven't explained where the indeterminacy in that particular theory comes from. You've just put it in at the beginning, and asserted that since reductionism (or simplexity) rules, and indeterminacy is important at the quantum level, then it must still be important at the level of the brain. The collapse of chaos by way of complicity implies that this need not be so. Our principle that large-scale emergent phenomena are insensitive to underlying fine structure would lead us to expect exactly the opposite: that if quantum-based brains can exhibit consciousness, then brains that obeyed classical—nonquantum—physics would also be able to. Not that anyone can build such brains to check whether we're right, but the point is that if anybody ever discovers the mechanism of consciousness, our bet is that it would work in a classical universe too.

We are surrounded by evidence that complicated systems possess features that cannot be traced back to the individual components. Do the molecules of a cat possess, in some rudimentary and minuscule form, an unusual attraction for the molecules of a mouse? Ridiculous! Inasmuch as such features come from anywhere in particular, they come from the manner in which the components, en masse, interact. The ability of conscious brains to make choices—assuming they actually do—must be a feature of large systems of neurons, or any comparable computational devices, that arises from the manner in which they interact. It is complicity, not quantum mechanics, that leads to consciousness. We can see why consciousness evolved: A good way to avoid predators or find mates or generally manage your life is to decide between alternatives rather than blindly following predictable rules.

NEWTON'S EARS

There are many ways in which we can detect the workings of complicity in the human mind. Here's one. Isaac Newton's mind came up with the rule (law of motion) that acceleration is proportional to force. One of the reasons why previous generations had failed to discover this simple rule is that one of its concepts, acceleration, is not something we perceive directly. It is a rate of change of a rate of change and as such is rather subtle. However, the other concept, force, is something that we feel all the time. Gravity pulls us down; we spend our lives fighting it. When we want to move a rock along the ground, we have to push it. Precisely this intuition led Aristotle to propose that velocity is proportional to force (because the faster you want the rock to move, the harder you have to push). The same intuitions about forces, combined with Galileo's experiments, led Newton to replace Aristotle's velocity by acceleration. The rest, as they say, is history.

However, there is another side to the story. Newton's experience of forces included such things as being pushed by other people. We humans detect such forces through sense organs—in this case, the inner ear. But the sensors in the inner ear do not actually detect forces. When something pushes us, it doesn't push the sensors directly. However, the force accelerates us, and the sensors in our ears respond to that acceleration. Newton's law of motion is an unwitting "deconstruction" of how his ears functioned. If Newton's law hadn't worked, Newton's ears wouldn't have worked either. (Aristotle presumably failed to deconstruct his own ears correctly.)

This bears on a major question, which we asked right at the start of the book and have not yet fully answered. Are the patterns we perceive in nature real, or are they figments of our imaginations? In *The Mind's Sky* Timothy Ferris uses the hourglass (Fig. 51a) as an image of matter/mind: The upper half of the glass is the material universe, the lower half is a human mind, and the sand trickling from one to the other is information from the material world entering the brain. But, as the story of Newton's ears shows, that image is far too symmetric; it puts the mind's view of the universe on an equal footing with the universe's view of the mind. Now, the sand inside the mind is indeed a model of the material world, and it is that model that we manipulate when we seek patterns. The material world is not part of our minds. But our minds *are* part of the material world. The lower half of the hourglass is inside the upper half (but *not* vice versa). A better image might be the mathematician's Klein bottle (Fig. 51b) which not only bends the lower half of the hourglass inside the upper, but does so in such a way that the combined surface has only one side. The "mind" half is turned inside out, so that its contents are not directly identified with any of the contents of the material half.

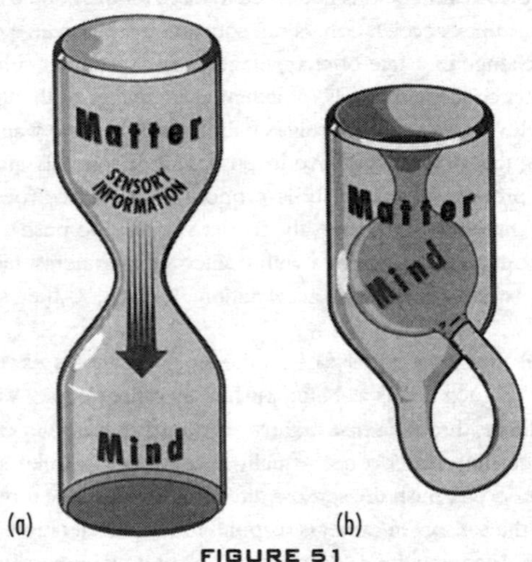

FIGURE 51

Two images of mind/matter.
(a) The hourglass, (b) the Klein bottle

This image emphasizes the fact that mind is a strange loop. "Information" about the material world can get into our minds by routes that do not pass through our sense organs. If somebody hits you on the head with a hammer and kills you, some rather nasty information about the material world has had a direct and major effect on your mind, by destroying it. Hallucinogenic drugs affect the mind through its chemistry but create images that seem to have come from sense organs. The drugs affect those parts of the brain machine that process sensory data. They have the same kind of "unexpected" effect as the "wrong" coin in the ticket machine, mentioned in chapter 1, which persuaded the machine to disgorge an entire roll of tickets. This possibility was implicit in the mechanics of the ticket machine but was not intended to be part of its function. The same goes for drugs and brains.

There is no hourglass symmetry between mind and matter. To repeat an image from chapter 1, reality may perhaps be a figment of our imagination, as some philosophers argue, but our imagination is definitely a figment of reality. Thus the concept "acceleration" in Newton's imagination was apparently derived from the sensory information "force." We put it this way because that is how his brain interpreted what was happening: "I feel a force pushing me." However, that sensory impression actually came about because his ears responded to acceleration.

Some brain puns are better than others; the best way to simplify the perceived universe is to tune into a collapse of chaos in the real universe. And that is exactly what Newton did: He tuned into a collapse of chaos in his ears, which produced the sense of a force from an acceleration. It was the exact opposite of what he thought he was doing, which was explaining acceleration in terms of a force.

■ GOD IN THE BIG BANG

Reductionist science can seem very inhuman, and many people of a spiritual turn of mind find this so disturbing that they reject science altogether. But science and the human spirit are not incompatible. Understanding the universe involves far more than just finding mechanistic explanations in terms of internal bits and pieces. The human mind structures the universe in terms of features, and we use those features to determine our place within nature. For example, we relate to a tree in terms of its features. We find its majesty so inspiring that we write poems about it and make paintings of it, we cut it down and use it to build things, we admire the beauty of its grain, and we associate its buds, blossoms, and falling leaves with the seasons. We compare its rough bark to our weatherbeaten skin, we talk of damage to a tree as a "wound." The feeling of unity with trees goes beyond mere brain puns: Both trees and humans use dead cells on the outside to help keep water in.

The universe also sees a tree in terms of its features. Birds evolved the ability to perch because trees possessed the feature "twig." Mycorrhiza and trees became symbiotic because their mutual features reinforced each other. On the astrophysical level, the sun "sees" Mars as a concentrated mass exerting a gravitational force, rather than as a collection of atoms and force vectors, and this is one way in which science sees Mars too.

Thus, even though it may be true that the inner workings of the universe, deep down the branching reductionist funnels, can be captured in simple equations, the universe itself often functions by operating upon a high-level structure *as* a high-level structure. So do we. This congruence between the way the universe works and the way we see ourselves within it offers satisfaction on a human level. There is a shared dynamic, the complicity of our own evolution within the universe, the fact that our perceptual systems are figments of reality. This goes well beyond the anthropic principle, which says that for purely logical reasons we necessarily find ourselves in a universe that is able to produce creatures like us. That may explain why we are in this universe rather than a different one, but it doesn't explain why we feel comfortable here. In contrast, the complicity of a shared dynamic implies that wherever we are we will find spiritual satisfaction from being in a universe that fits us. The dynamic leads us to see ourselves within the universe in a benevolent way—not as the arbitrary consequence of cold rules so far removed from ourselves that we cannot comprehend the connections, but as a natural part of the universe's high-level features. We belong here.

The mere possession of a romantic view of how wonderful the universe is, however, is not enough. It is too mindless and too dangerous. The universe rewards us for understanding it and punishes us for not understanding it. When we understand the universe, our plans work and we feel good. Conversely, if we try to fly by jumping off a cliff and flapping our arms the universe kills us, and if we release too many greenhouse gases into the atmosphere it overheats us. When we understand the universe we acquire spiritual equilibrium. This is what attracts most scientists into science, but they often forget it as they become enmeshed in the nuts and bolts of laboratory work and administration. Romanticism alone can seriously damage your mind, but reductionism alone can seriously damage your soul.

This is a very religious view, similar to the Spinozan position of being at one with the warp and woof of the universe. For Spinoza, the discovery of constraints gave "peace" (*shalom* = wholeness) and—usually—more freedom. This paradox of the perception of constraints leading to more freedom is reminiscent of Zen koans and Hindu parables. They make people's lives and thoughts conform to reality and make effective, balanced people. They contrast with Christianity, Taoism, and Confucianism, which (at least in their popular representations) present the unrealities of heaven and hell, fate, and ancient wisdom as guides to human thought and action.

Some scientists who have delved into the question of humanity in relation to the universe have ended up by adopting positions surprisingly close to conventional religion. In *A Brief History of Time* Stephen Hawking pointed his mental radio telescope at the remnants of the Big Bang, and what he saw was God. *We don't want to point you in that direction*. Whatever your personal persuasions on such questions are, there is one thing we want you to understand. A dynamic does not necessarily imply a purpose. Darwinian evolution has a dynamic, but organisms do not seek to evolve. The existence of attractors does not imply that dynamical systems are goal-seekers: on the contrary, they are goal-*finders,* which only recognize what the "goal" is when they have found it.

The fallacy of seeing God in the Big Bang is the leap of logic from "There is a dynamic in the universe that created us and makes us feel at home" to "The universe was set up *in order to* create us and make us feel at home." It may have been, but the discovery of a dynamic does not of itself imply any such thing. However, as with the anthropic principle, it is tempting to derive a sense of purpose from a dynamic and see it as a spiritual frame that we must use. "We are here because we are meant to be here." Well, we may be and we may not. But because of the shared dynamic, we feel at home here—whether we are a goal or an accidental byproduct—so the dynamic provides a spiritual frame that we find both comforting and awe-inspiring.

Romanticism alone is empty. Reductionism alone can provide a strong feeling of *sympathy* with the universe: We understand individual bits and pieces; we can kind of see how a tree works, how it gets its water into the topmost branches. But the shared dynamic goes much further: It creates a feeling of *empathy* with the universe. Put your cheek against a tree and feel the roughness of the bark against your skin.

SO WHAT *ARE* LAWS OF NATURE?

We can't tell you why nature has laws. We could try to argue that they are "all in the mind," but as we've just reminded you, minds are "all in reality," and it's hard to see how a lawless universe could contrive to evolve minds that had kidded themselves that their universe looked as if it obeyed laws when it really didn't. Lawless universes don't evolve organized structures like brains that think they can detect laws. Yes, that's another anthropic principle, but we think this one works, because it doesn't rely on exploring just a few limited axes in a space of the possible.

What we can do is try to sort out where the laws come from. The conventional picture—boiled down to essentials—is that there are "deep" laws that govern how nature really works and shallower laws that reflect mathematical consequences of them. The deep laws are things like quantum mechanics; the shallow ones are like aerodynamics, crystallography, and econometrics. This is not to suggest that our current versions of the deep laws are one hundred percent right, but their agreement with experiment is very good, so they must be pretty close to the "real truth." We could summarize this view by taking it to extremes: Laws of nature result from the true Theory of Everything through the process of simplexity. This is the reductionist Tree of Everything described earlier in Figure 44.

We think that the above view is far too neat and tidy. In particular, it assumes that the patterns our minds can use are the same as those nature uses. We compute the motion of particles, using mathematical rules that we have evolved by watching particles move in the real world, but that doesn't prove that the universe is a computer. Since the motion of a particle can also be described by sequences of words, you could equally well argue that the universe is a word processor. Information is our concept, the result of complicity in the evolution of the brain, but complicity works in terms of features, not in terms of detailed internal structure. Our mathematical models have the right features, true; but that doesn't entitle us to deduce that the internal structure of the universe is the same as that of the model, especially since the features may be immune to internal structure. Analogy and metaphor are not the same as identity; analogies have identical features but different substructure.

Our minds have been forced to evolve quick-and-dirty feature-recognition systems in order for us to survive in a hostile world. Quick-and-dirty methods cheat; they take shortcuts, such as "Anything orange and black is a tiger." Reality is more subtle and contains many orange-and-black things that are not tigers, but survival demands a rapid (and therefore imperfect) decision that errs on the side of safety, not a considered, reflective judgment. The mental computations involved in our survival techniques must be carried out in real time, and the pressure of evolutionary competition is so great that there may be room only for such flawed solutions. Having evolved those mental computational devices we now use them to look for other kinds of feature, too, although the computations do not have the same direct survival value and speed may no longer be of the essence. So a human brain presented with the elliptical orbit of Mars will (after considerably more thought) conclude, "Inverse-square law." And this perception may perhaps be equally quick-and-dirty in terms of its congruence with what nature is actually doing. Our mental models of nature are not so much faithful reflections of reality as cut-down models that focus on certain essential features.

The universe may be much closer to a glass menagerie, consisting of structures and processes that are more or less "transparent" (meaning that the details of their internal workings are irrelevant to how they behave) wreathed together by complicity. Our picture is more like Figure 52, into which we've thrown in a few strange loops to make the point that it is just a rough approximation. In this view, laws of nature are features. They are structured patterns that collapse an underlying sea of chaos, and they are conditioned and created by context. Even in fields such as physics, where reductionism seems to be king, we have argued that a great deal of mental selection goes into concepts such as "electron." An electron is a feature of a set of physical rules with a matching instance (detectable only by apparatus whose interpretation is based around those same rules). Fine structure, be it patterned or chaotic, implies features. Even a system with no obvious features would posses the feature "featureless"; it's a Catch-22.

Features arise because rules at one level of interpretation "simplexify" or "complicify" to give features one level higher.

Laws are thus the "grand designs" of the universe, the features that can be extracted from it as a consequence of context. From a theoretical point of view, organisms are features in an ecosystem—not wave functions, not DNA strings made flesh, not neural nets. Societies are features of the inter-

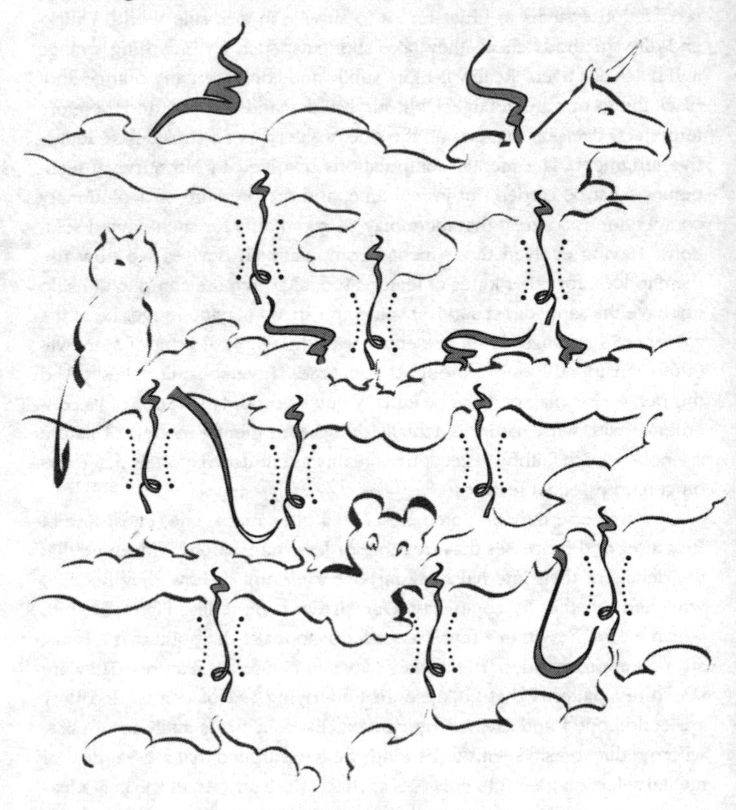

FIGURE 52

The complicit universe, or glass menagerie. When you look down mental funnels you see "insides": The animals are nested. But strange loops mean that sometimes A is inside B is inside A.

action of large numbers of human brains and bodies. Planets are features of collections of atoms wandering around in space under the influence of gravity. In the real world, we find matching instances for all these theoretical concepts.

You could invent bizarre features, such as "strings of atoms arranged in a line according to the square roots of prime numbers," and those too would obey "laws" of a kind. But they wouldn't interest our kind of brain, and for a good reason. Our brains do not just invent patterns at will. Because of complicity between physics, chemistry, and biology in the course of evolution, it is far more accurate to see our brains as figments of reality than to see reality as a figment of our imagination. The patterns that our brains perceive are accurate representations of large chunks of reality, because our brains and sense organs evolved that way. Similarly the hydras that survived were those that bent *toward* prey. When we select a feature, the selection process is not arbitrary. It works in terms of features of our brain mechanisms, and those mechanisms are based on physics, not imagination.

Reductionism seeks to explain all patterns in nature, obvious or hidden, as simplexities arising from underlying internal simplicities. We think that many patterns do not fit this description at all; they are complicities, arising from internal complexities and simplicities under the influence of external complexities and simplicities. Because our brains themselves evolved through complicity between their internal representation of reality and the external reality itself—between their content and their context—they can recognize features, analogies, and metaphors, and see patterns in them. Our complicit brains are both made with and aware of genuine complicities in the universe, and the ringmaster in our head tells himself stories about them.

Our prized laws of nature are not ultimate truths, just rather well-constructed Sherlock Holmes stories. But those stories have been scrubbed and polished, over the centuries, until they capture very significant features of the way the universe works.

That's what laws of nature are.

FORMALIZING EMERGENCE

We've argued that emergence is the rule rather than the exception, and that there are at least two distinct ways for high-level rules to emerge from low-level rules—simplexity and complicity. But we haven't offered any kind of formal structure within which they can be studied. In contrast, reductionist science has built up a huge store of theories, formal rules for the behavior of the natural world. Physicists can write down the equations of quantum mechanics and extract numbers from them that agree with experiments to an impressive degree. We have argued that reductionism is full of gaps, but what we offer toward filling them can be attacked as lacking the crisp precision of a system of formal rules. Can we write down the equations for emergence?

The short answer is no. There are several excuses. Conventional science has an enormous head start, and the Panda Principle applies. "Equation" is in any case the wrong image; the formulation of detailed laws is a reductionist concept, and the whole point about emergence is that it is not reductionist. However, there ought to be some kind of formal structure that captures the essence of emergence. A Nobel Prize awaits whoever finds one and makes it work, and we have no such aspirations; but we can make a stab in the general direction, enough to put up something of a case for the possibility that such a formal structure might one day be devised.

Essentially, what is needed is a mathematical justification for the belief that simple high-level rules not only can, but usually do, emerge from complex interactions of low-level rules. By "emerge" we mean that a detailed derivation of the high-level rules from the low-level ones would be so complicated that it could never be written down in full, let alone understood. We can certainly offer examples where something like this seems to be true, all drawn from chapter 6. Because the examples are mathematical, and mathematical proofs are reductionist, we will be talking only about simplexities. However, until we can formalize simplexity, there is little point in worrying about the far more elusive topic of complicity.

Dynamical systems have features that certainly seem to be emergent: attractors. Attractors are high-level features, but there is no known way to look at the equations for the system, the low-level rules, and see what kind of attractors the system will have. As we said in chapter 6, the attractors emerge when you follow the dynamics, and this is often the only way to find out what they are.

The same kind of thing happens for Langton's ant. Despite all its chaotic excursions, and no matter how chaotic the "environment" of black squares that it lives in, it eventually settles into a simple pattern of behavior: highway construction. Although we cannot yet prove that it always does this, we can check it in any particular case. We merely follow the ant's motion, using the simple rules that determine which direction it takes, until we find a particular sequence of 104 steps that moves it two squares diagonally into a region that contains no black squares. From that moment on it will build a highway. Even if there are no black squares initially, we have seen that the ant wanders for some ten thousand steps before it locks into the highway-building cycle. At the moment, the only way to see that this happens is to work each step out in turn. So, in order to reduce highway building to the original rules, we have to carry out a very long, somewhat messy chain of calculations.

This could change if someone is clever enough to discover a logical shortcut, which is what mathematicians instinctively look for. For instance, it can be proved in a few lines that the ant will always escape from any bounded region of the plane. But the proof does not say *how* it escapes. It looks as if we have found a simple high-level rule, "Eventually builds a highway," which is a consequence of the simple low-level rules about black and white squares but whose chain of deduction is some ten thousand steps long. The simplicity is not passed directly from low-level rules to high-level patterns. And the more complicated the initial environment of black squares is, the longer this type of derivation becomes. Moreover, the proof doesn't explain what the ant does; it just reaffirms that it is true.

More drastically, we know that Conway's game of Life can generate a programmable computer. The halting problem—"Does this program eventually stop?"—is known to be undecidable. You can take any program whose halting problem is undecidable and run it on the Life computer, arranging matters so that if the program stops then the entire structure annihilates itself by shooting itself down with gliders. So either it goes on forever, or it wipes itself out. The problem is that there's no way to tell which will happen. The long-term fate of this particular object, in Conway's simple rule-based game, cannot be determined in advance by any finite computation. We can only let the system run, and watch what it does. We have a mathematical guarantee that the answer to the simple question "Does the configuration eventually vanish?" is implicit in the rules, but cannot always be made explicit.

This kind of uncomputability occurs because the chain of logic that leads from a given initial configuration to its future state becomes longer and longer the further you look into the future, *and* there are no short cuts. (This is not how the mathematical proof goes, but is what it implies.) In other words, as we have told you many times before, the problem with reductionist explanations is that they can proliferate without limit—the Reductionist Nightmare.

These examples focus our attention on the link between low-level rules and high-level structure. Reductionism holds that the high-level structure is a logical consequence of the low-level rules, and we have no wish to dispute this here. The question is, What kind of consequence? If we wish to use reductionist rules to explain and understand the high-level structures, then we have to be able to follow the chain of deduction. If that chain becomes too long, our brains lose track of the explanation, and then it ceases to be one. But this is how emergent phenomena arise. They are not outside the low-level laws of nature; they follow from them in such a complicated manner that we can't see how.

The problem facing a formal theory of emergence is to pin this kind of effect down, to show that no currently undiscovered shortcut can provide a simple derivation. This leads to a central question. Given a system of low-level rules, must there necessarily exist simple high-level features that can *only* be deduced from the rules by enormously long chains of logic? This is a more subtle question, because the same mathematical statement can have many different proofs, some shorter than others. We can only deduce the behavior of Langton's ant by doing ten thousand calculations, but for all we know some brilliant intellect is at this moment putting the finishing touches to a one-page paper that invents a cunning shortcut. That, after all, is how mathematicians make their living.

What we need, then, is a formal proof that in any sufficiently rich rule-based system there exist simple true statements whose deduction from the rules is necessarily enormously long and complicated. (We say "sufficiently rich" because systems that are too simple might have, say, only three statements, each with a two-line proof.) Call this the Existence Theorem for Emergent Phenomena. Its proof is an exercise in metamathematics, the formal mathematics of formal mathematical systems. We don't know how to give such a proof, but we can deduce something along the right lines from Gödel's theorem. This celebrated discovery says that if you choose your favorite mathematical formalization of ordinary arithmetic, then it is impossible to devise a procedure within that formalization that will automatically work out whether any given statement is true or false. There is no "decision algorithm" for the truth or falsity of arithmetical statements. This doesn't keep some statements from being true (2 + 2 = 4) and some from being false (2 + 2 = 5); but it means that you can't write a computer program that will always be able to tell which is which.

We will use Gödel's theorem to show that rich formal systems must possess a kind of emergence. Our conclusions aren't as strong as we would like, but they're a start. The idea is to ask whether there is some general relationship between the length of a statement and the length of its proof. The length is just the total number of symbols, and it is a crude but effective measure of complexity. In this language the Existence Theorem for Emergent Phenomena would be "There exist true statements involving only a few symbols whose *shortest* proofs are enormously long." We can't prove that, but let's see how close we can get.

Suppose you knew that every true statement must have a proof that is at most a thousand times as long. Then you can decide—in principle—whether a given statement is true by counting how many symbols it contains, multiplying by a thousand, and then working your way systematically through all proofs of that length. Either one of them proves the statement (and it is true) or none do (and it is false). So such a system cannot obey Gödel's theorem. The same argument holds if all we know is that every true statement must have a proof that is at most a billion times as long—or indeed if we can give any method for putting a limit on the size of the proof that depends only on the size of the statement. But sufficiently rich formal systems do obey Gödel's theorem, so they must contain true statements whose proofs are a thousand times as long, or a billion times as long, as the statements themselves are. That is, such systems necessarily possess properties that are far simpler than any route by which you could establish them.

This is a formal proof of the existence of a kind of emergence. But it doesn't prove the Existence Theorem for Emergence, because it offers no control over how long the relevant statement is. Does a statement of length

1,000,000 whose shortest proof has length 1,000,000,000 count as being emergent? Probably not. What we want is more like a statement of length 100 whose shortest proof has length 1,000,000,000,000. *That* would be genuine emergence. Our argument from Gödel's theorem falls short of this target, but it gets close. In fact there are good reasons to believe that improvements are possible, but there's a Catch-22 element to the subject, which means that really strong improvements will be extremely elusive.

If the above approach is along the right lines, then it tells us that the Reductionist Nightmare will occur in any sufficiently rich formal system. A Theory of Everything is surely a rich enough system—after all, it is supposed to explain everything in the universe. So Theories of Everything inevitably lead to the Reductionist Nightmare. Not only are emergent phenomena possible; they are unavoidable. Moreover, they are necessarily inaccessible to reductionist explanations.

All of this argues that we must find a way to deal with emergent phenomena in their own terms, and on their own level. We can push them a few layers down the Reductionist Nightmare, but the deeper we push them, the messier the explanation becomes, and no comprehensible chain of deduction will ever connect them to the bottom. On the other hand, we have no wish to ignore the very real advances made by reductionist science. We have to find a way to combine content and context, reductionism and high-level features, into a seamless whole.

A neat trick, if it could be done. Is there any chance? There are a few areas in which something of the kind is already happening. One is the theory of dynamical systems, whose practitioners nowadays are happy to employ both the equations (reductionist derivations from the rules) and the geometric language of phase space and attractors (emergent features) as part of the same explanation. That's one reason why we've employed images from dynamical systems to explain the relationship between simplicity and complexity in nature. Complexity theory is heading in the same direction, but its emergent features are largely discovered in computer simulations rather than by any formal treatment of them *as high-level features*. Dynamical-systems theory and complexity theory are formal theories of simplexity; we don't yet have good ideas about a formal approach to complicity.

Although we can't recognize complicity from its internal details, we can recognize it by its universal meta-patterns. Complicity transcends its internal details, and there's a kind of scale of transcendence. At the bottom is reductionism. One level higher is simplexity, where a single space of the possible produces emergent features. One level higher still is complicity, where the interaction of several spaces of the possible leads to an explosion of the combined space and the emergence of features that can't in any sense be traced back to the components. Beyond complicity comes . . . what? Not the interaction of even more spaces of the possible—that's just a sort of complicit reductionism. No, the next layer has to transcend complicity, in the same qualitatively different manner that complicity transcends simplexity and simplexity transcends reductionism.

Over to you.

REDUCTIONISM PLUS?

We began by asking whether the universe is simple or complicated. The answer: It depends on the context you have in mind when asking the question and the kind of answer you want. Simplicity, complexity, simplexity, complicity, nothing is as simple—or as complex—as we thought when we began. Simple rules can breed simple behavior or complex; complex rules can breed simple behavior or complex. Contrary to common belief, complexity is one of the least conserved quantities in the universe. So are those things that go with it, such as information, meaning, organization, awareness. You *can* sometimes get something for nothing—or nothing for something.

How, then, should science approach its avowed goal of understanding the universe and humanity's place within it? The reductionist Tree of Everything is insufficient: There are huge gaps in its explanations. We think that DNA controls biological development, but we don't know how; we think that appropriately arranged neural networks generate consciousness, but we don't know how. We can see what's at the bottom of the reductionist funnels, but not how it rises to the top.

We need a different approach to those gaps, something that can deal with the big patterns of complex systems. No such approach yet exists, though we have tried to indicate some of the phenomena with which it must come to terms and some of the conceptual frameworks that it might adopt. We said earlier that many new think tanks, such as the Santa Fe Institute and the Center for Complex Systems, have recently been set up to study such questions. The main point we would make about their brainchild, complexity theory, is that it is a theory of simplexity; complicity isn't even in the picture.

We think that the key is to understand complicity, not as an incredibly complex reductionist network, but as the interaction of features within different spaces of the possible. That is, we must put the *dynamics* back into biological development, evolution, and brain function, with the emphasis being on qualitative forms and features. Think of DNA and organisms. We've already argued that the organism does not "see" the DNA code, but only those features of it that produce particular effects that matter to the organism. Similarly, DNA does not "see" organisms; all that matters to DNA is that the organism bearing it should survive to replicate it. This is the "selfish gene" image, but as one aspect of a double-edged process, not as the sole factor. Each system reacts only to the *features* of the other. So what we need is a theory of features, an understanding of how the geographies of spaces of the possible conspire to create new patterns and combined dynamics. Such a theory would see weather as the motion of cyclones and rain clouds, not as the motion of billions of tiny, indistinguishable particles of fluid. But it would treat those features in a precise way, unlike old-fashioned qualitative meteorology. It might well use computers, but they would run programs that understood the *meaning* of the large-scale features, rather than allowing them to emerge as a by-product of billions of bits of information. It might not predict weather any better than we can now, because our current understanding is that weather is inherently unpredictable, thanks to the butterfly effect, and features collapse the chaos that they *can* collapse, not the chaos that we would like them to collapse. But it would remove the butterfly effect from the path whereby we understand weather.

Reductionism is great for quantitative aspects of internal details. In contrast, our current understanding of external large-scale effects is mostly descriptive and qualitative, geometric rather than numerical. We can recognize a hurricane from satellite photos; but we can't tell what it's going to do. Somehow we must combine the best aspects of these two approaches. Richard Feynman speculated on the possibilities of devising such a theory; turn back to the preface and reread what he said. We must find a theory of mathematical complicity between the quantitative and the qualitative.

How? Lord knows. Look how long it's taken the human race to get just the reductionist part working. All we can do right now is point in some direction and claim that that's where science ought to be going. But there *are* big gaps in the reductionist picture, and we think there's a strong case that they can't be filled by yet more reductionism. We're not saying that the reductionist approach should be abandoned, and we're certainly not advocating replacing it by Just So Stories. But we think that too much of the emphasis currently placed on reductionism stems from the Panda Principle: It was there first and its devotees won't let anything else displace it. And we think that a lot more effort should be put into questions such as meaning, structure, and development, so that science can combine internals and externals into a single, coherent scheme.

■ THE INSTITUTE FOR SIMPLE SYSTEMS

We leave the last word to the Zarathustrans, who are equally interested in the question of understanding the octiverse and Zarathustranity's place within it. Characteristically, their approach is a trifle offbeat.

NEEPLPHUT: Welcome to the Institute for Simple Systems.
CAPTAIN ARTHUR: It's very impressive, Neeplphut. I've never seen such an enormous building! And all this incredible equipment—it's breathtaking!
NEEPLPHUT: Yes, I am sorry about that. It is a sign of our ignorance.
CAPTAIN ARTHUR: Pardon?
NEEPLPHUT [*ignoring him*]: Tell me something, Captain. Several times you have talked of "Sherlock Holmes stories." I am having considerable problems with this concept. What is a Sherlock Holme?
CAPTAIN ARTHUR: Sherlock Holmes is a fictional detective.
NEEPLPHUT: "Detective" I understand; but the translator is having problems with "fictional."
STANLEY: Fiction is stories that aren't true, but could be.
NEEPLPHUT: I believe my translator has gone on the blink.
CAPTAIN ARTHUR: Stanley is referring to a logically consistent sequence of events, compiled for purposes of entertainment.
NEEPLPHUT: Ah. You mean *history*.
CAPTAIN ARTHUR: No—these are imaginary events. They didn't really happen.
NEEPLPHUT: How can a sequence of events be logically consistent if it did not happen?

STANLEY: Take "The Hound of the Baskervilles." It is about an ancient curse on the Baskerville family and the sudden death of Sir Charles. Nearby are found the footprints of a gigantic hound, emerging from the great Grimpen Mire . . .

NEEPLPHUT: That does sound exciting. I am very fond of mud. But I am sorry to hear of the untimely death of Sir Charles. His family must be most distressed.

CAPTAIN ARTHUR: Well—Neeplphut, you have to understand that there is no actual Baskerville family, and no real hound. It's all made up.

NEEPLPHUT: Oh, you mean it is a *lie!* Lies, I understand. We use them for the instruction of the young.

STANLEY: I'm sorry, Neeplphut, but you're way off the mark. Sherlock Holmes stories aren't lies.

NEEPLPHUT: Do Terran records include this particular Baskerville family?

CAPTAIN ARTHUR: No. It's imaginary.

NEEPLPHUT: But the story says that it exists?

CAPTAIN ARTHUR: Yes.

NEEPLPHUT: My case rests.

CAPTAIN ARTHUR: No, look, you don't understand. Everybody knows the story isn't *true*. But a *realistic* story has to be logically and emotionally consistent. For instance, "The Gerbil of the Baskervilles" wouldn't carry much conviction.

NEEPLPHUT: Why not?

STANLEY: Nobody would be terribly upset about gerbil prints emerging from a bog.

NEEPLPHUT: I am having serious problems here. Readers know the story is false but require it to be convincing. It must be realistic but it cannot be real. It seems to me that since all falsehoods are mathematically equivalent, *any* lie is logically consistent! It is consistently untrue. So any lie makes a realistic story, and a story is just an extended lie.

CAPTAIN ARTHUR: No, no, no . . . it's just the internal logic of the story that has to be consistent. It doesn't have to agree totally with reality.

NEEPLPHUT: Mmmm, interesting. *Internal*. We Zarathustrans do not really think along those lines. But surely, if it is a matter of internal logic, then there is no difficulty in envisaging a world in which everybody is terrified of gerbils. The conflict with reality is not part of the story.

CAPTAIN ARTHUR: No, but it's part of the context.

NEEPLPHUT: Internal logic is part of the context? Are you *sure?*
CAPTAIN ARTHUR: No, I mean . . . well, every story involves a context employed by both writer and reader.
NEEPLPHUT: But that context is not reality? Even though you say a story must be "realistic"?
STANLEY: It's a selected part of reality. The *background* must be realistic. You couldn't have Sherlock Holmes watching television or flying in a jumbo jet, for instance. Wrong historical period. Reader and writer must tacitly agree on a context, before the story makes sense.
NEEPLPHUT: Ah! I begin to see! It is like a restricted system of axioms, but never stated explicitly. That is a wonderful new thought! *Very* exciting! Complicity between content and context, each artificially limited to an arbitrary subset of reality . . . Let me access my knowledge base . . . excellent. Am I right in thinking that "The Hound of the Fotheringay-Smythes" would be a perfectly acceptable variant?
CAPTAIN ARTHUR: Yes, though perhaps without the resonance. Just steer clear of gerbils.
NEEPLPHUT: And about halfway between "The Hound of the Baskervilles" and "The Red-headed League" you would find "The Red-headed Hound"?
STANLEY: Yes, but that would be a very poor story. Hounds can't copy out the *Encyclopaedia Britannica*.
NEEPLPHUT [*gets very excited*]: Aha! So there is a geography of Sherlock Holmes story space, governed by the dynamic of logical consistency within the agreed context. . . . Stories must maximize their degree of conviction, so they sit at peaks in the landscape. "The Red-headed Hound" sits in a valley, so it does not occur in practice. But that is exactly like characters and species and paradigms and functions and ecological niches . . . and anthropic principles! Credible Sherlock Holmes stories are the attractors, the peaks of credibility, in Sherlock Holmes story space. Spread about between them are the stories that do not work, whose logic falls apart, such as "The Gerbil of the Baskervilles" and "The Red-headed Hound." Change any one detail, and the entire structure collapses. But there are lots of Sherlock Holmes stories—actual ones, like "The Hound of the Baskervilles" and "The Red-headed League," or invented, like "The Hound of the Fotheringay-Smythes" and "The Bald-headed League." In the same way, just because a tiny change in Planck's constant destroys the special features of carbon that make life as we know it possible, that does not imply that

very similar kinds of life cannot occur in a universe with totally different laws! I must tell the librarian at once!

CAPTAIN ARTHUR: Why?

NEEPLPHUT: These analogies are a major simplification. We can amalgamate five wings of the library, pull down two buildings, and fire ten percent of the staff. Everybody *will* be pleased!

CAPTAIN ARTHUR: *Pleased?* To lose their jobs?

NEEPLPHUT: Of course. That is the overriding aim of the Institute for Simple Systems. To understand something is to simplify it. Theories destroy facts, metatheories destroy theories, and so on. The culmination of all that the institute stands for is to close itself down. What use is science if all it can do is *complicate* your view of the world? Every scientist should be trying to see the world in the simplest possible way. I am certain that your Terran scientists would agree/disagree.

Delete whichever is inapplicable.

NOTES

Here we give additional comments, references, and other material to clarify technical points or encourage further reading. The rule in the Notes is "Anything goes." Any books mentioned without full bibliographic details can be found in Further Reading, which follows these notes.

EPIGRAPH

v *The next great awakening of human intellect . . . :* Quoted from Richard P. Feynman, Robert B. Leighton, and Matthew Sands, *The Feynman Lectures on Physics II-41-12* (Reading, Mass.: Addison-Wesley, 1963). See also p. 436 of James Gleick, *Genius.*

1: SIMPLICITY AND COMPLEXITY

16 *. . . what Thomas Kuhn calls a paradigm shift:* See Thomas Kuhn, *The Structure of Scientific Revolutions.*

18 *consequences of its "program," its program being the laws of nature:* A leading advocate of this point of view is Ed Fredkin at MIT. See Julian Brown, "Is the Universe a Computer?" *New Scientist,* July 14, 1990, 37–39.

20 *"antichaos," in which complex causes produce simple effects . . . :* See Mitchell Waldrop, *Complexity,* and Roger Lewin, *Complexity.*

22 *This is the dream of the Theory of Everything:* See Steven Weinberg, *Dreams of a Final Theory,* and John Barrow, *Theories of Everything.*

23 *fine structure on all scales of magnification:* See Benoit Mandelbrot, *The Fractal Geometry of Nature.*

23 *it was pretty much impossible to notice it:* More accurately, the process of noticing it coevolves with the formulation of the general concept "fractal." You have to notice the unity in order to formulate the concept behind it, but you can't notice it in any significant way until you have that concept available. In *The Fractal Geometry of Nature*, Mandelbrot says that he studied what he now recognizes as various aspects of fractals for twenty years, without realizing that they had anything to do with each other.

23 *mimic a huge variety of plants and flowers:* See Przemyslaw Prusinkiewicz and Aristid Lindenmayer, *The Algorithmic Beauty of Plants.*

25 *even a square one:* This follows from Figure 53. Just arrange the speeds of the circles to make the planet follow the square. Which orbits can be obtained using (possibly an infinite number of) uniform-speed circles is a difficult question, related to Fourier analysis. Only mathematicians would want to know, and they can work it out for themselves.

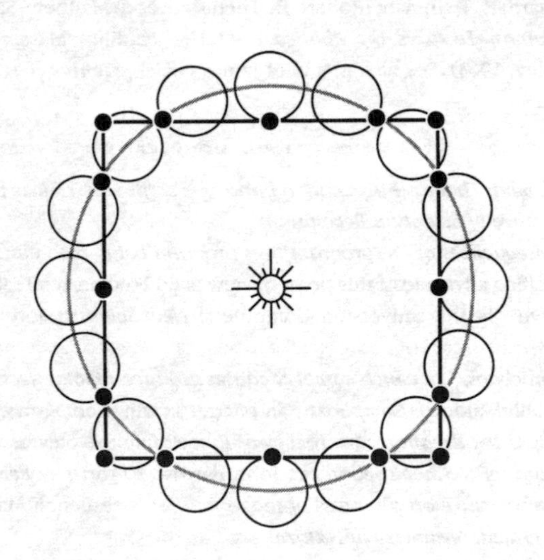

FIGURE 53

One variable-speed epicycle yields a square orbit.

26 *except when hit with a blunt instrument:* Bishop Wilberforce, a major critic of Darwin's theory of evolution, was killed when he fell off his horse and hit his head on a rock. Thomas Henry Huxley remarked that "for the first time in his life, reality and his brains came into contact—and the result was fatal."

27 The *Goon Show* of Science: The *Goon Show* was a BBC radio program that specialized in surreal comedy and WWII army jokes. It was born on May 28, 1951, under the title *Crazy People,* but was renamed in 1952 and came to dominate fifties British humor. It featured Spike Milligan, Peter Sellers, Harry Secombe, and Michael Bentine, playing characters such as Major Denis Bloodnok, Bluebottle, Eccles, and Hercules Grytpype-Thynne. *Goon* fans still recall classics such as "The Dreaded Batter-Pudding Hurler (of Bexhill-on-Sea)" and "The Affair of the Lone Banana." The final program, "The Last Smoking Seagoon," was broadcast in January 1960. See Spike Milligan, *The Goon Show Scripts* (London: Sphere Books, 1973), and *More Goon Show Scripts* (London: Sphere Books, 1974).

2: The Laws of Nature

31 *Chemistry leads to cooking, warfare, and fashion:* The Chinese character for "chemistry" means exactly "the study of change."

35 *1 knife, 200 forks, and 202 glasses—but it's not likely:* Unless the Arcturian ambassador is dating Miss Betelgeuse.

37 *Eight of these are now known to exist:* The other two do not; the values of various measurements used by Mendeleev have now changed, and we wouldn't expect his other two elements to exist in any case.

40 *Carbon, uniquely, can form covalent bonds with itself:* Well, almost uniquely; so, for instance, can hydrogen, whose one outer electron, in a shell that needs two to be full, counts equally as one pimple or one socket. But after you've joined two hydrogen atoms together, pimple to socket and socket to pimple, you've used up all the pimples and sockets and you can't join anything else on. With carbon, you can leave some spare pimples and sockets, and keep going.

42 *under just the right conditions:* See Jim Baggott, "Great Balls of Carbon," *New Scientist,* July 6, 1991, 34–38; Robert F. Curl and Richard E. Smalley, "Fullerenes," *Scientific American,* October 1991, 32–41.

43 *discovered lining cracks in rocks:* See Jeff Hecht, "Russian Rock Yields Natural Buckyballs," *New Scientist,* July 18, 1992, 18; Peter R. Buseck et al., *Science* 257, July 10, 1992, 215–217.

44 *"God throwing dice," . . . :* In a letter to Max Born, Einstein wrote: "You believe in a God who plays dice, and I in complete law and order in a world which objectively exists, and which I, in a wildly speculative way, am trying to capture. I firmly *believe,* but I hope that someone will discover a more realistic way, or rather a more tangible basis than it has been my lot to do. Even the great initial success of the quantum theory does not make me believe in the fundamental dice game, although I am well aware that your younger colleagues interpret this as a consequence of senility."

47 *COBE satellite has observed . . . :* These observations were at first considered controversial, because they lay at the extreme limits of sensitivity of the instruments employed. But independent observations now seem to confirm them. See Marcus Chown, *Afterglow of Creation.*

48 *Einstein's theory of relativity . . . :* General relativity, that is. There's nothing about bent space-time in special relativity. To avoid complicating the issue we'll generally omit the adjective "general."

49 *Planck's constant, whose value . . . :* This is the value in SI units of joule-seconds.

50 *It's fudges all the way down!:* In support we quote the great physicist Richard Feynman: "That is the same with all our other laws—they are not exact. There is always an edge of mystery, always a place where we have some fiddling around to do yet. This may or may not be a property of nature, but it certainly is common to all the laws as we know them today." See Richard P. Feynman, *The Character of Physical Law* (Cambridge, Mass. : MIT Press, 1965), p. 33; and James Gleick, *Genius,* p. 365.

51 *Zarathustra is inhabited by intelligent aliens:* "Not as harmless as you might think," to quote the *Good Galaxy Guide* (Zarathustra: Octopi Press).

3: The Organization of Development

57 *most organisms are also organism factories . . . :* Most bees aren't—only queen bees reproduce, and each swarm or hive of bees has only

one queen. But remove the queen, and the worker bees start turning fertile. They're inhibited chemically—as nuns are theologically—but have the potential to reproduce.

59 *mercifully abbreviated to DNA:* You will often see the acronym DNA interpreted as "deoxyribonucleic acid." But deoxyribose is a sugar, like glucose and fructose; the "-ose" isn't an adjectival ending. Ribose, another sugar, similarly lends its name to the chemical RNA.

61 *without oxygen, we wouldn't be here:* That's the usual story. However, there's an alternative explanation, in which most of the oxygen arose from geological disturbances. See D. J. Des Marais et al., "Carbon Isotope Evidence for the Stepwise Oxidation of the Proterozoic Environment," *Nature* 359, October 5, 1992, 605–609.

67 *Similarly for Mom's eggs:* The pairs in the egg don't split up until after the sperm has entered the egg. Discarded half-pairs go into so-called polar bodies.

69 *the secret of life:* See James Watson, *The Double Helix.*

73 *for right-hand threads, of course:* This story is told at greater length by Jack Cohen in *Reproduction*, chapter 8, and by Ian Stewart and Martin Golubitsky in *Fearful Symmetry: Is God a Geometer?*, chapter 7. Mind you, Steve Jones has just told us that although it's in all the textbooks, it's not entirely true! (Murphy's law strikes.) But something very similar must be.

75 *Lewis Wolpert has developed a theory . . . :* See Lewis Wolpert, *The Triumph of the Embryo.*

79 *anomalies and developmental oddities . . . :* Organization of embryos can usually be explained as the interaction of a "map" and a "book." The map tells the cell where it is in the organism; the book tells it what to do when it finds itself in that position. Thus the cell uses clues, perhaps chemical, from its surroundings to place itself on the map—metaphorically speaking—and then it looks in its DNA book to find out what genes should be turned on if it is in that position. For instance, any piece of skin from a newt tadpole makes balancers (a pair of spikes) when it is transplanted to the head of a frog tadpole; and any piece of skin from a frog tadpole makes horny teeth when transplanted to the head of a newt tadpole. Effectively, it is as if the frog DNA book says "*Position:* around mouth—*action:* make teeth," but the newt DNA book says "*Position:* around mouth—*action:* make balancers," while

the map is the same in both cases. Edward Koller showed that chick gum epithelium still gave "Make tooth" signals; chick gum tissue underneath it couldn't respond; but mouse mesenchyme from anywhere could. It made hen's teeth. See Stephen Jay Gould, *Hen's Teeth and Horses' Toes.*

85 *DNA code still controls the whole process:* While writing this chapter we asked ten geneticists whether Mendel's pea color is a maternal-effect gene. None of them knew, and none of them could understand why it was an interesting question, because the numbers come out right in any case.

86 *"slow" ones for when it's hot:* See Diala C. Amanze, "Regulation of the Early Embryonic Development in the Zebra Fish *(Brachydanio rerio),"* unpublished Ph.D. thesis Dis-S2-B86, University of Birmingham (England), 1986.

93 *and falling off the bottom edge:* Mathematicians may prefer to think of one ball bearing on a multidimensional sheet.

95 *which enzymes to use at what temperatures:* This last statement is contentious: The frog's egg regulates its development in the pond for a day or so before the phylotypic stage—that is, without using its own DNA program. Some maternal messenger RNA is labeled "Use if the temperature falls below 50°F," we suppose.

4: The Possibilities of Evolution

101 *there's no problem in explaining the origins of lots of DNA:* But see Richard Lewontin, "The Dream of the Human Genome," *New York Review of Books,* May 28, 1992, 31–40, and our chapter 9.

102 *But it takes extraordinarily long to do so: Hamlet* contains roughly 150,000 characters. A standard keyboard contains about 80. The odds *against* typing them in the correct order are roughly $80^{150,000}$ to 1, or about $10^{285,000}$ to 1 against. Not impossible, but *very* unlikely. The hemoglobin molecule is a protein containing 539 amino acids; the odds against "typing" that are rather better, 10^{619} to 1. See Isaac Asimov, *Hemoglobin and the Universe.* The age of the universe is of the order of 10^{10} years.

102 *we can speculate on what might have happened:* An alternative theory involves the use of copper atoms as a scaffolding; see John Emsley, "DNA's Ancestors Made Themselves Up," *New Scientist,* August 25, 1990, 27.

107 *his message wasn't widely appreciated:* Karl Marx got the message, though.

109 *Sewall Wright introduced the idea of "peaks of adaptedness" . . . :* See William B. Provine, *Sewall Wright and Evolutionary Biology* (Chicago: University of Chicago Press, 1986).

112 *"Just* which *ape do you claim to be descended from, Mr. Darwin?":* This is a paraphrase, not a verbatim quote. A similar remark was used in a Victorian debate on evolution, but there is no written record of the debate, only written recollections of witnesses.

119 *any particular base is as likely to mutate as any other:* So it is generally assumed—probably wrongly. There is experimental evidence that some parts of the genome seem to be more stable than others, either because they mutate less frequently or because any mutations in these areas are lethal, so are rarely discovered.

120 *Brian Goodwin had published a book about it . . . :* See Brian Goodwin, *Temporal Organization of Cells.*

122 *artificial organisms loose in the computer's memory:* This particular system was invented by Robin Jones. More sophisticated examples are described in Mitchell Waldrop, *Complexity*, and Roger Lewin, *Complexity.*

127 *many beautiful essays . . . :* See *Ever Since Darwin, The Panda's Thumb, An Urchin in the Storm, Hen's Teeth and Horses' Toes,* and *Bully for Brontosaurus.*

132 *if the K/T meteorite . . . :* The "K" in "K/T" is the first letter of the German for "Cretaceous," and T stands for "Tertiary," which has the same first letter in both German and English.

132 *killing off the dinosaurs . . . :* This theory is a bit controversial, but nowhere near as controversial as it was a short time ago. See the debate between Walter Alvarez and Frank Asaro, "An Extraterrestrial Impact," *Scientific American*, October 1990, 44–52; and Vincent E. Courtillot, "A Volcanic Eruption," *Scientific American*, October 1990, 53–60. Then take a look at Virgil R. Sharpton et al., "New Links Between the Chicxulub Impact Structure and the Cretaceous/Tertiary Boundary," *Nature* 359, October 29, 1992, 819–21.

132 *reconstructed upside down . . . :* See "Weird Wonders," *Scientific American*, June 1992, 12–14.

5: The Origins of Human Understanding

138 *is not unoccupied; it is nonexistent:* A New Zealand ground parrot, the kakapo, has become so rare that the last surviving dozen or so have been removed to a tiny offshore island. (Worries that they might all be male have recently proven unfounded.) On the kakapo lives a tiny parasite, specific to that bird. If the kakapo becomes extinct, then a niche will disappear, and so will the parasite.

139 *Darwin's finches are still evolving:* See Peter R. Grant, "Natural Selection and Darwin's Finches," *Scientific American*, October 1991, 60–65.

140 *bottom-grubbing cichlids died out:* The real world lacks textbook neatness. The tale is confused by the introduction of a second outsider, the Nile perch, into the Rift Valley lakes. It's a generalized predator, and it's well on the way toward wiping out *all* of the cichlids. And the catfish are starting to look worried.

145 *Neurons communicate by electric impulses . . . :* But a hot topic is nonspiking neurons, which send nondigital signals. Nerve cells don't "compute" digitally, but they do use discrete pulses to communicate over long distances. See William H. Calvin, *The Throwing Madonna.*

145 *roughly like messages being transmitted over a telephone line:* A more accurate image is the burning of a trail of gunpowder, in which any activity at a given position triggers activity farther along.

146 *the simple "threshold" internal mechanism is without doubt far too simple . . . :* The image may be *way* off. A neuron could resemble a microprocessor—an entire computer—rather than just a single transistor. If so, the brain would be far more massively parallel than most people think, and much further beyond the reach of present technology.

149 *Donald Hebb said that nerve cells that fire together grow together:* JC read Donald Hebb's book *The Organization of Behavior* (New York: Wiley, 1949) in 1951, and is still amazed by the staying power of his insights.

151 *a lion partially hidden behind a bush:* For years the Stewart household has always had a cat. One Christmas, in about 1973, an aunt gave us a concrete cat for the garden—very lifelike, in a stalking position, with shining plastic eyes. When the parcel was opened, the real cat, Seamus Android by name, became very alarmed. With fur standing on end, it edged cautiously toward the interloper, eyes locked in battle . . .

until their noses touched. It then stepped back, gave everyone a look as if to say, "Of course, I never imagined for a moment it was a *real* cat," and thereafter ignored the concrete ornament totally.

152 *to stop the electrical signal from escaping:* As always, nature is more complex. At intervals along the nerve axon are gaps in the insulation, called nodes of Ranvier. Why have gaps if the insulation speeds up the signal? The answer is that, just as in a transatlantic cable, the signal dies away as it propagates. The nodes of Ranvier are effectively repeaters, local amplifiers that pick up a faint signal and make it stronger.

152 *quick-and-dirty solutions can often be the most effective:* Quick-and-dirty processing makes brain puns more likely—indeed, it relies on them. See earlier note (two back) about the concrete cat, and Daniel Dennett, *Consciousness Explained.*

154 *which is what your eyes are filled with:* Bees don't fill their eyes with water, and can see into the ultraviolet. Plants exploit this by using ultraviolet billboards. They're called petals.

156 *information about the presence or absence of light:* See David H. Hubel and Torsten N. Wiesel, "Brain Mechanisms of Vision," *Scientific American,* September 1979, 130–45, and Charles R. Michael, "Retinal Processing of Visual Images," *Scientific American,* May 1969, 104–114.

156 *tell the brain that they've done so:* They also prevent the nerve cell from becoming "habituated" and unable to respond.

158 *those computations are inordinately complex:* See Daniel Dennett, *Consciousness Explained.*

160 *on the eighth step—hence the term "octave."* Septimists will argue, with numerical justification but lack of euphony, that it should be called a septave.

161 *at right angles to each other:* This may be a physiological reason why we like to think of three-dimensional space as having a rectangular coordinate system—forward/back, left/right, and up/down. We have coordinates in our ears. Mathematicians, who in recent years have emphasized the "coordinate-free" approach to space because it reflects intrinsic geometric structure rather than artificially imposed frames, are going against their own ears.

162 *Dr. Delius dreams in undiscovered colors:* One of our guinea pigs for an early draft came up with this phrase. We have no idea where it comes from, but we like it.

163 *feel the electrical activity of other creatures' nerves:* This is as close to genuine telepathy as nature seems to get.

167 *Searle's "Chinese room":* See John R. Searle, "Minds, Brains, and Programs," *Behavioral and Brain Sciences* 3 (1980), 417–58. It is reprinted in *The Mind's I,* edited by Douglas Hofstadter and Daniel Dennett.

168 *to* be *an intelligent entity:* Arguments between Searle and the Artificial Intelligentsia are so cross-paradigm that neither side will ever understand what the other is trying to say. Like Searle, we think the Artificial Intelligentsia are a bit naïve; but we much prefer their positive efforts to Searle's negative dismissals. Let's face it, a machine that *acted* as if it were intelligent, even if it really wasn't, would be a pretty neat gadget. Searle's style is too reminiscent of opposition to heavier-than-air flying machines. ("Can't you *see,* dammit—they're *heavier than air*!") But we don't think you'll see any Wright Brothers of AI for a while yet. For similar (and better) arguments see Daniel Dennett, *Consciousness Explained.*

172 *left-handed or ambidextrous . . . :* Or ambisinistrous, to avoid accusations of "handism."

172 *less symmetric than left-handers'; nobody knows why:* However, there is at least one plausible theory why right-handedness is more common, and once you can explain that, brain differences between right-handers and left-handers become more plausible too. It is described in William H. Calvin's *The Throwing Madonna.* In fact, it's the reason for the title. The idea is that the human heartbeat is strongest on the left side of the body, which is why mothers nearly always hold babies with their heads to that side. A baby in the womb gets used to its mother's beating heart, so this posture helps keep it pacified once it's born. Now, this leaves only the right arm free for throwing rocks, so that over a long enough period, evolution will favor the right hand for such functions and modify the brain accordingly.

Of course, you also need to explain why the heart is usually displaced toward the left side.

174 *attach a word as a label:* See Antonio R. Damasio and Hanna Damasio, "Brain and Language," *Scientific American,* September 1992, 63–71.

175 *I know that I am conscious.* "I" here refers to either of the two authors—we both agree on this bit.

175 *Well, sorry, but probably not:* See Marvin Minsky, *The Society of Mind*, and Daniel Dennett, *Consciousness Explained.*

176 *his conscious mind "makes the decision":* See Timothy Ferris, *The Mind's Sky*, and Daniel Dennett, *Consciousness Explained.*

176 *"I think, therefore I exist" . . . :* This is usually translated as "I think, therefore I *am*." But in the Latin original it is "*Cogito ergo sum*," and "*sum*" comes from the verb "*esse*," which, although it is usually translated as "to be," has the sense of "to exist." The use of "am" is traditional but archaic, whereas "exist" captures the sense more accurately.

177 *"I think I exist":* Compare "It is raining, therefore *it* exists." In Korean you say "Rain comes" and "Thought happens." The qualification "with reference to me" has to be added separately.

6: Systems of Interactive Behavior

181 *when the structure of the brain is represented as a DNA data string . . . :* Another possibility is that it's not *represented* as a DNA sequence at all.

182 *half the square of the number of objects:* With n objects there are $\frac{1}{2}n(n-1)$ distinct pairs, because each of the n objects can be paired with the remaining $n-1$, but then we count every pair twice, once for each member. And that's just *pairwise* interactions. What about triples, etc.?

182 *effective explanations in ecology, epidemiology, or economics:* There is an interesting study of how each of economics and ecology tends to support its theories by appealing to analogies with the other, instead of comparing them with experiment. See David J. Rapport, "Myths in the Foundations of Economics and Ecology," *Biological Journal of the Linnaean Society* 44 (1991), 185–202.

182 *The new area of complexity theory . . . :* See Mitchell Waldrop, *Complexity*, and Roger Lewin, *Complexity.*

182 *whose theories flatly contradict each other:* Monetarism, eagerly espoused by the British government as a sure road to economic salvation, has delivered two slumps within one decade, the second being the worst collapse since the 1930s. The same theories of supply-side economics have all but wrecked the economy of the United States, and with it the global one. But Nobels were awarded for supply-side eco-

nomics, and not so long ago, either. Maybe to get a Nobel Prize in economics you don't have to be right, you just have to look clever.

184 *They think that the Theory of Everything will turn out to be mathematical . . . :* See Steven Weinberg, *Dreams of a Final Theory*, and John Barrow, *Theories of Everything*.

184 *the theoretical prediction for a quantity known as Dirac's number . . . :* See Richard P. Feynman, *QED*.

185 *the dorsal/ventral distinction, and segmentation . . . :* See Brian K. Hall, *Evolutionary Developmental Biology*.

186 *10 percent of each generation:* J.B.S. Haldane, "The Cost of Natural Selection," *Journal of Genetics* 55 (1957), 511–24.

186 *based on nonsensical assumptions:* See Rodney Needham, *Against the Tranquility of Axioms*.

195 *legal/illegal, alive/dead, or male/female:* The legal profession thrives on the fuzziness of the first boundary; the second poses severe ethical problems for doctors with patients on life-support machines; and the third is complicated by chromosome abnormalities: There are at least seventeen different human sexes. (Unless you believe that sex is *only* chromosomal, or *only* behavioral, or *only* hormonal.)

196 *If it doesn't escape like that, color it black:* The whole process is explained in an extremely accessible way in Arthur C. Clarke, *The Colours of Infinity*.

197 *data needed to define the rule that generates it:* Gregory Chaitin has systematized data compression of this kind under the name "algorithmic information theory." See his book *Information, Randomness, and Incompleteness*.

198 *you have to contemplate* all possible rules *that might generate it:* See Michael Barnsley, *Fractals Everywhere*.

198 *something that we haven't told you to imagine:* If you've done that, try *not* to think of a hippo with an ear trumpet for the next five minutes. Concentrate very hard on not thinking about it.

207 The Legend of the Haggis: The modern haggis is a sheep's stomach stuffed with a mixture of offal, suet, oatmeal, and herbs, and boiled. It tastes great. But you may prefer the legend to the reality. If so, you may care to contemplate the possibility that the modern haggis is an artificial one, and the real haggis was driven to extinction by overhunting. Compare mock turtle soup.

213 *building a broad diagonal "highway":* See David Gale, "Mathematical Entertainments," *Mathematical Intelligencer 15* (1993), 54–55.

216 *using pulses of gliders . . . to carry and manipulate information:* See E. R. Berlekamp, J. H. Conway, and R. K. Guy, *Winning Ways for Your Mathematical Plays.*

7: Complexity and Simplicity

225 *prefer a false but beautiful theory to a correct but ugly one:* For this and similar pronouncements, see Helge Kragh, *Dirac: A Scientific Biography.*

226 *some genes* are *scattered all over the genome:* Most eukaryote genes are broken up into smaller segments (exons) by intrusive segments (introns) with no apparent relation to the gene concerned. Exons of some genes, like those for mammalian antibody molecules, are even on different chromosomes.

227 *Even an elliptical one:* Octagons have corners, but if you put enough octagons together, getting smaller and smaller, the corners cease to be visible, just as a circle on a computer screen looks circular, even though it is made from dots arranged on a square grid.

231 *the one that reveals a new simplicity:* Except in economics, apparently. But even economists *proclaim* new simplicities.

233 *"Does the Wind Possess a Velocity?":* See Lewis Fry Richardson, *Weather Prediction by Numerical Process* (Cambridge, England: Cambridge University Press, 1922).

234 *won't let them publish their paper:* There are honorable exceptions. An unproved conjecture known as the Riemann hypothesis is so important and useful that mathematicians often write papers that assume it to be true. In such cases the rule is that if you assume something, you have to say so. There is a rationale, here: If you do lots of work assuming the truth of the Riemann hypothesis, then you might blunder into something that contradicts what's already known, and if you do, you've disproved the Riemann hypothesis. But this is just an excuse. The real reason is that the number theorists are so keen to see what lies beyond the Riemann hypothesis that they're too impatient to wait for it to be proved. To be fair, what they see beyond it could be a new idea that leads to a proof. Sometimes the best way around an obstacle is to find out what's on the other side.

238 *how many times the explanations involve quarks":* At first sight our argument doesn't work very well for technological gadgets such as PET (positron emission tomography) scanners. Positrons are subatomic particles, and a PET scanner is not immune to subatomic details. But that's because it was designed to amplify certain features of the subatomic level. Technology does this kind of thing all the time, but it does it only in a simplified, selected setting. We're talking about explanations, not engineering, and explanations of natural phenomena hardly ever require details from the very deep levels.

241 *"down the road to the chemist":* Douglas Adams, *The Hitchhiker's Guide to the Galaxy*, start of chapter 8: " 'Space,' *it says,* 'is big. Really big. You just won't believe how vastly hugely mindbogglingly big it is. I mean you may think it's a long way down the road to the chemist, but that's just peanuts to space. Listen . . .' *and so on.*"

242 *there's just one of it:* We mean that there aren't millions of totally different molecules that do the same as DNA. Every creature's DNA has a different sequence of bases, so in a sense there are millions of different molecules. But not *that* different.

244 *it wouldn't tell us why:* An embryologist would answer the "why" question in the "how" form. JC is an embryologist.

244 *sensors at the side:* For purposes of illustration we rule out more exotic possibilities, such as a ring of eight eyes all the way around. Notice what's happened here: Even if our theory of Zarathustran insect eyes turns out to be wrong, it has caused us to ask a new and important question. Reductionists can now publish thousands of papers counting insect eyes, measuring the angles in which they point, and making esoteric statistical analyses.

8: The Nature of Laws

249 *a Swedish company managed to follow this route:* Michael W. Geis and John C. Angus, "Diamond Film Semiconductors," *Scientific American* 267 (October 1992), p. 64.

250 *Low-pressure synthesis of diamond . . . :* A physicist friend of JC's in Vienna was trying to find the best source of carbon for depositing diamond onto surfaces. Pure methane and carbon dioxide produced very

lumpy surfaces. Impure methyl alcohol was better. Then he found something that gave a beautifully even coating of diamond: Japanese whisky.

251 *a lukewarm gas with no interesting structure whatsoever:* This image, popular though it may be, is misleading. In 1979 Freeman Dyson published a paper (*Reviews of Modern Physics* 51, 447–60), showing that even in a universe heading relentlessly toward "heat death" interesting structures such as intelligent life could continue to survive. See page 114 of his book *Infinite in All Directions.*

256 *495 trillion . . . :* To be precise, 495,918,532,948,103.

259 *when previously combined systems are separated in this manner:* See P.C.W. Davies, *The Physics of Time Asymmetry.*

260 *if you time-reversed some (apparently isolated) subsystem of the universe:* We got this argument from David Ruelle, *Chance and Chaos.* See also Ilya Prigogine and Isabelle Stengers, *Order Out of Chaos,* and Peter Coveney and Roger Highfield, *The Arrow of Time.*

260 *the gravity of far galaxies, say:* Arrange twelve pool balls in a line, and hit the first one so that each collides with the next. In theory, the motion of a planet in a distant galaxy will change what happens by an appreciable amount. This is assuming classical mechanics; in quantum mechanics the same is true for just seven balls in a row.

261 *perhaps we just observe things directionally:* This is Stephen Hawking's conclusion in *A Brief History of Time.*

264 *how crucial that assumption is:* See Ilya Prigogine and Isabelle Stengers, *Order Out of Chaos.*

265 *If you ever get the opportunity,* do *try it yourself:* Art Winfree's simplification of Zhabotinskii's recipe works reliably even in front of students, and uses only four cheap components having a shelf life of years. It produces bromine, but not in dangerous amounts in an airy room. CAUTION: The mixture is moderately poisonous.

Make up four components:

1. 25g sodium bromate, 335ml water to dissolve, then 10ml concentrated sulfuric acid.
2. 10g sodium bromide, water to 100ml.
3. 10g malonic acid, water to 100ml.
4. 1, 10 phenanthroline ferrous complex (Fisons, Loughborough, UK).

Put 6ml of solution 1 in a glass beaker, then add 0.5ml of 2, then quickly mix in 1ml of 3. Leave the brown mixture to lose bromine (by an open window) until it is a pale straw color or colorless (2-3 minutes if agitated or in a flat dish). Add 1ml of the redox indicator 4, mix thoroughly, and pour into a 9cm glass or plastic petri dish on a white (preferably illuminated) background.

It will turn patchy blue, then clear to a brown-red. Foci of blue will appear (you may have to wait up to five minutes) and grow into a series of concentric rings, expanding slowly. If the dish is shaken to restore homogeneity, the patterns reappear. Otherwise do not jar or vibrate the dish. The effect lasts for twenty to twenty-five minutes. The experiment works very well on an overhead projector, for visibility in a classroom, but the rings may be fuzzy if the cooling fan is unbalanced.

267 *the sad fate of Schrödinger's cat . . . :* See John Gribbin, *In Search of Schrödinger's Cat.*

267 *as T. S. Eliot puts it . . . :* See T. S. Eliot, *Old Possum's Book of Practical Cats* (London: Faber and Faber, 1962).

268 *whether the atom has decayed . . . :* See John Horgan, "Quantum Philosophy," *Scientific American*, July 1992, 70–79.

271 *an effect that physicists call decoherence:* See Thanu Padamanabhan, "Bridge over the Quantum Universe," *New Scientist*, October 10, 1992, 27–29. Another peculiarity of the "collapse of the wave function" is that it need not *stay* collapsed. Raymond Chiao has found cases when it revives. Moreover, there is evidence that quantum devices known as SQUIDs can be caught in a superposed state, so their wave function doesn't collapse at all. See John Horgan, "Quantum Philosophy," *Scientific American*, July 1992, 70–79.

272 *he will never confess:* Eliot's cat was male. There is no record of the sex of Schrödinger's.

273 *quantum mechanics is reductionist, but relativity is contextual:* A more technical reason, but an important one, is that quantum mechanics is linear, whereas relativity is nonlinear.

274 *words in place of objects:* See John Casti, *Paradigms Lost.*

276 *both kinds of behavior at once:* It may be possible either to beat the Heisenberg uncertainty principle or to disprove Bohr's complementarity principle. See Jim Baggott, "Beating the Uncertainty Principle," *New Scientist*, February 15, 1992, 36–40.

279 *they follow the quickest path . . . :* Actually this should be the *extremal* path; slowest paths can also occur, for example. But we didn't want to digress on the meaning of "extremal," and we still don't. See Tim Poston and Ian Stewart, *Catastrophe Theory and Its Applications.*

279 *a totally different "physical" picture:* The same general point that equivalent mathematics gives different physical interpretations was made by Richard Feynman in a 1946 letter to his friend Theodore Welton: "There are so very few equations that I have found that many physical pictures give the same equations." See James Gleick, *Genius,* p. 326.

283 *to seed the formation of galaxies:* See Nigel Henbest, "Big Bang Echoes Through the Universe," *New Scientist,* May 2, 1992, 4–6; and "The Golden Age of Cosmology," *Scientific American,* July 1992, 9–12.

9: THE DEVELOPMENT OF ORGANIZATION

286 *Michael Crichton's novel and Steven Spielberg's blockbuster movie* Jurassic Park *. . . :* See Further Reading.

287 *sequencing DNA from an insect known as a nemonychid weevil.* See R. J. Cano, H. N. Poinar, N. J. Pieniazek, A. Acra, and G. O. Poinar, Jr., "Amplification and Sequencing of DNA from a 120–135 Million Year Old Weevil," *Nature* 363 (June 10, 1993), 536–38.

288 *sequencing the human genome:* Here's a typical example. The Institute of Biology is a major British scientific society. The program of its 1993 symposium on Recent Advances in Human Genetics includes: "The 'handbook of life' that eventually will be produced will include the identity and location of each of the 100,000 or so human genes, together with their nucleotide [base] sequence. Eventually it will contain the complete sequence of every chromosome. Such information will be the prime reference material for *all* biological and medical science and will increase enormously our potential for understanding the human body and for diagnosing and treating human disease." (Our italics.) Biologists will tell you that they don't say such naïve things. They do.

288 *Tom Easton's "gengineer" stories . . . :* See *The Magazine of Fantasy and Science Fiction,* March 1989.

293 *biological, chemical, and cultural processes . . . :* The human ability to grasp develops after birth and is influenced by an environment pro-

vided by other human beings. Kittens can recognize features of their surroundings, such as the edge of a deep hole, only if they are exposed to those features at a suitably early age.

293 *Development seems to involve dynamics . . . :* See Stephen Jay Gould, *Ontogeny and Phylogeny*, and Brian K. Hall, *Evolutionary Developmental Biology*.

293 *to stabilize a particular dynamic form:* See Brian Goodwin, "Development as a Robust Natural Process," preprint, Open University (Milton Keynes, England), 1992; Mae-wan Ho, "Where Does Biological Form Come From?" *Revista di Biologia* 77 (1984), 147–79.

294 *Richard Lewontin has pointed out that . . . :* See the brilliant multiple review-cum-essay by Richard Lewontin, "The Dream of the Human Genome," *New York Review of Books*, May 28, 1992, 31–40. Ignore the bit from the middle of page 38 onward—that's less brilliant.

301 *precisely 3,333,333,333 bases long:* This is not correct, but it's not far off (the true number is about 3,154,000,000) and it helps the dramatic line. See J. Weissenbach et al., "A Second-Generation Linkage Map of the Human Genome," *Nature* 359, no. 6398 (October 29, 1992), 794–800. Also we've assumed "the" human genome is unique, which of course it isn't. Though you wouldn't realize this from much of the current discussion about sequencing "it."

302 *at the 43,772,331st group!* Repeat: For the purposes of this dramatic sketch, we assume there's a unique genome.

304 *treadbare tires . . . :* This word doesn't exist, but it should.

305 *DNA sequences live in DNA space:* See Manfred Eigen, "Viral Quasispecies," *Scientific American*, July 1993, 32–39. He uses the term "sequence space" and defines it in detail. His main point is that viruses mutate so frequently that there is no meaningful concept of a virus "species," but that regions of sequence space can be used to provide a substitute.

306 *Geneticists who see the interaction this way:* More specifically, they suppose that x percent of a character can be traced to genes and $100-x$ percent to environment. Then they perform a series of mathematical incantations that they claim measures x. The technique is first to compare genetically identical creatures in different environments (twins that have been separated soon after birth, say); then to compare environ-

mentally identical creatures with different genes (siblings brought up in the same home). Some statistical jiggery-pokery then produces the value of x.

This approach is open to all kinds of criticisms. Neither experiment considers the "real" case, when *both* genetic and environmental differences are acting. The statistical method assumes that nature and nurture are independent influences and that their effect is simply added up—an assumption that may have made sense back in the dark ages of linear mathematics, but which strikes nasty discords in an era that has discovered the importance of nonlinearity. Finally, the statistical method will always give *some* value for x, whatever the system is doing. You could assume that the flying ability of the bumblebee is x percent due to sunspot activity and $100-x$ percent to the major league baseball scores, and the method would come up with a value for x. For an excellent technical critique see Douglas Wahlsten, "Insensitivity of the Analysis of Variance to Heredity-Environment Interaction," *Behavioral and Brain Sciences* 13 (1990), 109–161.

307 *They'd still be elephants:* See "If Not a Dinosaur, a Mammoth?" *Science* 253 (September 20, 1991), 1356.

10: The Evolution of Possibilities

312 *no other valid viewpoint exists:* This is the central "simplex, complex, multiplex" theme of the unusual science fiction novel *Empire Star*, by Samuel R. Delany.

322 *Stephen Jay Gould calls this the Panda Principle . . . :* See page 61 of his book *Bully for Brontosaurus* (Further Reading). For the full story of the panda's thumb, see his book *The Panda's Thumb.* Actually, the panda's thumb isn't really an instance of the Panda Principle. It's not that a bad thumb got there first and kept better ones out, it's that the possibility of a better one got used up.

324 *Ray calls his fake world Tierra . . . :* See Ed Regis, *Great Mambo Chicken and the Transhuman Condition;* Roger Lewin, *Complexity;* and Mitchell Waldrop, *Complexity.*

328 *dinosaurian Cohens and Stewarts would argue:* See Jack Cohen and Ian Stewart, "Chaos, Contingency, and Convergence," *Nonlinear Science Today* 1, no. 2 (1991), 9–13.

335 *Phrases such as "evolutionary dead end" . . . :* Tim Poston sent us a piece of catastrophe geometry that makes this point very vividly. Suppose the dynamics on creature space and DNA space are related by a multidimensional surface, shown schematically in Figure 54. The DNA sequence D_1 and its corresponding creature C_1 are "trapped" in a dead end because C_1 is at a local optimum of the creature dynamic. On the other hand the DNA sequence D_2, corresponding to the

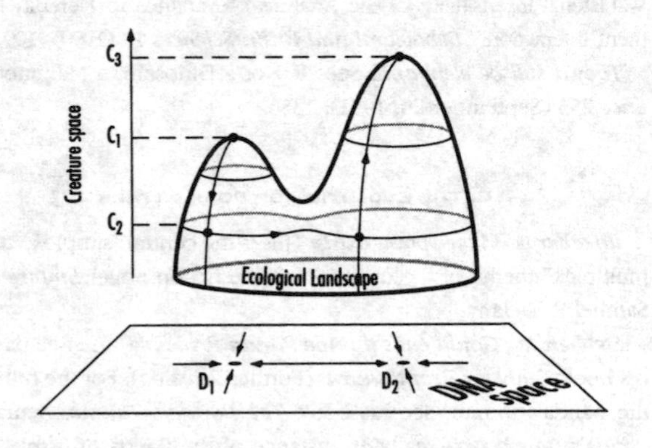

FIGURE 54

Catastrophe surface for an evolutionary dead end

less optimal creature C_2, can move, without change of phenotype, by cryptic mutations whose statistics look just like "no selection at all." When the DNA sequence reaches D_2 there is an easy path to the creature C_3, with no change in DNA sequence, and C_3 is "fitter" than C_1. The usual analysis by evolutionary theorists considers motion in either DNA space or in creature space, but fails to understand the new possibilities that arise when both spaces operate in combination.

336 *to make a bridge that led to it:* We repeat that there's an alternative explanation, in which most of the oxygen arose from geological disturbances. See D. J. Des Marais et al., "Carbon Isotope Evidence for the Stepwise Oxidation of the Proterozoic Environment," *Nature* 359 (October 5, 1992), 605–609.

11: The Understanding of Human Origins

343 *no serious chance of survival on their own before the age of ten or twelve . . . :* In Rio de Janeiro street kids survive without parental help at a much younger age. But they, too, are in a "privileged" situation—in our sense. The streets provide them with nongenetic means of survival—discarded food, materials like cardboard or wood, and weapons.

344 *which means brains:* Indeed, as computer scientists and the Artificial Intelligentsia have come to recognize, sophisticated vision requires enormous computing power—and more. A *big* brain is not enough; vision requires processing methods so clever that nobody as yet has any idea how the brain does the job. Our silicon-based machines have enough trouble recognizing an object of fixed shape and size when it can vary in orientation; indeed, they even have trouble working out where the edges of the object are. Our own visual system can identify an object at different distances, when partially concealed by others, and from unfamiliar viewpoints. It attaches labels to features of our surroundings, which persuade us that we see full color, three dimensions, and "real-time" motion. See Daniel Dennett, *Consciousness Explained.* Compared to this, the biggest supercomputer is no more than an abacus.

346 *theft has obvious advantages over honest toil . . . :* As Bertrand Russell once remarked about a rather erudite mathematical question: the role of axioms.

346 *beta-carotene, found in plants:* This is the source of a wartime rumor, spread by British military intelligence and still widely believed, that carrots are good for night vision. What was actually responsible for the high success rate of British night fighters, and what military intelligence didn't want the enemy to discover, was radar. Of course, if you have a deficiency of beta-carotene, the rumor is true.

351 *the biblical tale of Noah . . . :* The story of Noah was an adaptation of an already extant Babylonian flood tale, the Epic of Gilgamesh, which was polytheistic. The Israelites gave their version one God, so it is a story that evolved according to its context.

351 *Just So stories . . . :* And often Not So Stories.

356 *a very perceptive contextual thought:* The science-fiction writer Orson Scott Card had the same idea. "I was trying to show the borderline between human and animal, the exact comma in the punctuational model of evolution that marked the transition between nonhuman and human. For me, that borderline is the human universal of storytelling; that is what joins a community together across time; that is what preserves a human identity after death and defines it in life." See Orson Scott Card, *Maps in a Mirror*, vol. 1 (London: Arrow Books, 1992). He is writing about the meaning of his story "The Originist," in that collection.

356 *the tools needed to interact with our culture:* "Our very identity is a collection of the stories we have come to believe about ourselves. We are bombarded with the stories of others about us; even our memories of our own lives are filtered through the stories we have constructed to interpret those past events." Orson Scott Card, introduction to *Maps in a Mirror*, vol. 2 (London: Arrow Books, 1992).

356 *we learn what "timid" is from mouse stories . . . :* See Paul Shepard, *Thinking Animals.*

356 *or for a bigger lexicon:* See Steven Pinker and Paul Bloom, "Natural Language and Natural Selection," *Behavioral and Brain Sciences* 13 (1990), 707–784.

361 *to develop resistant strains:* This process has been going on for centuries; we're not just thinking of fancy "genetically engineered" strains. But overspecialized breeding has faults of its own: After a short period of success, the "resistant" strains often succumb to disease, because ge-

netic diversity has been eliminated. The benefits of science are great, but few things are *wholly* beneficial.

365 *A shrimp net yields shrimp, a tuna net tuna . . . :* And sometimes dolphins—but that's because the tuna net is really a dolphin net, too, even though it wasn't intended to be.

12: The Behavior of Interactive Systems

371 *if different choices are made:* Mae-wan Ho, "How Rational Can Rational Morphology Be?" *Revista di Biologia* 81 (1988), 11–55, argues against the basic assumption of phylogenetic systematics: that classification on the basis of form should be consistent with genealogy. She shows that you can set up a branching tree of forms by a process you *know,* as a thought experiment, but that the usual methods fail to reconstruct the tree correctly.

376 *balloonism probably can't evolve in our present ecosystem:* In technology, the balloon evolved first and winged flight came later. The two have evolved to exploit distinct niches, avoiding direct competition.

377 *as most of the selachians—skates, rays, and sharks—do today.* The cephalopods tried this and failed. *Nautilus,* an osmotic "balloon," implodes at a depth greater than six hundred feet because its shell is hard.

380 *They'd need to develop terraforming apes to do that.* See Ian Stewart, "The Ape That Ate the Universe," *Analog,* July 1993, 100–121.

381 *We might easily destroy about 99.99 percent of the eukaryote species on this planet . . . :* See Jared Diamond, *The Third Chimpanzee.*

383 *there are many kinds:* There's a lovely story told by the parasitologist Miriam Rothschild, whose daughter was given an essay to write on the topic "Let there be light." The story she wrote went something like this: "Animals, in order to see, need pigments that only plants can make. Animals see the plants using those pigments, and then eat the plants; they breathe out carbon dioxide, which the plants turn back into oxygen. At this stage the animals are just exploiting the plants. But now that animals can see, plants can exploit animals by making flowers to tempt them with nectar. As soon as the first bee saw the first flower, *then* there was Light!" JC recalls hearing this story in a lecture. A similar story exists in print. See Miriam Rothschild, "Remarks on Carotenoids in the

Evolution of Signals," in *Coevolution in Animals and Plants* (eds. Lawrence E. Gilbert and Peter H. Raven), (Austin: University of Texas Press, 1975), 20–37.

389 *a theory recently announced by Lee Smolin . . . :* See John Gribbin, "Evolution of the Universe by Natural Selection?" *New Scientist*, February 1, 1992, 22; and Lee Smolin, "Did the Universe Evolve?" *Classical and Quantum Gravity* 9 (1992), 173–91.

389 *in which the speed of light is six hundred miles per hour.* See Terry Pratchett's Discworld series, of which the most recent (as we write!) is *Witches Abroad* (New York: ROC Books, 1993).

389 *It seems generally accepted . . . :* See Martin Rees, "Black Holes at Galactic Centers," *Scientific American*, November 1990, 26–33; Simon Mitton, " 'Searchlights' Reveal Black Hole in Action," *New Scientist*, June 13, 1992, 15; John Gribbin, "Black Holes Reveal Themselves," *New Scientist*, October 3, 1992, 32–35. The evidence is still controversial and some astronomers discount it. At the other extreme, Edward van den Heuvel has calculated that there ought to be at least a hundred million black holes in our galaxy alone; see Govert Schilling, "Black Holes by the Million Litter the Galaxy," *New Scientist*, July 11, 1992, 16.

392 *One of his research students—her name was Baugenphyme . . . :* Baugenphyme is a Zarathustran Rosalind Franklin. See chapter 1, and James Watson's *The Double Helix.*

391 *invented a process of edge doubling . . . :* Actually the process was one of edge octupling (effectively repeating the edge-doubling process three times over), but to make the ideas more congenial to nonoctimists and bring out the analogy with the Feigenbaum number, we have slightly rewritten Zarathustran history.

393 *the Principle of Octimality was thoroughly accepted by all but the most die-hard septimists . . . :* The classic work on the history of the ideological battle between Octimism and Septimism is *The Rise and Fall of the Septimist Heresy: A Brief History of Misbelief in an Octimal Universe*, in eight volumes, by Dawkingjay Steephun, published by Octopoid Press. Dawkingjay traces the rise of septimism to a cryptic tendency to conflate the interval between one to eight with the difference between those two numbers, namely seven.

394 *let us call it a furcle . . . :* This is *not* a standard term.

13: Complicity and Simplexity

400 Causality and Contingency: See Jack Cohen and Ian Stewart, "Chaos, Contingency, and Convergence," *Nonlinear Science Today* 1, no. 2 (1991), 9–13.

401 *perhaps even mantis shrimps . . . :* See Jack Cohen, "How to Design an Alien," *New Scientist*, December 21, 1991, 18–21.

402 *the Reductionist Nightmare, branching forever . . . :* In an interview on British television in 1981 Richard Feynman made a similar point when he said: "If it turns out there is a simple ultimate law which explains everything, so be it—that would be very nice to discover. If it turns out it's like an onion with millions of layers . . . that's the way it is." See James Gleick, *Genius*, p. 432.

406 *solar system for the next million years:* One thing they found was chaos, the butterfly effect—so those predictions must be taken with a pinch of salt. That's why we say "very nearly in practice." The laws do predict the future motion, but they may not predict it correctly. However, the chaos is surprisingly limited. The precise positions of the planets can change from one run of the calculation to the next, but the general character of the motion remains within strict limits. No planet wanders off into the interstellar void or falls into the sun. The time-honored question of the mathematical stability of the Newtonian solar system may yet be given a satisfactory answer—but with a modified concept of stability.

412 *Stuart Kaufmann's concept of antichaos . . . :* See Mitchell Waldrop, *Complexity*, and Roger Lewin, *Complexity*.

412 *conveniently moves in a near ellipse:* We choose to ignore quantum effects because there's no evidence that these affect the conclusions about Mars's orbit. They do however destroy the particular simplexity "rigid sphere" that we use in classical mechanics to compute that orbit: If we go quantum, the whole argument falls to bits. An appeal to decoherence, converting quantum mechanics into a good approximation to classical, might save it.

415 *Complicity enlarges it:* Margaret Boden makes a similar distinction in her book *The Creative Mind*. She distinguishes between P-creativity, which explores a fixed conceptual space, and H-creativity, which expands the conceptual space itself. ("P" stands for "psychological" and

refers to ideas that are new to their creator; "H" stands for "historical" and refers to ideas that are new with respect to the whole of human history.)

418 *the person in the toll booth:* What happens in the mind of the maintenance engineers is another story—but in their minds "bridge" has a very different meaning. They still deal with it as a feature, but that feature has different mental associations from those in the mind of the user. They "see" it differently. Features are context-dependent.

418 *in insects, pterodactyls, bats, birds:* Or more if you include almost-wings, such as thistledown, sycamore seeds, and the thread used by spiderlings to sail on the breeze.

419 *The two types of aircraft have similar wing designs . . . :* We do realize that there exist balloons, helicopters, aircraft that land on skis, but the basic point works; you just have to narrow the context enough.

421 *Feedback between spaces with different geographies . . . :* Actually, feedback between spaces with the *same* geography can also produce new types of behavior, as soon as the dynamics becomes more complicated than steady states.

421 *In place of compromise we find conflict . . . :* Halfway between N and X in the alphabet is S. So the compromise between conteNt and conteXt is conteSt.

423 *We call such interactions strange loops . . . :* We borrowed this term from Douglas Hofstadter, *Gödel, Escher, Bach.*

424 *adult eels that look different:* This story is a bit controversial, but there's no doubt that the European eels never make it. So even if our story is false, something very like it must be true.

424 *hit the earth about 130 years from now:* See Jeff Hecht, "Will We Catch a Falling Star?" *New Scientist*, September 7, 1991, 48–53.

425 *predicted to hit Jupiter on July 25, 1994:* See Jeff Hecht, "Comet on Collision Course with Jupiter," *New Scientist*, June 5, 1993, 14.

425 *into the bottom of the biology one:* To cap it all off, Jupiter also flings asteroids in toward the earth's orbit, by way of Mars, thanks to chaos. See Ian Stewart, *Does God Play Dice?*

425 *What needs explaining is not the self-referential nature of consciousness . . . :* See Nicholas Humphrey, *A History of the Mind*, and Daniel Dennett, *Consciousness Explained.*

429 ***a direct and major effect on your mind . . . :*** See the Huxley quotation in the note for ***except when hit with a blunt instrument,*** chapter 1.

436 ***we have no such aspirations . . . :*** Put it this way: We wouldn't actually turn a Nobel down.

FURTHER READING

We thought this section would be a lot more useful if we annotated each entry with a mini-review. To preserve modesty, each of us reviewed the other's books.

Vladimir I. Arnold. *Catastrophe Theory*. 2nd ed. New York: Springer-Verlag, 1986. A compact account emphasizing the mathematics, brilliantly written, rather scathing about applications to the social sciences, sometimes unfairly. Very much a Russian view.

Isaac Asimov. *Hemoglobin and the Universe*. London: Collier, 1962. The first of Isaac's collections of essays, published once a month in *The Magazine of Fantasy and Science Fiction*. All thoughtful and informative.

Ralph Baierlein. *Newton to Einstein: The Trail of Light*. Cambridge, England: Cambridge University Press, 1992. Good biographical-scientific account of relativity and what led up to it.

Michael Barnsley. *Fractals Everywhere*. Boston: Academic Press, 1988. Advanced mathematics text with a huge number of pictures; explains fractal image-compression.

John Barrow. *Theories of Everything*. New York: Fawcett, 1992. A very readable account which dismisses Theories of Everything for reasons similar to ours.

Gregory Bateson. *Steps to an Ecology of Mind*. Northvale, N.J.: Jason Aronson, 1988. Essays by a very intelligent scientist who emphasizes context.

Graham Bell. *The Masterpiece of Nature: The Evolution and Genetics of Sexuality*. Berkeley: University of California Press, 1982. A very original professional biological discussion of the function and evolutionary patterns of sex.

Elwyn R. Berlekamp, John H. Conway, and Richard K. Guy. *Winning Ways for Your Mathematical Plays*. 2 vols. New York: Academic Press, 1982. Hugely original compendium of mathematical games. Full of puns. *The* source for Life.

R. J. Berry. *Neo-Darwinism*. London: Edward Arnold, 1982. A short, excellent, straight account.

Margaret A. Boden. *The Creative Mind: Myths and Mechanisms*. New York: Basic Books, 1992. Delightful study of mind and creativity; emphasizes the "creative space" and distinguishes between exploring it and expanding it.

S. Brenner, J. D. Murray, and L. Wolpert. *Theories of Biological Pattern Formation*. Port Washington, N.Y.: Scholium International, 1982. A series of examples of how to see patterns in a reductionist universe.

Don Brothwell, ed. *Biosocial Man*. London: Institute of Biology, 1977. An old but useful collection—biology sees social problems as *content*.

A. G. Cairns-Smith. *Seven Clues to the Origin of Life*. Cambridge, England: Cambridge University Press, 1990. A more accessible "origin of life" than his *Genetic Takeover*. Our inspiration for "Sherlock Holmes stories."

William H. Calvin. *The Throwing Madonna: Essays on the Brain*. New York: Bantam, 1991. Stunning series of essays by a neurobiologist with a solid grasp of the science but an unconventional turn of mind. Why left-handedness is rare.

John L. Casti. *Paradigms Lost*. New York: William Morrow, 1989. Informed and informative survey of changes in the scientific worldview.

———. *Searching for Certainty: What Scientists Can Learn about the Future*. New York: William Morrow, 1991. A well-written overview of modern reductionist science, with many indications of incompleteness and contextuality.

Gregory J. Chaitin. *Information, Randomness, and Incompleteness*. Singapore: World Scientific, 1992. Penetrating technical papers on the meaning of "random" and the information-theoretic cost of computations. Insightful view of uncomputability and undecidability. Mostly for professionals, but some articles are accessible to the lay reader.

Marcus Chown. *Afterglow of Creation*. London: Arrow Books, 1993. The science behind the COBE satellite observations that hit the headlines as "ripples of the edge of time" and "seeing the face of God."

Arthur C. Clarke. *The Colours of Infinity*. London: Strange Attractions, 1992. Everything you want to know about the Mandelbrot set; gentle but thorough description of the mathematics.

Jack Cohen. "Maternal Constraints in Development," in *Maternal Effects in Development* (D. R. Newth and M. Balls, eds.). Cambridge, England: Cambridge University Press, 1979. For technical details on nongenetic heredity in an orthodox biological context.

———. *Reproduction*. London: Butterworths, 1977. A classic student text collecting "all" aspects of biological reproduction between two covers. Highly illustrated.

———. *The Privileged Ape: Cultural Capital in the Making of Man.* Carnforth, England: Parthenon, 1989. An account of human evolution, especially the evolution of multicultures, emphasizing the role of privilege. Sometimes turgid.

Jack Cohen and Brendan Massey. *Animal Reproduction: Parents Making Parents.* London: Edward Arnold, 1984. Highly simplified version of Cohen's *Reproduction.*

Peter Coveney and Roger Highfield. *The Arrow of Time.* New York: Fawcett, 1992. Well-written but inconclusive account of time reversal and chaos.

Michael Crawford and David Marsh. *The Driving Force.* New York: HarperCollins, 1989. Substantial and well-presented setting for human evolution, getting brains on the seashore. Seafood means success.

Michael Crichton. *Jurassic Park.* New York: Ballantine, 1993. Build-a-dinosaur, mini-review of contents superfluous, except to say that the basic premise is *silly* but well established as a public myth. Gripping yarn with one of the best fictional mathematicians ever.

Charles Darwin. *The Origin of Species.* Harmondsworth, England: Penguin, 1985. Still very worth reading.

Paul Davies. *Other Worlds.* Harmondsworth, England: Penguin, 1988. Stimulating account of the nature of reality in a quantum world.

P. C. W. Davies. *The Physics of Time Asymmetry.* Leighton Buzzard, England: Surrey University Press, 1974. Technical backup for our discussion of why the second law of thermodynamics is irrelevant to time reversibility.

Richard Dawkins. *The Blind Watchmaker.* New York: Norton, 1987. A beautifully written account of modern evolutionary theory—marred only by its basic premise that genes map to characters.

———. *The Extended Phenotype.* New York: Oxford University Press, 1989. Sparkling and extremely clever backtrack on and amplification of *The Selfish Gene.*

———. *The Selfish Gene.* 2nd ed. New York: Oxford University Press, 1990. Elegant account of neo-Darwinism; argues the view that DNA rules. Superb, well worth reading, but we don't believe it.

Samuel R. Delany. *Empire Star.* New York: Ace Books, 1966. Remarkable science fiction story, with many wise comments on simplex, complex, and multiplex worldviews.

Daniel C. Dennett. *Consciousness Explained.* Boston: Little, Brown, 1991. Utterly brilliant.

Jared Diamond. *The Third Chimpanzee: The Evolution and Future of the Human Animal.* New York: HarperCollins, 1992. Why are humans so different from chimps when they share 98 percent of the same DNA? Argues, compellingly and in fascinating detail, that most differences have precursors elsewhere in the animal kingdom. Tends to forget that in DNA it's *quality,* not quantity, that counts.

Freeman Dyson. *Disturbing the Universe.* New York: Basic Books, 1981. Deep thoughts about the universe by one of the wisest leading physicists.

———. *Infinite in All Directions.* New York: Basic Books, 1988. More of the same. Explains why the "heat death of the universe" is a misleading image.

Timothy Ferris. *The Mind's Sky.* New York: Bantam, 1992. Thought-provoking essays on mind, matter, and the universe. Sees reality as a figment of the imagination, less aware of mind as a figment of reality.

Richard P. Feynman. *QED: The Strange Theory of Light and Matter.* Princeton, N.J.: Princeton University Press, 1985. Crystal-clear explanation of what quantum mechanics has to say about the universe; discussion of the Dirac number. Dates from the period when he didn't worry much about what quantum theory *meant.*

Michael J. Field and Martin Golubitsky. *Symmetry in Chaos.* New York: Oxford University Press, 1992. Amazing picture book of what chaos looks like in the presence of symmetry, with possible applications to pattern formation. State-of-the-art research mathematics that shows how chaos and stable form might coexist.

R. Foley. *Another Unique Species: Patterns in Human Evolutionary Ecology.* New York: Wiley, 1987. A professional, but contextual and ecological, view of human evolution—savannah version.

Alan Garfinkel. *Forms of Explanation.* New Haven, Conn.: Yale University Press, 1981. What do we mean by "explanation"? Carefully and sensibly argued. We've reinvented some of his wheels.

Ronald N. Giere. *Explaining Science.* Chicago: University of Chicago Press, 1988. Philosophy of scientific theories, very close to our features/instances approach.

James Gleick. *Chaos: Making a New Science.* New York: Viking, 1987. Brilliant on personalities, but not much solid science.

———. *Genius: The Life and Science of Richard Feynman.* New York: Pantheon, 1993. Excellent biography that misses out a little on the playful side of one of the great physicists of this century, but illuminates his thinking on the deep issues of physics.

Brian Goodwin. *Temporal Organization of Cells: A Dynamic Theory of Cellular Processes.* London: Academic Press, 1963. Marvelous for its time, but dated now because nonlinear dynamics has moved on. However, most biologists have still not received the message.

Andrew Goudie. *The Nature of the Environment.* 3rd ed. London: Blackwell, 1993. Well-explained ecology book, serious science but accessible to nonspecialists.

Stephen Jay Gould. *An Urchin in the Storm.* New York: Norton, 1988. Witty, pithy, memorable essays. You can learn better biology from them than from most textbooks. Ditto for his other books listed here.

———. *Bully for Brontosaurus.* New York: Norton, 1992.

———. *Ever Since Darwin.* New York: Norton, 1992.

———. *Hen's Teeth and Horses' Toes.* New York: Norton, 1993.

———. *The Mismeasure of Man.* New York: Norton, 1993.

———. *Ontogeny and Phylogeny.* Cambridge, Mass.: Harvard University Press, 1977. His most professional book, one of the great biology books of the century.

———. *The Panda's Thumb.* New York: Norton, 1992.

———. *Time's Arrow, Time's Cycle.* Cambridge, Mass.: Harvard University Press, 1987.

———. *Wonderful Life: The Burgess Shale and the Nature of History.* New York: Norton, 1990. The famous account of the soft-bodied creatures of the Burgess shale, and the celebration of contingency in evolution.

John Gribbin. *In Search of Schrödinger's Cat.* New York: Bantam, 1984. Good popular book on the meaning of quantum mechanics.

Brian K. Hall. *Evolutionary Developmental Biology.* London: Chapman and Hall, 1992. Superb textbook.

Nina Hall, ed. *Exploring Chaos: A Guide to the New Science of Disorder.* New York: Norton, 1993. Collection of articles by experts for the general reader. One of the best introductions to the scientific content of chaos.

A. H. Halsey, ed. *Heredity and Environment.* London: Methuen, 1977. Very interesting collection of essays: puzzled, "sure," stupid, and elegant.

M. Harris. *Cows, Pigs, Wars, and Witches: The Riddle of Culture.* New York: Random House, 1974. Enormously entertaining contextual pop-anthropology book.

Douglas R. Hofstadter. *Gödel, Escher, Bach: An Eternal Golden Braid.* Harmondsworth, England: Penguin, 1980. Classic mind-expanding cult book of the 1980s—funny, infuriating, and enormously illuminating unless you think science must be discussed solemnly. Excellent for distinction between information and meaning, and accessible source for Gödel's theorem. Often tough but rewards effort.

Douglas R. Hofstadter and Daniel C. Dennett. *The Mind's I: Fantasies and Reflections on Self and Soul.* New York: Bantam, 1982. Essays with punch: content and context in studies of self.

Nicholas Humphrey. *A History of the Mind: Evolution and the Birth of Consciousness*. New York: Simon and Schuster, 1992. A very elegant and readable book that explains awareness by the persistence of sensory data caused by time delays in the brain's processing activities.

Richard M. Jackson and Philip A. Mason. *Mycorrhiza*. London: Edward Arnold, 1984. Short book for more details about this symbiosis. Much has happened since.

Helge S. Kragh. *Dirac: A Scientific Biography*. Cambridge, England: Cambridge University Press, 1990. Excellent biography, examines Dirac's penchant for beautiful falsehood in preference to ugly truth.

Thomas Kuhn. *The Structure of Scientific Revolutions*. Chicago: University of Chicago Press, 1962. The book that presented science as a collection of shifting paradigms and let social scientists stop feeling guilty that they weren't achieving the same success as the physical scientists.

Bernd-Olaf Küppers. *Information and the Origin of Life*. Cambridge, Mass.: MIT Press, 1990. A very Germanic account. Some very interesting and sophisticated philosophy of information.

David Layzer. *Cosmogenesis: The Growth of Order in the Universe*. Oxford, England: Oxford University Press, 1990. Marvelous account of modern cosmology and what it means for us.

Roger Lewin. *Complexity: Life on the Edge of Chaos*. New York: Macmillan, 1992. A people-based description of the work of the Santa Fe Institute.

Konrad Z. Lorenz. *The Foundations of Ethology*. New York: Springer-Verlag, 1981. Classic, still worth reading.

James Lovelock. *The Ages of Gaia: A Biography of Our Living Earth*. New York: Norton, 1988. Gaia from the horse's mouth.

Jean-Pierre Luminet. *Black Holes*. Cambridge, England: Cambridge University Press, 1992. Brilliant popular account of relativity and black holes; explains the science in depth but is always comprehensible.

Benoit Mandelbrot. *The Fractal Geometry of Nature*. 2nd ed. San Francisco: W. H. Freeman, 1982. Fractals from the horse's mouth. Penetrating, elegant, infuriating, obscure.

John Maynard-Smith. *Evolution and the Theory of Games*. Cambridge, England: Cambridge University Press, 1982. Much better than *The Evolution of Sex*; contextual.

———. *The Evolution of Sex*. Cambridge, England: Cambridge University Press, 1982. A reductionist attempt to explain sex, which fails heroically.

Marvin Minsky. *The Society of Mind*. New York: Simon and Schuster, 1985. Up-to-date view of human intelligence.

Elaine Morgan. *The Aquatic Ape.* Lanham, Md.: University Press of America, 1984. Evocation of Alister Hardy's theory, well presented and well argued.

Rodney Needham. *Against the Tranquility of Axioms.* Berkeley: University of California Press, 1983. Takes anthropological "certainties" apart with a deft hand.

Susan Oyama. *The Ontogeny of Information.* Cambridge, England: Cambridge University Press, 1984. The very best nature/nurture destruction; a witty, wise, illuminating book with an *awful* title.

Roger Penrose. *The Emperor's New Mind.* New York: Viking Penguin, 1991. Definitive source for "quantum uncertainty, therefore free will." Fun, unafraid of mathematical formulas, and brilliant on almost everything except its central theme.

H. C. Plotkin, ed. *Learning, Development and Culture.* Chichester, England: Wiley, 1982. Essays on content and context with regard to culture.

Tim Poston and Ian Stewart. *Catastrophe Theory and Its Applications.* London: Pitman, 1978. Timely when published but now dated, especially as regards applications; a very good introduction nevertheless.

Terry Pratchett. *Witches Abroad: A Fantasy Novel.* New York: NAL/Dutton, 1993. One of many Discworld novels. Fantasy parody with a deeper wisdom. A world in which light speed is 600 mph; inspiration for Zarathustran obsession with 8.

Ilya Prigogine and Isabelle Stengers. *Order Out of Chaos: Man's New Dialogue with Nature.* New York: Bantam, 1984. Nonequilibrium thermodynamics and the emergence of structure.

Przemyslaw Prusinkiewicz and Aristid Lindenmayer. *The Algorithmic Beauty of Plants.* New York: Springer-Verlag, 1990. Beautiful picture book showing how the mathematical structures of L-systems can reproduce the branching patterns of plants. Secret order amid nature's complexity.

Ed Regis. *Great Mambo Chicken and the Transhuman Condition: Science Slightly Over the Edge.* Reading, Mass.: Addison-Wesley, 1990. Wild, wacky, unputdownable collection of unorthodox science, including artificial life. The story of the maverick gurus.

Michael Riordan and David N. Schramm. *The Shadows of Creation: Dark Matter and the Structure of the Universe.* New York: W. H. Freeman, 1991. Very good account of modern cosmology as cosmologists see it. Reductionist, uncritical.

Rudy Rucker. *Mind Tools: The Five Levels of Mathematical Reality.* Boston: Houghton Mifflin, 1988. Good, exciting, but occasionally frenzied pop science, over the top on conservation of information.

David Ruelle. *Chance and Chaos.* Princeton, N.J.: Princeton University Press, 1991. Wise words from one of the mathematical founders of chaos theory; very good on time reversibility, and astonishingly readable.

Erwin Schrödinger. *What Is Life?* Cambridge, England: Cambridge University Press, 1944. Asked the same question that we do, fifty years before we did. We stole some of his answers. Classic, with many brilliant insights.

Paul Shepard. *Thinking Animals.* New York: Viking, 1978. How animal images affect our thinking and our culture.

Philip Steadman. *The Evolution of Designs.* Cambridge, England: Cambridge University Press, 1979. Analogies between evolution of technology and evolution of organisms. Good, critical view.

Ian Stewart. *Does God Play Dice?* Oxford, England: Blackwell, 1989. Best general account of chaos, against a background of modern science.

———. *The Problems of Mathematics.* 2nd ed. Oxford, England: Oxford University Press, 1992. Readable overview of current state of the subject; includes fractals, dynamical systems, catastrophes, chaos.

Ian Stewart and Martin Golubitsky. *Fearful Symmetry: Is God a Geometer?* New York: Viking Penguin, 1993. A whole new way of looking at pattern, complexity, and the generation of order in nature.

René Thom. *Structural Stability and Morphogenesis.* Reading, Mass.: Addison-Wesley, 1975. Catastrophes from the horse's mouth. Penetrating, elegant, infuriating, obscure.

D'Arcy Wentworth Thompson. *On Growth and Form.* 2 vols. Cambridge, England: Cambridge University Press, 1961. Wonderful, full of thoughtful and thought-provoking examples.

———. *On Growth and Form.* J. T. Bonner, ed. Cambridge, England: Cambridge University Press, 1961. Abridged edition, which conveys the same message in more digestible form.

C. H. Waddington. *The Evolution of an Evolutionist.* Edinburgh: Edinburgh University Press, 1975. The most accessible account of canalization.

Mitchell Waldrop. *Complexity: The Emerging Science at the Edge of Order and Chaos.* New York: Simon and Schuster, 1992. How emergence is becoming respectable; a detailed look at the Santa Fe Institute and the theories that it is developing.

James Watson. *The Double Helix.* New York: Signet, 1968. Insider's warts-and-all view of the discovery of the secret of life. Necessarily biased.

Steven Weinberg. *Dreams of a Final Theory: The Search for the Fundamental Laws of Nature.* New York: Pantheon, 1992. A leading advocate explains what he means by a Theory of Everything. Thoughtful, fascinating, but tacitly assumes that "fundamental" in the sense of "ultimate bits and pieces" is the same as "fundamental" in the sense of "foundation for everything else."

Lancelot Law-Whyte. *Internal Factors in Evolution.* London: Social Science Paperbacks, 1965. A totally confused, very peculiar book that nonetheless contains many valuable evolutionary insights, especially contextual ones.

Arthur T. Winfree. *The Geometry of Biological Time.* 2nd ed. New York: Springer-Verlag, 1990. Difficult but rewarding account of temporal structures in biology. Intended for professionals but worth dipping into.

Lewis Wolpert. *The Triumph of the Embryo.* Oxford, England: Oxford University Press, 1991. A leading biologist discusses the problem of biological development in terms of positional information.

Alexander Woodcock and Monte Davis. *Catastrophe Theory: The Revolutionary New Way of Understanding How Things Change.* New York: Dutton, 1978. Brief, simple account of catastrophe theory and some of the issues that surround it, told as it seemed at the time.

E. C. Zeeman. *Catastrophe Theory: Selected Papers 1972–77.* Reading, Mass.: Addison-Wesley, 1977. Geometric models of sudden change in a dozen branches of science. Enormously interesting even when wrong, and right far more often than many people think, even now.

INDEX

Main references are given in bold type.

READ MORE IN PENGUIN

SCIENCE AND MATHEMATICS

Six Easy Pieces Richard P. Feynman

Drawn from his celebrated and landmark text *Lectures on Physics*, this collection of essays introduces the essentials of physics to the general reader. 'If one book was all that could be passed on to the next generation of scientists it would undoubtedly have to be *Six Easy Pieces*' John Gribbin, *New Scientist*

A Mathematician Reads the Newspapers John Allen Paulos

In this book, John Allen Paulos continues his liberating campaign against mathematical illiteracy. 'Mathematics is all around you. And it's a great defence against the sharks, cowboys and liars who want your vote, your money or your life' Ian Stewart

Dinosaur in a Haystack Stephen Jay Gould

'Today we have many outstanding science writers . . . but, whether he is writing about pandas or Jurassic Park, none grabs you so powerfully and personally as Stephen Jay Gould . . . he is not merely a pleasure but an education and a chronicler of the times' *Observer*

Does God Play Dice? Ian Stewart

As Ian Stewart shows in this stimulating and accessible account, the key to this unpredictable world can be found in the concept of chaos, one of the most exciting breakthroughs in recent decades. 'A fine introduction to a complex subject' *Daily Telegraph*

About Time Paul Davies

'With his usual clarity and flair, Davies argues that time in the twentieth century is Einstein's time and sets out on a fascinating discussion of why Einstein's can't be the last word on the subject' *Independent on Sunday*